ENGINEERS OF VICTORY

RANDOM HOUSE | NEW YORK

PAUL KENNEDY

ENGINEERS OF VICTORY

The Problem Solvers
Who Turned the Tide in the
Second World War

Published in the United States by Random House,
an imprint of The Random House Publishing Group,
a division of Random House, Inc., New York.

RANDOM HOUSE and colophon are registered trademarks of Random House, Inc.

LIBRARY OF CONGRESS CATALOGING-IN-PUBLICATION DATA
Kennedy, Paul M., 1945.
Engineers of victory: the problem solvers who turned the tide in the Second World
War / Paul Kennedy.—1st ed.
    p. cm.
Includes bibliographical references and index.
    ISBN 978-1-4000-6761-9 (alk. paper)—ISBN 978-1-58836-898-0 (ebook)
1. World War, 1939–1945—Campaigns. 2. World War, 1939–1945—Naval
operations. 3. Naval convoys—Atlantic Ocean—History—20th century. 4. World
War, 1939–1945—Aerial operations. 5. Bombing, Aerial—History—20th century.
6. Germany—Armed Forces—Organization. 7. Germany—Armed Forces—
History—World War, 1939–1945. 8. World War, 1939–1945—Amphibious
operations. 9. Amphibious warfare—History—20th century. 10. World War,
1939–1945—Campaigns—Pacific Area. I. Title.
D743.K425 2013
940.54—dc23   2012024284

Title-page image credit copyright © iStockphoto / © Todd Headington

Printed in the United States of America on acid-free paper

www.atrandom.com

9 8 7 6 5 4 3

Book design by Mary A. Wirth

To Cynthia

The young Alexander conquered India.
On his own?

Caesar defeated the Gauls.
Did he not even have a cook with him?

— Excerpt from Bertolt Brecht's 1935 poem "Fragen eines
    lesenden Arbeiters," in which Brecht imagines a young
    German worker beginning to read a lot of history
    books and being puzzled that they are chiefly histories
    of great men

Contents

Maps and Tables

Introduction

This is a book about the Second World War that attempts a new way of treating that epic conflict. It is not another general history of the war; it does not focus upon a single campaign, nor upon a single war leader. It focuses instead upon problem solving and problem solvers, and chooses to concentrate upon the middle years of the conflict, from roughly the end of 1942 to roughly the high summer of 1944.

In a book as complex as this one, it is best to state at the beginning what it is *not* about and what it does *not* claim. It resists all efforts at reductionism, such as that the winning of the war can be explained solely by brute force, or by some wonder weapon, or by some magical decrypting system. Claims that the war was won by Royal Air Force (RAF) bombers, the Red Army's T-34 tanks, or the U.S. Marine Corps's amphibious warfare doctrine are treated with respect and care in the pages below, but none of these explanations dominates the book. Nor should they. The Second World War was so infinitely more complex, and fought out across so many theaters and by so many different means, that the intelligent scholar simply has to go for a multicausal explanation as to why the Allies won.

This complexity is reflected in the five large chapters below. Each chapter tells a story of how small groups of individuals and institutions, both civilian and military, succeeded in enabling their political masters to achieve victory in the critical middle years of the Second World War. It is about what the military-operational problems were, who the problem solvers were, how they got things done, and thus why their work constitutes an important field of study. The story begins at the Casablanca Conference of January 1943, when the earlier Allied strategic thinking was brought together into a much more cohesive and wider-ranging blueprint for the defeat of the Axis powers, and it ends around seventeen months later, in June/July 1944, when, remarkably, all five of those operational challenges either had been overcome or were headed for success. It is an analysis of how grand strategy is achieved in practice, with the explicit claim that victories cannot be understood without a recognition of how those successes were engineered, and by whom. In this sense, the word *engineers* here means not strictly people possessing a B.Sc. or Ph.D. in engineering (although the founder of the Seabees, Admiral Ben Moreell, and the inventor of the mine detector, Józef Kosacki, certainly did) but those falling under *Webster's* wider definition: "a person who carries through an enterprise through skillful or artful contrivance." The book's potential transferability to large non-military organizations will seem obvious.

Of course, the five individual chapters themselves do not, and cannot, begin in January 1943, for in each case there is an antecedent tale to help the reader understand the background and contours of the analysis that follows. Still, there is not a simple, mechanistic structure to every chapter. Convoying merchant ships across the oceans (chapter 1) and landing on an enemy-held shore (chapter 4) were such long-standing military challenges, and drew upon so many lessons and principles from past fighting, that those chapters deserve a lengthier historical introduction. By contrast, grappling with the Wehrmacht's armored-warfare techniques (chapter 3) and being shot out of the sky by enemy fighters (chapter 2) were so recent an experience that those two chapters begin with anecdotes of clashes in 1943 itself. Chapter 5 sits somewhere in between. Trying to figure out how to move large forces across the Pacific after 1941 certainly demanded new weapons and organizations, but the operational challenge had been pondered for a full two decades beforehand and needs its own introduction.

By contrast, each chapter falls away quite rapidly after it reaches June/July 1944. There is brief coverage of how fighting led all the way to Berlin and Hiroshima, but the arguments in this book are complete by around July 1944. The tide really did turn in those critical eighteen months of the war, and no desperate actions by Berlin or Tokyo could block the oncoming waves.

Authors come to write the books they do for many reasons. In my case, a long detour in research and writing during the 1990s, to help compose a study to improve the effectiveness of the United Nations, probably was the reason I became more interested in the idea of problem solvers in history.[1] Later, a class in grand strategy that I taught each year at Yale further spurred this intellectual interest. That class is a remarkable twelve-month course that examines the great classics (Sun Tzu, Thucydides, Machiavelli, Clausewitz) together with a number of historical examples of grand strategies that went right or wrong, and then concludes with an analysis of contemporary world problems.[2] The pedagogical justification for such a course is a strong one: if we are teaching talented future leaders in the realms of politics, the military, business, and education, the period of their lives when they are advanced undergraduates and graduate students is probably the optimal time for them to grapple intellectually with enduring writings and historic case studies. Very few prime ministers or CEOs have much time to study Thucydides at the age of fifty or sixty!

But the teaching of grand strategy has, by its very nature, to address strategy and politics from the top. Therefore, what transpires at the middle level, or the level of the practical implementation of those policies, is often taken for granted. Great world leaders order something to be done, and lo, it is accomplished; or lo, it stumbles. We rarely inquire deeply into the mechanics and dynamics of strategic success and failure, yet it is a very important realm of inquiry, though still rather neglected.[3] To give but a few examples: historians of Europe know that for a staggering eighty years, Philip II of Spain and his successors sought to quell the Dutch Protestant Revolt, far to the north of Madrid and separated by Europe's many rivers and mountain chains, but we rarely inquire into just how that military campaign was pursued so successfully and impressively along the "Spanish Road." Scholars know also that the Elizabethan navy outmaneuvered and outshot the far larger Spanish Armada of 1588, but rarely are they aware that only Sir John Hawkins's drastic redesign of the queen's galleons a decade earlier gave those vessels the necessary speed and firepower to do just that. The astounding growth of the British Empire in the course of the great eighteenth-century wars is recorded in many a book, but usually without explaining the degree to which it was financed by the merchants of Amsterdam and other European capital centers. When Britain declared war on Germany in August 1914, historians will tell us that that same empire was put immediately upon military alert across the globe, but they say little of the astonishing undersea cable communication system that executed that order.[4] Grand strategists, leaders and professors alike, take a lot of things for granted.

By the same token, historians of the Second World War also know that in January 1943, following the successful North African landings, Winston Churchill, Franklin Delano Roosevelt, and the Combined Chiefs of Staff met at Casablanca to decide upon the future ordering of the war; and that from those intense debates emerged both the political and the operational guidelines to future Anglo-American grand strategy. Politically, the enemy would have to offer unconditional surrender. With Germany recognized as the most formidable of their enemies, victory in Europe would make the first claim upon resources, but Fleet Admiral Ernest J. King ensured that this ruling should not exclude comeback operations in the Pacific and Far East at the same time, however ambitious that may have seemed. The Russian ally would have to be given all possible help in resisting the Nazi blitzkrieg, even if that help couldn't include direct battlefield assistance on the Eastern Front. More immediately, the Western navies, air forces, and armies would have to figure out how to achieve their triple operational mission: (1) win control of the Atlantic sea-lanes, so that the convoys to Britain could get through safely; (2) attain command of the air over all of west-central Europe, so that the United Kingdom could act not only as the launching pad for the invasion of the continent but also as the platform for the systematic aerial destruction of the Third Reich; and (3) force their way across Axis-held beaches and carry the fight to the European heartland. With all this agreed upon, the U.S. president and British prime minister could pose for the conference photographs, approve these strategic directives, and fly home.[5]

We also know that, little more than a year later, all of those operational aims were either accomplished or close to being realized (the "unconditional surrender" part would take another year). North Africa was taken, then Sicily, then all of Italy. The policy of unconditional surrender was maintained, except for pulling Mussolini's collapsing empire out of the war and neutralizing Italy. The "Germany first" principle was kept intact, and, as hoped, the United States showed that it was also able to commit such enormous military resources to the war in the Pacific that Japan's surrender followed a mere three months after the fall of the Third Reich. The Atlantic sea-lanes were made safe. Aerial dominance over Europe was established, and with it came the increased strategic bombing campaign against German industry, cities, and people. Russia was given further aid, though its own fortitude and resources were by far the greatest reason for its ultimate victory on the Eastern Front. American forces surged across the Pacific. France was finally invaded in June 1944, and less than a year later the Allied armies met along the Elbe River to celebrate their mutual, hard-fought victory in Europe. What was ordained at Casablanca had really come about. This book attempts to explain how and why.

As so often, appearances are misleading. No straight causal line connects the confident Casablanca statement of Allied war strategies and their realization. For the plain truth was that at the beginning of 1943 the Grand Alliance was in no position to carry out these declared aims. Indeed, in many of the fields of war, and especially in the critical struggles for command of the sea and command of the air, things deteriorated in the months following the Casablanca conference. The ultimate wartime victory of 1945 has all but erased this truth, much as the final victories over Philip II of Spain and Napoleon tended to obscure how difficult things seemed—and were—for their opponents in the middle years of those conflicts.

In the battle for control of the Atlantic sea-lanes, the campaign that Churchill confessed gave him more cause for worry than any other during the entire war, merchant ship losses intensified in the months after Casablanca. In March 1943, for example, Admiral Karl Doenitz's U-boats sank 108 Allied vessels totaling 627,000 tons, a rate of loss that horrified the Admiralty's planners, especially as they knew they would face even larger numbers of German submarines in the summer ahead. Thus, far from the convoys easily providing massive amounts of men and munitions for a second front, there were fears of Britain not getting enough commercial bunker fuel to survive. Unless and until this danger was beaten off, there could be no question of an invasion of Europe.

As 1943 unfolded, things also went from bad to worse in the Allied strategic bombing campaign of Germany. Under Albert Speer's extraordinary reorganization of German war industries, the Luftwaffe doubled its number of night fighters. Air Marshal Arthur "Bomber" Harris's famous thousand-bomber raids dealt a few dramatic blows (Cologne, Hamburg) to German industry, but so many of the RAF's bombers were destroyed when they moved on to attack more distant Berlin that the force came to the brink of paralysis. In the sixteen massive aerial attacks upon the Nazi capital between November 1943 and March 1944, Bomber Command lost 1,047 planes and sustained damage to another 1,682. The daylight raids of the United States Army Air Forces (USAAF) led to an even higher attrition rate per operation. In the famous raid of October 14, 1943, for example, 60 of the 291 Flying Fortresses attacking the vital ball-bearing plants at Schweinfurt were shot down and a further 138 were damaged. Both air forces had to wrestle with the blunt fact that the interwar saying "The bomber will always get through" was wrong. In consequence, Allied command of the air had become as illusory as command of the sea. But without both, the defeat of Germany was impossible.

In any case, the Western Allies hadn't worked out how to achieve their third military task—how to land on an enemy-held coastline possessing the defensive capacities of the "Atlantic Wall," how to repel the inevitable and massive Wehrmacht armored counterattacks against the bridgeheads, and how to push two to three million soldiers from the Channel beaches to the heart of Germany. The North African landings that had preceded Casablanca were relatively easy, since the Vichy French naval and political opposition there was negligible, which perhaps ironically contributed to the general confidence exuded by Roosevelt and Churchill at Casablanca (less so by canny practitioners such as Alanbrooke and Dwight Eisenhower). But cracking the German fortifications along the Atlantic shore was a totally different matter, as the Chiefs of Staff must have known, since the one exploratory venture to test those defenses—the catastrophic Dieppe Raid of August 1942—resulted in the death or capture of the majority of Canadian troops deployed. In consequence, the conclusion drawn by Allied planners from that raid was that it would be virtually impossible to take a well-defended enemy harbor. But if that was the case, where exactly could one land millions of men and thousands of ships? On an open beach, battered by the usual Atlantic storms? That also seemed impractical. So how, then, was the West successfully to invade France—or, for that matter, to invade Japan amid turbulent Pacific tides?

The challenge of defeating German counterattacks upon the beachheads brings up a further large question: how does one stop a blitzkrieg? For particular historical and operational/technical reasons, the German armed services in the late 1930s and early 1940s had hit upon a form of mixed-weapons warfare (shock troops, mobile small arms, motorized infantry units, tanks, tactical support aircraft) that swiftly carved through their opponents' defenses. The Polish, Belgian, French, Danish, Norwegian, Yugoslav, and Greek armies were rent asunder. In 1940–41 the proud British Army was tumbled out of Europe (Norway, France and Belgium, Greece and Crete) in a fashion that had not happened since Mary Tudor lost Calais.

By the time of the Casablanca conference, there was at least some good news regarding this form of combat. At the far western borders of Cairo, around El Alamein, British-led armies had stopped the charismatic General Erwin Rommel's advance, damaged his most important military units, and begun to roll back the German forces along the North African shores. At almost the same time, the Red Army's counteroffensive in the southern sector of the Eastern Front had mired the brutal and imposing German offensive in Stalingrad, retaken that city house by house, and captured General Friedrich Paulus's entire Sixth Army.

Yet, shocked by these twin defeats on land, the Third Reich threw off its complacency and reorganized itself. Its armaments production in 1943 was well over twice that of 1941; its output of aircraft in that earlier year had been about half of Britain's, but by 1943 it was surging ahead again. The various German armed services were receiving better aircraft, better tanks, better submarines. Hitler's worried reaction to the Anglo-American landings in North Africa (November 8, 1942) was to seize control of all of southern (Vichy) France and pour crack divisions into Tunisia. As the Allied leaders were flying home from Casablanca, Rommel's newly arrived forces were punishing the inexperienced U.S.

units in the Kasserine Pass. After Stalingrad the Red Army's frontline forces had run out of steam, and as early as February and March 1943, Erich von Manstein's reinforced panzer armies had blunted the Russians' offensive, had retaken Kharkov, and were assembling a vast armored force for their own summer assault toward Kursk. If, in addition, Berlin really could continue to interrupt the Atlantic convoys, destroy the Western aerial offensive, and deny the Anglo-American armies entry into France, then presumably it could concentrate more of its massive forces on the Eastern Front, until perhaps even Joseph Stalin would admit a compromise settlement.

Another major operational challenge was the task of ensuring the defeat of Japan. This was clearly going to be an American enterprise, if not exclusively then overwhelmingly so. To be sure, British and British-Indian troops would attempt the recovery of Burma, Thailand, and Malaya, and Australian divisions would join Douglas MacArthur in taking New Guinea and pushing on to the Philippines. Yet the most sensible operational route was actually to avoid the jungles of New Guinea, Burma, and Indochina and instead to hop across the Central Pacific directly westward from Hawaii to the Philippines, then China, then Japan. Innovative U.S. officers had toyed with this "War Plan Orange" throughout the interwar years, and on paper it seemed most promising; it was, after all, the only campaign plan that didn't have to be tossed away or severely amended in consequence of the Axis's great successes of 1939–42.

The problem once again, as in the case of invading France, was always the practical one. How exactly did one land on a coral atoll, its inshore waters strewn with mines and obstacles, the beaches infested with dynamited booby traps, the enemy holed up in deep bunkers? As late as November 1943, Central Pacific Command began its long-awaited offensive with an assault in overwhelming force against the Japanese garrison holding Tarawa in the Gilbert Islands. The result was not in doubt, because Imperial General Headquarters had decided that the Gilberts lay outside its "absolute national defense sphere" in the Pacific, and the garrison numbered a mere 3,000 men, but the losses among the American marines stranded under heavy fire on the outlying coral reefs shocked the public back home. Whichever way one came across the Pacific, the indicators were grim. It was all very well for General MacArthur, with his flair for publicity, to promise "I shall return" to the Philippines when he left early in 1942, but the Japanese garrison in those islands now totaled 270,000 men, none of whom would surrender. How long, then, would it take to get to Japan's own shores? Five years? And at what cost, if the enemy garrisons in the Philippines were twenty or fifty times as large as those on Tarawa?

There was, then, a truly daunting list of difficulties to be overcome by the Grand Alliance, a list made all the more formidable because almost all of these challenges were not fully separate but depended upon gains being made elsewhere. Hopping across the Pacific islands, for example, first required gaining command of the sea, and that in turn depended upon command of the air, and then upon building giant bases on top of meager coral islands—and a great disaster in the Atlantic or Europe would have produced urgent calls for a relocation of U.S. resources from the Pacific to those theaters instead (and a furious row among the Chiefs of Staff). Invading France was impossible until the German U-boat menace to the Atlantic convoys had been defeated. Only when Allied shipyards could produce enough of the new, odd-looking assault vessels to surmount obstacles and fight their way onshore could a maritime invasion take place in any theater. Although Stalin would never admit it, the Red Army's successes in the field were helped significantly by the fact that the Anglo-American strategic bombing campaign compelled Germany to allocate enormous amounts of manpower to the antiaircraft, civil control, and emergency rebuilding programs required to keep the Third Reich in action. The vital Dodge and Studebaker trucks, the workhorses for the Soviet divisions in their westward advance, could not be transported from America to Russia unless maritime lines of communication were preserved by the Royal Navy. Conversely, it is difficult to see how the Anglo-American armies in the west could have made much progress at all had not scores of battle-hardened Wehrmacht divisions been pinned down (and decimated) in the east. In short, whereas one advantage gained by the Allies could help campaign(s) elsewhere, a serious defeat could damage the chances of the other aims being achieved.

Remarkably, all five separate though interconnected challenges were overcome between early 1943 and the summer of 1944—roughly, between Casablanca and the quadruple successes of Normandy, the fall of Rome, the Marianas landings in the Central Pacific, and Operation Bagration on the Eastern Front. Some strategic problems (aerial control over Germany, island-hopping across the Pacific) took longer than others (controlling the Atlantic sea-lanes, blunting the blitzkrieg), but in the space of something like seventeen months, the tide turned in the greatest conflict known to history.

Why was that so, and how did it happen? One reply lies close to hand, in the sense that the Fascist aggressor nations were rash enough to attack the rest of the world. Because of their earlier arms buildups in the 1930s, the Axis powers gained wide and stunning successes, but they could not succeed in defeating any one of their three major enemies. When the rest of the world recovered from those batterings, it steadily applied its far greater resources, fought its way back, and achieved final victory.*

Yet there is another equally important point to be pursued, namely, exactly *how* did the Allies recover and fight their way back? The relative productive capacities held by each side by 1943–44 do indeed point to the likely winners. But what if the U-boats had not been defeated in summer 1943, or if the Luftwaffe had not been destroyed early in 1944, or if the Red Army had not found ways to blunt German panzers? What if the legendary "turnaround" weapons such as the long-range fighter and miniaturized radar—whose arrival on the battlefields in 1943–44 most historians seem to take as a given—had not come into play at the time they did, or had not been developed at all?

At the very least, all this suggests that the "inevitable" Allied victories could have occurred much later than in May and August 1945, and that they would have been accompanied by far higher losses in the field. The story of the second half of the Second World War presumably would have looked very different than it does to us today. What we explore here, then, is a common conundrum: how does one achieve one's strategic aims when one possesses considerable resources but does not, or at least not yet, have the instruments and organizations at hand?

---

*This is, after all, the argument in my 1988 book *The Rise and Fall of the Great Powers*, 347–57.

This is the story, then, of that strategic, operational, and tactical turn-around from early 1943 to mid-1944. From the beginning, it was obvious that the investigation proposed here had to move downward from the Combined Chiefs of Staff's declarations to a detailed analysis of the process of carrying out the proclaimed missions. It was one thing to assert that defeating the German U-boat threat was paramount. How, actually, did one go about defeating it? Again, it was no doubt fine (and politic) for Washington and London to assure an irritated Stalin throughout 1942 and 1943 that a second front would soon be launched in France. But how? Certain individuals and certain organizations had to answer those questions; it was they who must solve these problems and thus make feasible the efforts of the millions of Allied soldiers, sailors, and airmen.

This is where the present work may set itself apart, because it seeks to tell the story of such individuals and organizations, not just in some anecdotal or romantic or military-history-buff sort of way, but as a key part of understanding the larger epic of how the tide was turned. Many readers will have some knowledge of the story of the Ultra decryption team at Bletchley Park and their equivalents in the Pacific. Some will know the story—as portrayed in the film *The Dam Busters*—of how Barnes Wallis invented the bouncing bombs that blew up the Ruhr Valley dams. Only a few are familiar with the eccentric Percy Hobart's creation of the weird tanks that could push right through D-Day coastal minefields and barbed wire, or with the individuals who devised the Mulberry harbors. Very few Western readers will have an inkling of the renowned T-34 tank's truly pathetic capacities, or know instead of the extremely important role of the Red Army's antitank weaponry. And only some will understand the significance of the cavity magnetron, or why putting a Rolls-Royce Merlin 61 engine into a rather limp P-51 Mustang fighter was such a critical step, or the significance of the extraordinary career of the founder of the American Navy's Seabees. Much of that World War II folklore is fascinating in itself to those who do know those stories. But the point being made here is that we have rarely if ever stepped back and understood how their work surfaced, was cultivated, and then was connected to the problems at hand, or appreciated how these various, eccentric pieces of the jigsaw puzzle fitted into the whole.[6]

This book seeks, through the chapters that follow, to make a contribution to such understanding. In many ways, it is a return to the research and writing done some forty years ago for Sir Basil Liddell Hart's own *History of the Second World War,* although this time I hope I am seeing through the glass more clearly than in that earlier time.* Yet the present study is intended not as a form of personal pilgrimage but rather as an effort to widen the debate about decision making and problem solving in history. It seems a story worth telling. And if that is true, it is a mode of inquiry that may be worth applying elsewhere.

---

*See the acknowledgments for further details.

ENGINEERS OF VICTORY

There is a tide in the affairs of men.
Which, taken at the flood, leads on to fortune:
Omitted, all the voyage of their life
Is bound in shallows and in miseries.
On such a full sea are we now afloat,
And we must take the current when it serves,
Or lose our ventures.

—WILLIAM SHAKESPEARE, *Julius Caesar,* Act IV, Scene 2

CHAPTER ONE

HOW TO GET CONVOYS
SAFELY ACROSS THE ATLANTIC

Thus was the stage set for Germany to fling into the Atlantic struggle the greatest possible strength. . . . It was plain to both sides that the U-boats and the convoy escorts would shortly be locked in a deadly, ruthless series of fights, in which no mercy would be expected and little shown. Nor would one battle, or a week's or a month's fighting, decide the issue. It would be decided by which side could endure the longer; by whether the stamina and strength of purpose of the crews of the Allied escort vessels and aircraft, watching and listening all the time for the hidden enemy, outlasted the will-power of the U-boat crews, lurking in the darkness or the depths, fearing the relentless tap of the asdic, the unseen eye of the radar and the crash of the depth charges. It depended on whether the men of the Merchant Navy, themselves almost powerless to defend their precious cargoes of fuel, munitions and food, could stand the strain of waiting day after day and night after night throughout the long, slow passages for the rending detonation of the torpedoes, which could send their ships to the bottom in a matter of seconds, or explode their cargoes in a searing sheet of flame from which there could be no escape. It was a battle between *men*, aided certainly by all the instruments and devices which science could provide, but still one that would be decided by the skill and endurance of men, and by the intensity of the moral purpose which inspired them. In all the long history of sea warfare there has been no parallel to this battle, whose field was thousands of square miles of ocean, and to which no limits in time or space could be set.

—S. W. ROSKILL, *The War at Sea, 1939–1945*

As Churchill and Roosevelt journeyed to and from Casablanca in January 1943, the weather in the North Atlantic had become as violent as any experienced sailor could remember. For much of December and January, huge storms at sea cramped naval and air activity. Merchant ships, pounded by giant waves, had heavy cargoes breaking loose and sliding around inside their hulls. Smaller warship escorts such as corvettes were tossed around like corks. Warships with heavier upperworks and gun turrets rolled from side to side. German U-boats, when they surfaced, could see nothing across the hundred-foot-high waves and were happy to submerge into quieter waters or to head south. Hundreds, perhaps thousands, of sailors were hurt, and not a few killed by accidents or washed overboard. In some extreme cases, the commander of a convoy was forced to order a return to base, or at least to send damaged ships back. General Sir Alan Brooke (later Viscount Alanbrooke) records in his diary that he, the other British Chiefs of Staff, and Churchill himself had their flight (or surface sailing) plans from London to Casablanca changed time and again.*

The result, quite naturally, was that convoy activity upon the storm-whipped North Atlantic routes was much less than normal in these midwinter months. Quite separate from this physical interruption, there was a second and cheerier reason the regular convoy traffic fell away at this time. Operation Torch itself demanded a vast number of escorts to assist the occupation of the Vichy states of Morocco and Algeria—the Royal Navy contributed 160 warships of various types to it—and in consequence the Gibraltar, Sierra Leone, and Arctic convoys had to be temporarily suspended.[1] Since ships carrying Allied troops, landing equipment, and immediate supplies were bound to get the

---

*The British practice of allowing someone elevated to the peerage to alter or conflate his name has been confusing here. For much of the war he was General (later Field Marshal) Sir Alan Brooke—or, simply, Brooke. When elevated to the peerage in 1945, he became Viscount Alanbrooke—or, simply, Alanbrooke. Thus for the war he was really Brooke, though so many later historians—Danchev, Bryant, Roberts—have preferred Alanbrooke that it is difficult to resist using the better-known name.

highest level of naval protection, and the Axis was ill-prepared for the Operation Torch invasion because of its obsessions with the Eastern Front and Egypt, it is scarcely surprising that the invading forces met with little or no U-boat opposition on North African shores.

The other, understandable consequence was that Allied losses to enemy submarine attacks fell dramatically during midwinter. If there were fewer Atlantic convoys at sea in the first place, those that did sail, while battered by storms, were often protected by the same lousy weather conditions. Some were routed far to the north in a sort of great circle, trading physical damage by ice floes for distance from the wolf packs. The Admiralty's monthly toll of shipping losses captured this dramatic decline well. For example, in November 1942 the Allies lost 119 merchant ships totaling 729,000 tons, and while many of these vessels were sunk in more distant waters, off South America, those supply lines were also part of an integrated effort aimed at sustaining and enhancing Anglo-American power in the British Isles. The sinking of British oil tankers coming from Trinidad would, as a consequence, hurt the buildup of the U.S. bomber groups in East Anglia; everything hung on command of the seas.

Because of the rough weather, U-boat sinkings in December and January fell to a mere 200,000 tons of Allied shipping, most of which also occurred on more southern routes (e.g., Trinidad to Gibraltar, to supply the Operation Torch armies). But the tallies for those months were exceptional, for the reasons explained above, and thus when the prime minister and president met at Casablanca they were under no illusion as to how serious the crisis over shipping losses had become. The Allied merchant fleets had lost a staggering 7.8 million tons in the course of 1942, almost 6.3 million of which had been sunk by that most formidable weapon, the U-boat. The amazing American mass production shipyards were still gearing up to full strength, but even their output in 1942 (around 7 million tons) meant that total available Allied shipping capacity had declined in absolute terms, and now had to compete with the even greater demands of the Pacific War. By early 1943, therefore, British imports were one-third less than those in 1939, and U.S. Army trucks and box-bound aircraft now competed with colonial foodstuffs, ores, and petroleum for space on the endangered merchant vessels. This grim fact imperiled *everything* in the European war

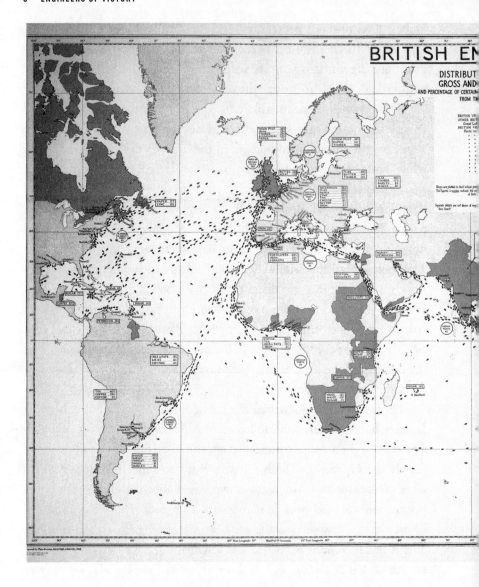

strategy. It threatened the British war effort; if things got worse, it threatened mass malnutrition for the islanders. The heavy losses suffered by oil tankers meant that only two to three months of fuel remained in Britain's storage tanks, but how could the country fight, or live, without fuel? This crisis also threatened the Arctic convoys to aid Russia, and the Mediterranean convoys to aid Malta and Egypt. It threatened, by extension, the entire Egyptian campaign, for Britain could scarcely send military reinforcements via Sierra Leone and the

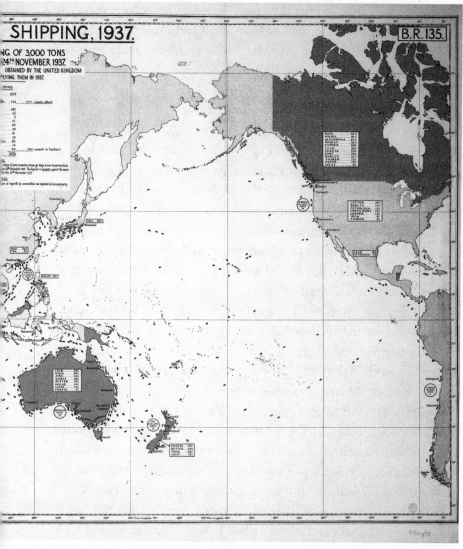

LOCATION OF MERCHANT SHIPS OF THE BRITISH EMPIRE, NOVEMBER 1937
A wonderful scattergraph map showing the sheer enormity and vulnerability of the British Empire's merchant shipping across the entire globe. Note the significance of the Mediterranean route, the massive trades from the Caribbean and South America, and the strategic significance of Freetown, Sierra Leone, and Gibraltar. This free-ranging system could not sustain the ravages of constant U-boat and aerial attack.

Cape of Good Hope to Suez if its own lifelines were being crushed. It even threatened unrest in parts of East Africa and India that had come to rely upon seaborne food imports. And it absolutely threatened the assumptions behind Operation Bolero (later renamed Overlord), which called for a rapid and massive buildup of the U.S. Army and the Army Air Forces in the British Isles in preparation for a second front in Europe; it would have been ironic to have sent 2,000 American heavy bombers and millions of GIs to England only to find that they had no fuel. Churchill later stated in his memoirs that, of all struggles of the war, it was the Battle of the Atlantic that he most worried about; if it was lost, so too might be Britain's gamble to fight on in 1940.

In addition, although the Admiralty did not possess an exact tally of the enemy's U-boats, there did seem to be an awful lot more of them. In the course of 1942, Allied warships and aircraft had destroyed eighty-seven German and twenty-two Italian submarines. But the Third Reich was also gearing up its war production and had added seventeen new U-boats *each month* during that year. By the end of 1942, therefore, Doenitz commanded a total of 212 submarines that were operational (out of a grand total of 393, for many were working up, training new crews, or receiving new equipment), very significantly more than the 91 operational craft he had had (out of 249) at the beginning of that year.[2] Although victory in the Second World War was critically affected by each side's inventiveness, technology, and organization, not just by sheer numbers, the blunt fact was that numbers *did* count. And by the time of the Casablanca conference it seemed that the Germans were having greater success at sinking Allied merchantmen than the Anglo-American forces were in sinking U-boats. Worse still, more and more U-boats were entering the fray.

In the months that followed, therefore, the prime minister's nightmare appeared to be coming true. As March and April 1943 arrived and the convoy traffic to the British Isles resumed at a higher rate, so too did merchant ship losses. February's total doubled that of the previous month, and in March the Allies lost 108 ships totaling 627,000 tons, making it the third-worst month on record during the war. What was more, nearly two-thirds of those ships were sunk in convoy; one was no longer talking here of the happy U-boat pickings of individual merchant ships off the well-lit shores of America early in 1942, or of the

almost equally easy raids upon Allied shipping routes in the South Atlantic. What was also truly alarming was that the losses had occurred
chiefly along the single most important convoy route of all, that between New York and Halifax and the receiving ports of Glasgow and
Liverpool. Between March 16 and 20 the greatest encounter in the entire Battle of the Atlantic saw Doenitz throw no fewer than forty
U-boats against the two eastbound convoys HX 229 and SC 122. This
epic fight will be analyzed in more detail below, but the result was awful
for the Allies: twenty-one merchant ships totaling 141,000 tons were
sunk, with the loss of only one U-boat. In the Admiralty's own later
account, "The Germans never came so near to disrupting communications between the New World and the Old as in the first twenty days of
March, 1943."[3] These ever-rising losses suggested, among other things,
that the whole principle of convoy as the best means of defending maritime commerce was now in doubt.

The Strategic and Operational Context

The British Admiralty's problems were nothing new in the annals of
naval warfare. The protection of ships carrying goods at sea from hostile attack is one of the oldest problems in the history of war and peace.
Even at the height of the Roman Empire merchants and consuls in Sicily and North Africa complained about the depredations of pirates
against the grain, wine, and olive oil trades. Fifteen hundred years later
Spanish commanders fumed at the plundering by Dutch and English
raiders of their galleons bearing silver and precious spices; only a generation or two afterward the Dutch found their long-haul trade with
the East Indies under French and English assault at sea. The age of
European expansion and then of the Commercial Revolution (sixteenth
to eighteenth centuries) had moved ever greater shares of national
wealth onto precarious maritime routes. In the age of Charlemagne, the
dependence of rulers and peoples upon command of the sea was negligible. By the time of, say, the Seven Years' War (1756–63) it was critical
in both the Atlantic and Indian Oceans, at least for all advanced economies. If a west European nation lost control of the trade routes, it was
most likely also going to lose—or at least not win—the war itself. This
was the message of that classic work, *The Influence of Sea Power upon*

History (1890), composed by the American naval author Alfred Thayer Mahan.

Mahan's ideas influenced the admiralties of Britain, Germany, Japan, the United States, and many lesser navies. The key belief was that the only way to gain command of the sea was to have the most powerful battle fleet afloat, one that would crush all rivals. Lesser forms of naval warfare, such as commerce raiding and cruiser and torpedo-boat operations—*la guerre de course*—didn't count for much, because they didn't win wars. It was true that during the Napoleonic Wars French predators had seized many independently sailing British merchantmen, but once the latter were organized in convoys and given an escort of warships, the sea routes were secure behind Nelson's assembled fleets. The same truth revealed itself, albeit at great cost, during the First World War. For three years, and even though the Grand Fleet had command of the sea, Allied merchant ships steaming on their own were picked off in increasing numbers by German U-boats. After the Admiralty was compelled by the British cabinet to return to the convoy system in 1917, losses to enemy submarines dropped dramatically. Within a short while, moreover, the Allied warships would possess asdic (sonar), so for the first time ever they could detect a solid object under water. Provided one had command of the sea on the surface, it was argued, one would also control the waters below. A submarine would thus be as recognizable as the sails of a French frigate 150 years earlier. Such was the prevailing assumption of naval staffs in the years following the Treaty of Versailles, 1919. Convoys, plus sonar, worked.[4]

Before we examine how and why that assumption was challenged by the renewed German U-boat threat during the first half of the Second World War, a couple of very important, though clashing, strategic-operational assumptions also need to be considered. The first of these, rarely articulated, is that one really didn't need to sink surface commerce raiders or submarines to win the maritime war. So long as the Royal Navy shepherded without loss a group of fifty merchantmen from, say, Halifax to Liverpool, it had won. The larger Allied strategy was to keep Britain in the fight and then to make it the springboard for an enormous invasion of western Europe. Thus, if every transatlantic (and South American, Sierra Leonean, and South African) convoy got to port safely without ever encountering U-boats, the war was being won, ship by ship, cargo by cargo. Even if the convoy escorts had to face a serious submarine attack but could beat off the predators, all would still be well. The task of the shepherd was to safeguard the sheep, not to kill the wolves.

The opposite argument was that killing the wolves had to be the essence of Allied maritime strategy. It too had its own logic: if the threat to the sea-lanes was forcibly removed, all would be fine and one of the Casablanca war aims could at last be implemented. In today's language, the prevailing authorities cannot wait for terrorists to attack the international system but have to go and root out the terrorists. In maritime terms, therefore, a navy charged with protecting its merchant ships would either go on a submarine hunt or, an even bolder tactic, simply drive its convoys through U-boat-infested waters and force the submarines to fight—and be killed.

The first of these two convoy strategies was clearly defensive; the second (whether submarine hunting or forcing the convoys through) was equally clearly offensive. Both visions, it is worth noting, involved a tricky, interdependent three-way relationship between the merchant ships, the U-boats, and the naval and aerial escorts, not unlike the children's rock-paper-scissors game. If the convoys could avoid an encounter or have the U-boats beaten off, fine for them; if the U-boats could get at the convoys without destruction from the escorts, fine for them; and if the escorts could destroy enough submarines, fine for them.

In the harsh world of the North Atlantic between 1939 and 1943, however, neither an Allied defensive operational strategy nor an offensive one was possible on its own. The way forward had to be achieved by a combination of both options, depending on the ups and downs of what turned out to be the longest campaign of the entire Second World War. And this route was, geopolitically, the most important maritime journey in the world. Of course the other Allied trade routes mattered, and all faced the same operational and logistical difficulties, or sometimes (as with the Arctic convoys) even greater ones. But maritime security across the Atlantic was the foundation stone of all Anglo-American grand strategy in the European theater. With a look forward to the remaining chapters of this book, it is worth restating the many interconnections. Winning this Atlantic battle preserved Britain's own very large military-industrial base. Britain was also the unsinkable aircraft carrier for the Allied strategic bombing campaigns, and the springboard for the eventual invasion of western Europe. Britain was the port of departure for most of the convoys to northern Russia and to the Mediterranean; it was the source for the many troopship convoys that Churchill dispatched, via the Cape, to Montgomery in Egypt and the Middle East. Controlling the Atlantic was the sheet anchor of the West's plans to defeat Italy and Germany.

The top-to-bottom logic chain of Allied grand strategy here is also unusually clear, a fine example of Millett and Murray's concept of the multilevel nature of "military effectiveness."[5] The *political* aim was the unconditional defeat of the enemy, and the return to a world of peace and order. The *strategy* to achieve that purpose was to take the war to the enemy by all the means that were available: aerial, land-based, naval, economic, and diplomatic. This required decisive successes at the *operational* level, and in all the areas covered in the chapters of the present book. It would be foolish to argue about which of those operational regions was more important than others (even if the Combined Chiefs of Staff had to do so as they wrangled over allocating resources); they all were part of Allied grand strategy. What is incontestable, however, is that if the British, the Americans, and their smaller allies were to reconquer Europe from fascism, they first of all had to have command of the Atlantic waters.

Yet control of that vital route was itself determined, in the last resort, by a number of key technical and tactical factors. In other words, there is also in this story a clear example of a bottom-to-top logic chain. Every individual merchant ship that was preserved and every individual U-boat that was sunk by Allied escorts directly contributed to the relative success rate of each convoy. The tactical success rates of each single convoy contributed to the all-important monthly tonnage totals, and those monthly tonnage totals were the barometer to the winning or losing of the Battle of the Atlantic. That operational battle, as we have argued, was key to victory in western Europe and the Mediterranean. And winning in the West was a part of the strategic tripos—victory in the West, victory in the East (Eastern Front), and victory in the Pacific/Far East.

The Battle of the Atlantic was an operational and tactical contest that hung upon many factors. The first of these, from which all the others flowed, was the possession of efficient and authoritative organization. This was so basic a point that it is often taken for granted, yet on brief reflection it is clear how important were the structures of command, the lines of information, and the integration of war-fighting systems. Both sides benefited greatly, of course, from the experiences of the epic campaign in the Atlantic during the First World War, and by the post-1919 lessons drawn from them. In terms of simplicity of command, Doenitz had it easier, for the U-boat service was separate from the German surface navy, and it became easier still for him when the failure of a squadron of heavy ships to destroy an Arctic convoy in the last days of 1942 led to an explosion of rage on Hitler's part and to Grand Admiral Erich Raeder being replaced as commander in chief of the entire navy by Doenitz himself at the end of January 1943. Doenitz decided to remain commander of U-boats, so as to keep control of submarine operations, and it is evident that he found it much easier to obtain the Fuehrer's backing than his predecessor had. This did not mean that he had no organizational fights. There was a constant struggle to gain the necessary share of war materials (steel, ball bearings, electrical parts, antiaircraft weaponry) against the enormous demands of the Wehrmacht and the Luftwaffe. And, as we shall see, Doenitz had the greatest difficulty in getting aerial support for his boats. Nonetheless, it was an enormous advantage to have a single and very experienced authority directing the entire U-boat campaign.

The organizational story on the Allied side was a lot less clear-cut. During the years of American neutrality, it actually had been simpler. Ensuring command of the seas was the traditional responsibility of the Admiralty in Whitehall, which then devolved defense of the Atlantic convoys to a particular authority, the Western Approaches Command, based in Liverpool. By the time our analysis begins, its chief was the formidable Admiral Sir Max Horton, like Doenitz a highly experienced submarine commander twenty-five years previously. The much smaller Royal Canadian Navy, operating out of ports such as St. John's and Halifax, could, as before, fit under what was essentially a British imperial command structure. This was certainly not true of the U.S. Navy when it entered the conflict in December 1941. Admiral King was known for his keen sensitivities toward the British, and while it might be thought that the United States had so much to do in the gigantic

Pacific War that it might find it agreeable simply to leave some of its warships to operate under an Anglo-Canadian command structure in the Atlantic, this did not happen easily. In any case, for much of 1942 the greatest U-boat challenge had occurred in America's own waters, off its eastern seaboard, and further south, in the Caribbean routes; the U.S. Navy clearly had to be centrally involved here. So, for all those Naval War College lectures about the benefits of having an integrated "command of the sea," the Allied navies had decided to settle for three zones with identified handover points.

This position improved significantly in early March 1943 following the Atlantic Convoy Conference in Washington. What could have been a serious crisis among the Allies—King for a while wanted to withdraw *all* U.S. warships from the North Atlantic in order to protect the military supply routes to American forces fighting in Tunisia—ended in a sensible compromise. The U.S. Navy would have primary responsibility for the convoys to Gibraltar and North Africa and would also protect all Caribbean convoys, while the British and Canadian navies assumed responsibility for the major routes to the United Kingdom. More important still was that King agreed to lend some naval forces (including a new escort carrier) to the North Atlantic theater, and did not oppose increasing the numbers of squadrons of very long-range B-24 aircraft to RAF Coastal Command and to the fast-growing but overstretched Royal Canadian Air Force. The last additions, as we shall see, came just in time.

The necessarily important factor of intelligence and counterintelligence fitted well into these larger command structures. Older forms of gaining information about the enemy's forces and possible intentions still operated in this war, and the British in particular used aerial reconnaissance, reports from their agents and anti-German resistance movements, and technical analysis of captured weapons systems to add to their miscellany of acquired knowledge. Both sides also developed some very sophisticated bureaus of operational research, whose analysts studied runs of data to figure out how best to utilize one's own resources and dilute the foe's. But it was in the 1939–45 struggles that signals intelligence, or "sigint," took a major lead over human intelligence, or "humint," in the great game of knowing one's enemy. Nowhere did this seem more important than in the Battle of the Atlantic. For the U-boats to know where the convoys were, or for the Allied navies to glean the disposition of the U-boats, made an utterly critical difference. Small wonder that the code breakers at Bletchley Park included a substantial naval intelligence section reporting directly to the Admiralty, or that Doenitz relied so heavily upon his invaluable B-Dienst.

Nonetheless, the warding off of a submarine attack and the destruction of the attackers had to be done through technology, that is, by defensive and offensive weapons platforms. It was true, obviously, in all theaters of war and at all times, but it is astonishing how much the exigencies of total war, and the terrible importance of winning the Battle of the Atlantic to both sides, led in the single year of 1943 to a staggering increase in the number of new ways of detecting an enemy and of deploying new weapons to kill him. This conflict was, more than any other battle for the seas, a scientists' war.

But the acquisition of newer technologies to detect and beat off, or to pursue and destroy, called in turn for their most efficient application—for significant improvements in tactics and training, both by individual submarines, surface escorts, and aircraft and (especially) by groups of them working together. Here the U-boats had an early advantage. They had all-volunteer crews, with some of their commanders remarkably young yet very capable, and saw themselves as an elite branch. They had a single operational task—to sink, and keep sinking, Allied merchantmen, and then avoid being sunk themselves. For a long time they enjoyed the tactical benefit of Doenitz's switching their mode of attack to nighttime surface encounters. They also possessed a very robust wireless communications system, so if one U-boat spotted a convoy, the other members of the wolf pack would very swiftly know about it and adjust their locations accordingly. Finally, the targets in question were usually very slow-moving and thus offered repeated chances for attack, so even if the submarines waited until the convoys were in the mid-Atlantic air gap, they still had lots of time.

Against such a strong hand the Allies had initially very few trumps. What limited number of aircraft carriers the Royal Navy possessed (often as few as only three or four) had to be deployed for aerial cover of the battle fleets and the Mediterranean convoys. Those same high-profile tasks also consumed the energies of the flotillas of the speedy fleet destroyers. The convoys were thus protected by a tiny group of smaller, slower, and often nearly obsolete craft, lacking aerial protection in the middle stretches, most of them lacking detection equipment, and armed with what were essentially weapons of the First World War. A U-boat could actually outrun most of the early Allied escorts, at least on the surface, if its commander was willing to accept the risk of being spotted—though of course neither hunter nor hunted could go at full speed in the massive Atlantic storms. Much-improved equipment was promised, and some was in the pipeline, but could it be gotten to Liverpool and Halifax and the air squadrons in time?

Two other inestimably important factors at play in battles of such strain, anxiety, and loss as the convoy shoot-outs were leadership and morale. As we shall see below, the element of morale did turn significantly to one side's favor later in the Atlantic campaign, after mid-1943, but on the whole the bodies of combatants were fairly evenly matched in this dimension. Doenitz and Horton were worthy opponents, and the latter received a great reinforcement when Air Marshal Sir John Slessor took over as C in C of RAF Coastal Command in February 1943: Slessor was dedicated to defeating the U-boats and a firm advocate of air-sea cooperation even if it involved frequent fights with Harris at Bomber Command over the allocation of planes. Doenitz had an extremely competent deputy, Rear Admiral Godt, for the day-to-day management of the U-boats, though very few middle-ranking staff. An enormous responsibility was thus placed upon the U-boat commanders themselves, many of whom became legends, not unlike the First World War fighter aces, with an instinct for both killing and surviving despite the very high loss rate—until the strain became too great. As we shall see, there was little evidence that Allied morale, whether of captains or crews, Royal Navy personnel or merchant sailors, ever sagged even when losses and general conditions were at their worst. Moreover, the Western Allies simply possessed far more trained officers of the captain and commander rank, not to mention naval reserve officers who could be thrown into any breach.

The remaining factor has to be that of relative force strengths and endurance. It was not just the measure of crews' physical and mental stamina during a fourteen-day convoy struggle; it was also a matter of sustaining each side's campaign through reinforcement, sending fresh numbers to replace those lost, and steadily building up the fighting punch of one's service. This was total, industrialized war, measured most clearly by the flows of new U-boats vis-à-vis Allied warships and aircraft, of merchant vessels, and of fresh crews. Here again, one might have assumed that by early 1943 the odds were tilting in Doenitz's favor; certainly the steady buildup in U-boat numbers pointed to that conclusion. Moreover, while German shipbuilding production could concentrate ever more narrowly on submarines and lighter attack craft (E-boats), British yards were being stretched to produce light fleet carriers for future Far East operations, newer classes of cruisers and destroyers, landing gear, and the Royal Navy's own submarines. Had it not been for the stupendous gearing up of American war industries as 1942 unfolded into 1943, this could have been a very one-sided production battle. In any event, at the time of Casablanca no one on the Allied side was very cheery about how the naval odds looked in the Atlantic.

The Battle at Sea, and the U-boats' Triumph

As Hitler's attack upon Poland in September 1939 led to the Anglo-French declarations of war, the strategic situation in the Atlantic basin and across western Europe was eerily similar to that of a quarter century earlier, when the Entente Cordiale and its empires had gone to war in response to Germany's invasion of Belgium. A small British Expeditionary Force (BEF) once again crossed the Channel to stand with the French armies. The other countries of Europe remained neutral, as did the United States, due to congressional fiat. Most of the British dominions (that is, Australia, Canada, Newfoundland, New Zealand, and South Africa, but not, unhappily, de Valera's Eire) joined the struggle, as did the dependent parts of the Anglo-French empires. The Royal Navy assembled its surface fleets at Scapa Flow and Dover to close the two egresses from the North Sea, except of course to a small number of German commerce raiders already in the wider oceans. The broad impression that history was indeed repeating itself was fully captured, symbolically and physically, by Churchill's return to the revered cabinet position of First Lord of the Admiralty, the position he had occupied in 1914. "Winston is back!" went out the message to the fleet.

The strategic position at sea could not have been worse for the German navy, headed by Raeder. His service did indeed have plans (the famous Z Plan) for a massive transoceanic battle fleet, with giant battleships, aircraft carriers, and all, but even Nazi Germany's formidable production capacities could not produce such a force—or even a quarter of it—by 1939. Another four or five years at least were needed, and Raeder had believed the Fuehrer would keep out of a major war for that long. He, like a lot of Wehrmacht generals, was badly mistaken. The German navy thus went to war with a force completely inadequate to match the Allied navies, with a service starved of the resources allocated to the German army and the Luftwaffe. Even its U-boat arm was weak, small in numbers, short in range, and forced to go all the way around northern Scotland to reach the broad Atlantic. The overall odds looked hopeless.

Those odds changed, in the most dramatic ways possible, during May and June 1940. The collapse of France and Belgium, and the escape of the battered BEF through Dunkirk, meant there was no longer a Western Front. Worse still, the Luftwaffe could now operate against England out of forward bases in Pas-de-Calais, while the German navy could steam in and out of Brest and the Gironde. To compound these disasters, there was the staggeringly fast Nazi takeover of the Netherlands, Denmark, and Norway, with all the consequent strategic and geopolitical implications. Now all the waters beyond the Channel and North Sea were open to German surface ships and submarines. The odds tilted further when Mussolini, cagily neutral in September 1939, opportunistically declared war on the British Empire and a falling France on June 10, 1940. An entire new navy, including one of the world's largest fleets of submarines, entered the war on Berlin's side, just as most French warships were abandoning hostilities and anchoring in Toulon and the North African ports.

The result was that—after the Battle of Britain and the survival of the island nation itself in 1940—the Battle of the Atlantic became the center of the western struggle. Once the immediate German invasion threat diminished, the various British countermoves—such as the maritime relief of Gibraltar, Malta, and Cairo, the preservation of the Cape routes to the East, the military buildups (including dominion and empire forces) in Egypt, Iraq, and India, and the development of the early strategic bombing offensive against the Third Reich—were, while critically important, impossible to sustain unless a constant flow of food-stuffs, fuel, and munitions reached the home islands from across the seas and new British divisions and weaponry were carried from the home islands to Africa and India. That simple strategic fact did not change when Germany attacked the USSR in June 1941, or when Ja-pan's attack on Pearl Harbor in December 1941 and Hitler's reckless declaration of war against the United States turned a European war into a global conflict. Indeed, the importance of winning the Battle of the Atlantic was only reinforced by the Anglo-American decision to build up a force of millions of men in the United Kingdom for the future invasion of western Europe.

Thus the struggle to defend the sea-lanes simply to preserve the British Isles now became a gigantic fight by the Allied navies as the first step in ensuring Germany and Italy's unconditional surrender. Fortu-nately for the British, the German surface threat could never reach full fruition (just as Raeder had warned). The sinking of the giant battle-ship *Bismarck* in May 1941 eliminated the greatest single danger, and the "Channel Dash" of the other German heavy ships from Brest back to Germany in February 1942, though highly embarrassing to the Royal Navy's pride, put those warships once again into constricted wa-ters, to be continually screened by the Home Fleet at Scapa and bombed repeatedly by the Royal Air Force. Futile and rather halfhearted sallies against Arctic convoys offered no challenge to the Allies' command of the sea. Only the submarines could do that.

And that they did very well. As the number of U-boats available to Doenitz rose steadily during 1942, their crews also grew in experience, their detection equipment became more reliable, their range was in-creased by the introduction of "milk cow" refueling subs, and they were masterfully coordinated by Doenitz. America's entry into the conflict gave them fabulous opportunities against a new enemy and his mer-chant fleet, almost completely unprepared for this type of warfare, all the way along the still floodlit eastern coast. Longer-range submarines were sent farther afield to pick off rich targets near Sierra Leone, in the eastern Caribbean, off Buenos Aires, and off the Cape. But Doenitz never stopped hammering away on the crucial North Atlantic routes.

Total Allied shipping losses had jumped from about 750,000 dead-weight tons in 1939 to an awful 3.9 million tons in the chaotic year of

1940, increased in 1941 to 4.3 million tons, and then soared again in 1942, as we have seen, to a colossal 7.8 million tons. Of course, there were heavy losses in other regions—off Dunkirk, in the South Atlantic, in the Mediterranean, and by 1941–42 in the Far East—but the heaviest losses (for example, 5.4 million tons of the 1942 grand total) occurred in the North Atlantic. By comparison, Doenitz's U-boat losses were moderate in those years: around twelve in 1939 and around thirty-five in 1941, increasing to eighty-seven in 1942. These U-boat losses were completely sustainable; the merchant ship losses—and equally the losses of their experienced crews—were much less so.[6]

It is therefore not surprising that the shipping losses of March 1943 scared Churchill and the Admiralty. If Doenitz's wolf packs could inflict that much damage during dark and stormy conditions—the main attacks ceased only after March 20, when already rough waters were joined by what was essentially a massive Atlantic hurricane—the planners worried how their Allied convoys could survive in lighter and calmer times, with the moon shining across the waters. Would the loss rate double again by May and June? And were the submarines becoming harder and harder to trace and to sink? The jubilant U-boat crews and their determined admiral must have hoped so.

And yet the one-sided results of the March battles were never repeated again. In fact, they turned out to be the high point of the submarine offensive against Allied shipping, a momentary peak that then fell away so precipitously that each side was stunned by the transformation. It is hard to think of any other change in the fortunes of war that was both so swift and so decisive in its longer-term implications.[7]

Precisely because it was so, it is important to take a closer look at the epic convoy battles of March to May 1943, when the balance of advantage swung so decisively from U-boat triumph to U-boat disaster. Fortunately, the sources for this story are excellent, down to the hour-by-hour tracking of virtually every submarine's movements and the turn of every convoy.[8]

The month of March began badly for the Allies. While the American, British, and Canadian naval authorities were at their Atlantic Convoy Conference, hammering out decisions regarding zones of control, reinforcements, and the rest, a confident Doenitz was dispatching more and more submarines to join each of the four large wolf packs he had established in the central Atlantic, usually two in the center and one each on the northern and southern flanks. Moreover, at this stage in the intelligence/decryption conflict the Germans very much possessed the upper hand; B-Dienst was providing its chief with extraordinarily complete descriptions of the time and course of the Allied convoys, sometimes even before they left harbor. By contrast, the code breakers at Bletchley Park and at the Admiralty were having difficulty reading German messages days after they were sent. In sum, the shepherds, though gallant, were more than normally disadvantaged, weaker than usual, groping in the dark. The wolf packs were ready to pounce.

Thus their slaughter of Convoy SC 121, which sailed from New York to various British ports on March 5, 1943. Despite the weather being perfectly foul, some U-boats not picking up signals, and a late rush of a few Allied escort reinforcements, the odds were overwhelmingly in Doenitz's favor. The great expert on these March convoy battles, Juergen Rohwer, provides us with a meticulous order of battle:

> The SC.121 . . . consisted originally of 59 ships and . . . was escorted by the Escort Group A 3 under Capt Heineman, USN, with the US coastguard cutter *Spencer,* the US destroyer *Greer,* the Canadian corvettes *Rosthern* and *Trillium* and the British corvette *Dianthus.* The Commander U-boats [i.e., Doenitz] deployed against this convoy the *Westmark* group comprising U 405, U 409, U 591, U230, U 228, U 566, U 616, U 448, U 526, U 634, U 527, U 659, U 523, U 709, U 359, U 332 and U 432. At the same time he ordered U 229, U 665, U 641, U 447, U 190, U 439, U 530, U 618 and U 642 . . . to form another patrol line, *Ostmark,* on the suspected convoy route.[9]

So there were fifty-nine vulnerable and slow merchantmen, with initially only five escorts, against twenty-six U-boats, and with no air cover for the convoy until the third day of the fight—and what was air cover anyway if the submarines attacked chiefly at night? The result was an ordeal from March 7 until March 10, when Doenitz called off his boats. Thirteen merchant ships totaling 62,000 tons had been sunk, but not a single submarine had been lost. It was perhaps the most disproportionate, one-sided battle of the entire war—and was deeply satisfying to Hitler, to whom Doenitz regularly reported.

There were, however, another couple of early March convoys across the Atlantic that also command attention. Convoy ON 170, for example, was ably directed away from all of these deadly mid-Atlantic battlefields and thus steamed across the northern waters without a loss and without (so far as we can tell) an encounter with a U-boat. Here was the case for shepherds and sheep simply taking the high Alpine passes and avoiding the wolf-strewn valleys below. Many an Allied convoy, in fact, survived the crossing unscathed, either because of clever routing or simply because Doenitz had directed all his boats to go after a different one.

A more mixed story is that of Convoy HX 228, which fought its way across the Atlantic between March 7 and 14 in a craziness and confusion that might remind naval historians of Nelson's entanglements with the French fleet at the Battle of the Nile (1798). At one stage in this battle the destroyer HMS *Harvester*, having rammed U 444, got its propeller shaft entangled in the latter's rudder and was only released by the French frigate *Aconit* ramming and sinking the submarine. The *Harvester* was torpedoed the next day, but the *Aconit* promptly sank the submarine that had done so, U 432. At the end of it all, HX 228 lost only four merchantmen plus the destroyer, while the escorts—joined, limpingly, by the first escort carrier, USS *Bogue*, a harbinger of things to come—kept a good account of themselves throughout. The U-boat crews were composed of formidable and intrepid men, but the British, American, and Canadian sailors—a small number of old hands and a vast recruitment of new officers and crew to their navies and merchant marines—showed themselves on this occasion to be equally resourceful.

However suggestive of a possible Allied recovery at sea, these hints were swept away by the achievements of the U-boats against convoys HX 229 and SC 122 between March 16 and 20. This was the most frightening moment, and not just for the fate of those two groups of merchantmen but for the overall convoy strategy as well.

Unlike a classic land battle (between Greeks and Spartans, or Wellington and Napoleon), where each opponent was roughly similar in composition, the two sides' forces in the Atlantic struggle were very different. Doenitz's U-boats were, roughly, all the same; the captain and crew of an older Type VII submarine were no doubt envious of those in the faster, larger, and better-equipped Type IXs—unnecessarily, as it turned out—but all of them could reach out far into the Atlantic, fire their deadly torpedoes, and dive fast, away from counterattack.[10] By contrast, the Allied convoys contained a motley assembly of ancient tramp steamers, ore carriers, oil tankers, mail and passenger ships, and refrigerated ships.*The cargoes they carried were equally heterogeneous— grain, linseed, meat, army supplies, aircraft fuel, sugar, bauxite (for aluminum), steel, tobacco, "African produce" (probably vegetable oil and hardwoods), and everything else needed to keep a nation of forty million people at war. British and American merchant ships were reinforced by boats flying the Panamanian, Norwegian, Greek, Polish, and Dutch flags. One of the unintended consequences of Hitler's aggressions was that the island state's limited fighting resources were boosted by considerable numbers of foreign merchant ships and crews, foreign fighter and bomber pilots, and foreign infantrymen—and Britain was happy to take them all.

The story of convoys HX 229 and SC 122 confirmed that the Royal Navy faced one of the greatest logistical challenges in all of military history. There were thousands of Allied merchant vessels on the high seas at any given time—probably up to twenty convoys, plus hundreds of independently sailing boats. From Trinidad to New Jersey, and from Adelaide to the Cape, the lines stretched out, with most of this produce ultimately destined for the critical North Atlantic passageway. As the maritime war unfolded, the convoys would necessarily become larger and larger, which was no bad thing in itself. During the Casablanca discussions on the shipping crisis, P. M. S. Blackett, the Admiralty's chief of operational research, had impressed listeners with an analysis showing that a convoy of sixty or even ninety ships was a more efficient way of getting goods across the Atlantic than a convoy of thirty; the number of escorts remained roughly the same, limited more by shipbuilding production and other duties (Operation Torch) than

---

*The increasing flood of U.S. and Canadian troops to Britain was transported by an entirely different method—the great liners of Cunard, which, when stripped inside to the bone, could each carry 15,000 GIs at a speed that even a fleet destroyer couldn't keep up with, let alone a U-boat. But, to repeat an earlier question, what would two to three million fresh soldiers do in the United Kingdom if they lacked food, fuel, and munitions?

anything else, and the U-boats only had a limited number of days and hours in which to attack, and a limited number of torpedoes, too. That mathematical analysis reinforced the planners' conviction that the convoy system was the best one to pursue, but it still left a practical problem: how on earth did one get such a large and heterogeneous bunch of merchant ships from one side of the ocean to the other, especially when the Allied warships were themselves such a mixed bag of destroyers, frigates, corvettes, cutters, trawlers, and others?

One response, which went back to almost the beginning of the war, was to make a simple division between "fast convoys" and "slow convoys." Much flowed from this, including the different nomenclature ("SC" was a slow convoy, "HX" a faster one, the latter usually coming out of the great harbor of Halifax, but also from New York itself). These convoys could leave from separate ports and be timed to arrive in the United Kingdom (or, on their return journeys, into East Coast harbors) on different days. Slower escorts such as sloops and armed trawlers were assigned more often to the slower convoys. The schedule of aerial patrols could be arranged accordingly. Allied warships dispatched to join one convoy in midocean might be instructed to help another one if it came under heavier attack. To be sure, and to the fury of every escort commander, all convoys, whether fast and slow, would have their stragglers—how could it not be, with forty, fifty, or sixty oddly assorted ships in a single group? Overall, dispatching large convoys and dividing the merchantmen into fast and slow concentrations made a lot of sense. Yet what would happen if the number of U-boats was simply too great?

Already by March 13, 1943, B-Dienst had evidence that the slow convoy SC 122 (fifty-one assorted vessels, with four or five close escorts) had set off from New York. It was to be followed a few days later by the fast convoy HX 229 (forty-one vessels, with four escorts), from the same port. The latter fact was not clear to the Germans at this time, but the intelligence nevertheless gave Doenitz plenty of time to order his western patrol group to ready for an attack, while also instructing additional U-boat groups to move westward, toward the key zone in the middle of the ocean. An examination of the route charts, losses, and reports of the convoys (there actually was a further fast convoy of twenty-five ships, HX 229A, well to the north, off Greenland, at this time), and especially of Rohwer's reconstruction of the maneuvers of the individual U-boats, leaves the reader with a sense not only of the complexity of this contest but also of its enormous size. Correlli Barnett has it right: "It could be said that for the first time an encounter in submarine warfare attained the scale and decisive character of the great fleet battles of the past."[11] In fact, one would probably have to go back to the grim multiday fights of the mid-seventeenth century between the Dutch and British navies to find a good historical equivalent.

The German attack upon these two convoys seemed to unfold like clockwork, although serendipity played a role, too. Because of engine trouble, U 653 was slowly heading westward to a relay point when it saw the approaching convoy HX 229 on the horizon, steaming toward British harbors. Its captain, the bemused though resourceful Lieutenant Commander Feiler, dived under for a long while as the entire convoy passed over it. When he resurfaced, the ocean was clear; the convoy had steamed on, and U 653 could send the critical message to Doenitz's headquarters, which then took immediate action. Twenty-one submarines responded to that news, a clear testimony to the way that electronic communication was changing the art of war.[12]

The seas were terribly rough, but the wolf packs pressed in, sensing that this was a great opportunity. The clear, moonlit night of March 16–17 truly was a night of the hunter, for which many merchantmen were to pay the price. The attackers were also advantaged by the fact that on that same night the HX 229 escort commanders decided to slow the convoy in order to pick up stragglers and, by doing so, unwittingly bumped into the first cluster of U-boats. Had that decision to slow down not been taken, the submarines "would certainly have passed by to the stern."[13] But once the convoy had been detected, there was little that its escort commander, Lieutenant Commander Luther in the destroyer HMS *Volunteer,* could have done to lead his flock to safety. His direction finder had located two U-boats 20 miles away and closing, so he had dispatched another escort to drive them away. But if one group of U-boats had missed their target, others surely would not. By this time, both the American and British admiralties were picking up U-boat signals from all around the convoy, so alterations in course to avoid one cluster of attackers simply brought HX 229 closer to another.

Thus, around ten o'clock on the moonlit night of March 16, the captain of U 603 found himself watching an entire Allied convoy slowly steaming by, with its remaining four escorts widely separated. By this stage Doenitz had his submarines equipped with the deadly FAT torpedoes, which ran straight at 30 knots over a given course, and then zigzagged in order to counter the opponent's evasive actions. Since the horizon was filled with targets and U 603 was able to come as close as 2 miles (3,000 meters), the zigzagging capacity was probably unnecessary on this occasion. The submarine fired its three front torpedoes, then its rear torpedo, and had the satisfaction of hearing a large detonation before slipping back under the waves; the freighter *Elin K.* sank within four minutes. With the convoy escorts now distracted by picking up the ship's survivors and looking for U 603, the convoy's northern flank was virtually uncovered. This gave Captain Lieutenant Mansek in U 758 the freedom to fire off torpedoes like a cowboy in a Western saloon. "At 2323 hrs Kptlt Mansek fired a FAT torpedo at a freighter of 6000 tons in the starboard column, one minute later a G 7e [torpedo] at a freighter of 7000 tons, at 2325 hrs a FAT torpedo at a tanker of 8000 tons behind her and at 2332 hrs a G 7e at a freighter of 4000 tons": four separate torpedo launches in nine minutes.[14] The Dutch freighter *Zaaland* and its American neighbor *James Oglethorpe* were hit, the first settling in the sea and the second, controlling an onboard fire, temporarily limping on (it would be finished off by U 91 later that night).

The Allied merchant ships were, in effect, running a gauntlet, and even those steaming in the inner columns were likely to be hit if a torpedo passed between the outer lines. What did this mean more generally? There were the two lost merchant ships themselves, of course, and thus two fewer in the limited tally of hulls that could go back and forth across the Atlantic. And then there were the crews, although many of them were picked up by the destroyer HMS *Beverly* and the corvette HMS *Pennywort*—at one stage in this chaotic night, incidentally, there was only *one* escort with the main convoy for a while. But perhaps the real point to notice was that the *Zaaland* was carrying a cargo of frozen wheat, textiles, and zinc, and the *James Oglethorpe* was transporting steel, cotton, and food in its hull, and aircraft, tractors, and trucks for the U.S. Army on its deck.

It is difficult, probably impossible, sixty-five years later to enter the mental world of the commanders of those four naval escorts, who were looking after such an immense responsibility (they were about to be joined that day by a fifth, HMS *Mansfield*). They did not occupy the mental space of, say, Alanbrooke, experiencing the larger worries and frustrations that filled his diaries at Casablanca and in these later desperate months.[15] Nor did they occupy the world of the common soldier, sailor, and airman, almost all of whom had been culled from civilian life into a new existence of danger, hardship, and terror in this seemingly everlasting war. The convoy escort commanders operated at the middle level—just like the German U-boat commanders—and had enormous obligations to fulfill, setbacks to deal with, and losses to swallow.* Yet it was upon this middle level that the fate of the war now depended.

At half past midnight on March 17, U 435 put a torpedo into the American freighter *William Eustis*, which stopped immediately and began to list. The convoy commander, Lieutenant Commander Luther, having just returned to the main body and aware that his fellow escorts were picking up survivors several miles behind, swept his destroyer, HMS *Volunteer*, to the rear of the convoy. There he found the wrecked *William Eustis*, with its lifeboats launched but with many of the crew swimming in the water in different directions, crying for help. This was one of those occasions when there were no obvious, good solutions— only bad ones and even worse ones. Rejecting the pleas of the master and chief engineer of the merchant ship (who had scrambled on board the destroyer) that a rescue attempt be made at dawn, Luther took on as many survivors as he could find and then, fearing a U-boat team might board the beleaguered vessels and seize vital codes and papers, he depth-charged the *William Eustis* and steamed back to the convoy.

It was still only 2:50 in the morning, and the overworked destroyer had just caught up once more with the convoy when it saw the distant

---

*One returns, then, with wonderment and humility to that classic black-and-white movie *The Cruel Sea* (based on a book written by the great novelist Nicholas Montserrat, with a screenplay by the American crime writer Eric Ambler), in which the actor Jack Hawkins portrays a convoy commander, in his little *Flower*-class corvette, enduring the attacks upon convoy after convoy across the Atlantic, and watching so many merchant ships blow up and their crews drown without being able to do much about it. Montserrat captained Atlantic escorts himself.

blast of another freighter going down. Two hours later Captain Lieutenant Zurmuehlen of U 600 achieved one of the most accomplished torpedo attacks of the Second World War, firing a salvo of four FAT torpedoes from the bow, and then one from the rear tube, at the totally unprotected starboard flank of the convoy. Within minutes the British freighter *Irena* and the whaling depot ship *Southern Princess* had each been hit by one torpedo, and the American freighter *Irene du Pont* by two. HMS *Mansfield* rescued survivors. As dawn broke, all that really could be done was finish off the abandoned, wrecked merchant ships. The corvette HMS *Anemone* did half that job, while U 91, arriving on the scene an hour or two later, completed the messy task. The British escorts were now full to overflowing with rescued crewmen.

Farther east, the convoy SC 122 had begun its own encounter with the furies. The fifty-one vessels enjoyed the protection of seven escorts, including the American destroyer USS *Upshur* and a specially equipped rescue ship, the *Zamalek*. At around 2:00 a.m. on this same dreadful night U 338 was racing westward to join the attack upon HX 229 when its commander, Captain Lieutenant Kinzel, saw a mass of ships on the horizon, a full 12 miles away; a hunter's moon, indeed. Kinzel's first two torpedoes mortally damaged the British freighters *Kingsbury* and *King Gruffydd*. The second salvo (also two torpedoes) tore into the Dutch four-masted freighter *Alderamin* and broke it into three pieces; it was gone in two minutes. In turning away from this mayhem, U 338 fired her stern torpedo at another nearby ship. The torpedo missed its intended victim but drove on through the middle of the convoy to the far end, where it blew a great hole in the side of the British merchantman *Fort Cedar Lake*. Five torpedoes, four ships.

As the next morning unfolded, the Allied naval authorities became aware that they were facing a double disaster: the gutting of two of their vital supply convoys plus the prospect that Doenitz was dispatching more and more U-boats into this great contest. The two convoys had not yet joined and would not do so until the night of March 17–18, but the overall picture was obvious. A great quantity of Allied merchantmen, seventy or eighty or more, and their cargoes were caught in the middle of the Atlantic, with a completely insufficient number of escorts, and with the enemy submarines massing for the coup de grâce.

But with the dawn came light, and with the light came aircraft.

And there was nothing that terrified a U-boat commander so much as the sight, or even the noise, of an approaching Allied plane. It really was the rock-paper-scissors game. Merchant ships were horribly vulnerable to U-boats and their torpedoes, but U-boats, even those that decided to stay on the surface and fight, were seriously outgunned against aircraft. On many occasions, the submarine hardly had more than a minute or two to fire, because the aircraft came in so fast; the only thing the sub's commander could do was dive, dive, dive . . . and wait for the depth charges. Catalinas, Liberators, Sunderlands, and Wellingtons were to them the scariest of the Allied weapons system, the Ringwraiths of the sky. This, after all, is why the U-boats waited to attack Allied merchant vessels in the mid-Atlantic air gap.

On March 17, 1943, that gap began to be closed. At the urgings of Western Approaches Command, the first few very long-range (VLR) B-24 Liberators were flown out of their base near Londonderry, Northern Ireland. For a number of reasons, all of RAF Coastal Command's squadrons (in Northern Ireland, Iceland, the Hebrides, and Scotland) were at this time significantly understrength, but the Liberators could at least reach the closest convoy, SC 122. During the course of the day these aircraft sighted U-boats on numerous occasions and made six attacks on them before they had to return to base. No submarines were destroyed or even damaged, but they were repeatedly forced to dive. In fact, during that whole day the only ship in the convoy to be sunk (again, by the indomitable Kinzel, in U 338) was the Panamanian freighter *Granville,* carrying supplies for the U.S. Army.

By contrast, convoy HX 229 was not within the reach of aerial protection, and the few escorts actually with the main body of the convoy, or not catching up after rescuing survivors or chasing a sonar trace, were completely overworked. This allowed the attackers to destroy the Dutch *Terkoelei* and the British *Coracero.* As nightfall came, and with fine visibility, the U-boats moved in on the convoy once more. But then a B-24 appeared and in fairly quick succession sighted and depth-charged first a pair of submarines and then another group of three, and then—completely out of depth charges—machine-gunned a sixth. None of the subs was damaged, but all were unnerved. The Liberator had stayed eighteen hours in the air, two hours longer than the normal recommended time. However, quietness was not given to SC 122, which lost another two valuable ships, the *Zouave* (to a direct hit from U 305) and the *Port Auckland* (wrecked by the same submarine but probably finished off by Kinzel).

These daylight sinkings of two more merchantmen had sent HMS *Volunteer* racing to the starboard of the convoy, but its own direction finders had become defective, and the other few escorts were picking up survivors. One guesses that Luther had either great courage, great obstinacy, or both, but it was only with the sinking of the *Terkoelei* and *Coracero* that the convoy commander "thought the time had come to ask for help." (One of the wonderful side benefits of reading the messages of these escort commanders, or a single ship's captain, is the chance to marvel at their tendency to understatement.) It was at this point that one of the faster American merchantmen, the *Mathew Luckenbach,* decided to break free from the convoy and steam ahead, despite repeated messages from the escorts and the convoy commodore; she was sunk, with few survivors, two days later.

Even before Luther's message for help, both admiralties decided to commit more resources to the battle. This was much easier for Doenitz, since he had fresh U-boats already in the Atlantic. The British planners faced a much tougher task, simply because there were a lot of other valuable convoys in the North Atlantic at exactly the same time. The two Gibraltar convoys KMS 11 (sixty-two merchantmen, with nine escorts) and KMF 11 (nine transports, with two escorts plus destroyers pulled from other duties) would probably have to fight their way across the Bay of Biscay, where they might face not only another cluster of U-boats but also the possibility of long-range German bombers; none of their escorts could be spared. Convoy ON 172 (seventeen merchantmen, with six escorts), south of Cape Farewell, was not only too far away but also in an area where another U-boat group was forming. And convoy ON 173 (thirty-nine ships, with six escorts), steaming far to the north of the battle, was very weak itself. Reinforcing corvettes dispatched from St. John's were not fast enough to catch up until later, and the two U.S. destroyers ordered from Iceland were damaged by the heavy seas on March 17 and 18.

The same was true of Allied aircraft, so critically important and yet so few. Also, possessing at this time poor ship-to-plane contact, the Liberators, Sunderlands, and the rest often failed to link up with the convoys. So further Allied shipping losses occurred on March 18. One of them, with brutal irony, was the *Walter Q. Gresham,* among the very first of the new American-built Liberty ships whose mass production would eventually solve the shipping shortages; the second was another modern vessel, the Canadian *Star.* By the time the vastly overworked corvette HMS *Anemone* had picked up survivors from these two ships to add to her earlier rescues, she had 163 exhausted civilians on board, including six women and two children. Then she rejoined station and spent all the following night driving away U-boat attacks on HX 229. Sometimes the asdic worked, sometimes not; the Hedgehog grenades failed twice; and while the depth charges worked, by the time they had reached their preset depth the submarines had slipped away. A multitude of survivors slept on the *Anemone* that night; Convoy SC 122 was luckier in that it possessed a proper recovery ship.

By March 19 the great plot map of the North Atlantic that covered an entire wall of the Admiralty's control room presented an extraordinary scene. Signals from enemy U-boats were being picked up from virtually every quadrant of the ocean. The Gibraltar convoys were coming under heavy attack from submarines and aircraft as they passed the northwest corner of Spain. To the far north HX 229A had avoided the U-boats but crashed into a sea of icebergs. One extremely large Danish whaling ship (the *Svend Foyn*), carrying numerous civilian passengers, was settling in the icy seas, not too far from where the *Titanic* had gone down. The only consolation was that HX 229/SC 122 had at last reached the 600-mile range limit from Allied air bases in Northern Ireland and Iceland, and the battered crews of the merchantmen and their surface escorts could watch Liberators, Fortresses, Sunderlands, and Catalinas guiding the warships to the U-boats and joining in the attack. Some of the U-boat captains still persisted, however, including Kinzel, whose vessel set course for home only after it had been badly damaged by aerial depth charges. Two hundred miles to his rear, the captain of the less experienced U 384 was caught on the surface by a low-flying Sunderland and blown to pieces early on March 20. This was the only submarine lost in the entire, prolonged battle. By evening Doenitz had recalled his patrols.

By the following day the convoy commanders were starting to send some of the escort vessels ahead to safe harbor. The first ones, properly enough, were warships such as the *Anemone, Pennywort,* and *Volunteer,* carrying their hundreds of civilian survivors. All on board had been pushed to the extremes of human endurance. Forty-two of the sixty ships in SC 122 reached their destinations, as did twenty-seven of the forty ships of HX 122. The commodore (i.e., chief of the merchantmen) of the slower convoy laconically reported that it had been a normal Atlantic voyage "apart from" the U-boats.

Nevertheless, though the heroism and fortitude displayed by the Allied crews had been of the highest order throughout, the fact remained that their navies had taken a tremendous beating. The Battle of the Atlantic was being lost.

The Allies' Many Weaknesses

Why did the battle of the convoys and escorts versus the U-boats go so badly for the Allies in these critical weeks? It will be clear to the reader that this field of war was so complex that no single factor was decisive. Indeed, some historians of this great campaign, including such a central participant in the convoy battles as Sir Peter Gretton, feel uneasy at suggesting that any single factor could be described as the major reason for its eventual outcome.[16] To this author, however, some causes clearly justify a greater emphasis than others.

To begin with, there was the sheer imbalance in numbers in early 1943. No doubt the British, Canadian, and American navies had many other very pressing calls upon their stretched naval resources in these weeks, but if this really was the critical theater for the Western Allies in 1943, then to allocate to a North Atlantic convoy of fifty slow merchantmen an escort of only four or five warships, and at a time when it was known that more and more U-boats were becoming operational, was taking a great risk. As one examines the charts of the ships' dispositions during the hours and days of attacks upon the convoys, it is clear that an escort commander such as Lieutenant Commander Luther faced a totally impossible task—how on earth could his ship possibly try to drive away submarines bunching together on one flank of the convoy, pick up survivors from sunk or sinking merchantmen in the receding waters, and protect the rest of his flock? In the rock-paper-scissors game, all three sides are equally strong and weak. This was simply not true of the battle between U-boat, convoy, and escort that was played with HX 229 and SC 122. Numbers were decisive. There were far too few escorts, and a late rush of aerial and surface support was indeed too late.

Then there was the imbalance in intelligence, for this was probably *the* time when B-Dienst was at the height of its competence, reading Admiralty messages at an impressive rate. Of course, individual submarines missed messages intended for them and steamed in the wrong direction, though sometimes with a favorable, unexpected sinking to follow. But, taken all together, those boats were being directed by Doenitz in a rigorous, demanding fashion—it was not just that he was in charge of one orchestra, but rather that he was directing, hustling, energizing *four* orchestras, four separate U-boat groups, sometimes more than a thousand miles from each other. By contrast, at this particular time Allied intelligence about German intentions seems to have been horribly behindhand. Even when Enigma decrypts came through to Western Approaches Command, they were likely to be late—and such delays came at a high price in the North Atlantic.

Another important element, as noted above, was that of air cover, or, rather, the lack of it. If anything confirmed the geopolitical significance of the mid-Atlantic air gap, it was the *location* of the losses of Allied merchantmen in the awful months when 1942 unfolded into 1943. Of course, U-boats were sinking Allied cargoes off Trinidad, Buenos Aires, and the Cape; the Gibraltar route, being so close to German naval and aerial bases in western France, was always under attack. But when one turns to the vital sea line of communications between North America and Britain, the evidence is overwhelming. Captain Roskill's official history gives us a map of the sinkings of the merchantmen: every one was in the air gap.[17]

This was the place where the three-way game with the convoy, the surface escorts, and the U-boats was played out. Even a weakly escorted convoy with some daylight aerial coverage, such as SC 122, had a better

THE NORTH ATLANTIC AIR GAP AND CONVOYS
During the first half of the war, almost all Allied merchantmen, whether in convoy or not, were sunk by U-boats in this mid-Atlantic gap; it was not to be covered until the coming of very long-range Allied aircraft in mid-1943.

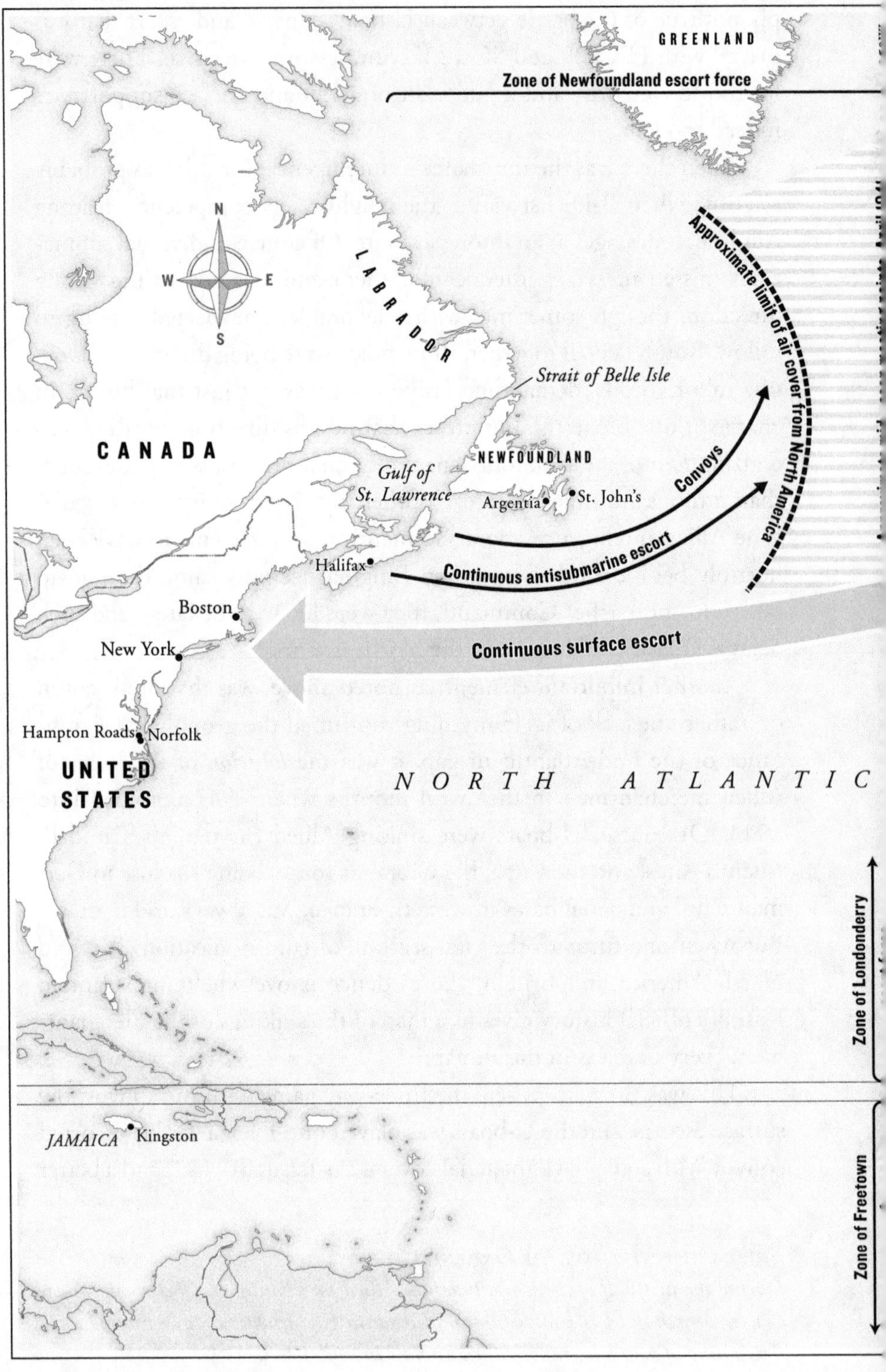

GREENLAND

Zone of Newfoundland escort force

Approximate limit of air cover from North America

N
W   E
S

L A B R A D O R

Strait of Belle Isle

CANADA

Gulf of
St. Lawrence

NEWFOUNDLAND

Argentia   St. John's

Convoys

Continuous antisubmarine escort

Halifax

Boston

New York

Continuous surface escort

Hampton Roads   Norfolk

UNITED
STATES

N O R T H   A T L A N T I C

Zone of Londonderry

JAMAICA   Kingston

Zone of Freetown

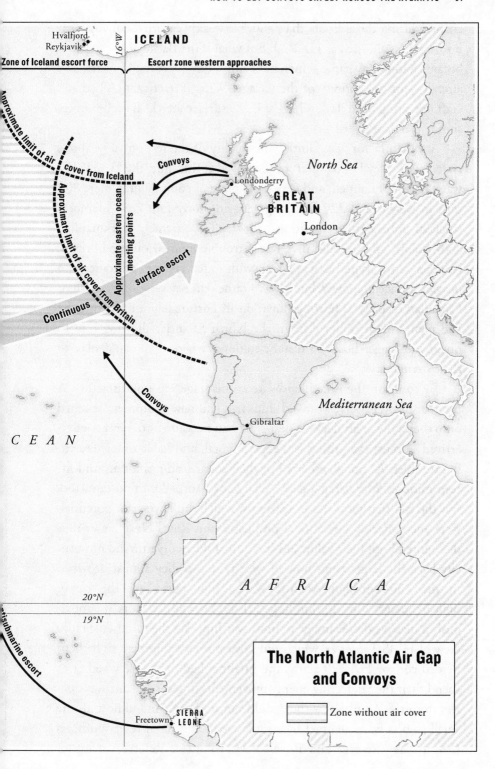

ICELAND

Hvalfjord
Reykjavik

19°W.

Zone of Iceland escort force

Escort zone western approaches

Approximate limit of air

cover from Iceland

Approximate limit of air cover from Britain

Approximate eastern ocean
meeting points

Convoys

Londonderry

North Sea

GREAT
BRITAIN

London

Continuous

surface escort

Convoys

Gibraltar

Mediterranean Sea

CEAN

AFRICA

20°N

19°N

submarine escort

Freetown

SIERRA
LEONE

**The North Atlantic Air Gap
and Convoys**

Zone without air cover

chance during those hours than another weakly escorted convoy with no aircraft at all, such as HX 229. But what if the Battle of the Atlantic became a four-way game—merchant ships, naval escorts, U-boats, *and* aircraft—for the entirety of the voyage? What if, tactically, Allied aircraft operating both day and night made surface attacks by submarines simply too dangerous?

A fourth factor was, simply, the quality of the respective weapons systems. The key U-boat that was deployed at this time, the Type VIIC, was a narrow, cramped, and very basic piece of equipment, a mere 800 tons in weight and 220 feet long, carrying forty-four crew members stuck in unbelievable conditions. It was also extremely workmanlike and reliable, and it could dive fast if trouble loomed on the horizon. Another series, Type IX, was larger, with much greater endurance and heavier surface firepower, but it was a rather clumsy vessel if a British or Canadian escort was bearing down on it. Fortunately for Doenitz, he had plenty of the older class in the North Atlantic, allowing him to divert the bigger boats to distant stations, where they could play to their strengths.

By contrast, the Allied hardware at this stage was unsatisfactory. A number of twenty-five-year-old ships had had new equipment stitched onto their bow or aft decks, a desperate half measure until newer vessels arrived. Almost everything was experimental, and liable to have teething troubles. As noted, escorts lost their sonar readings for around fifteen minutes after firing a depth charge or a torpedo. Important, too, was the fact that most newer weapons, which did have great potential, were encountering operational problems, and Royal Navy crews thus discounted both their value and their use. Older corvettes did not usually have the larger front bow of later types, so they almost drowned in the 150-foot-high Atlantic waves. Not possessing the newer high-frequency direction-finding (HF-DF) radio detectors was obviously a massive disadvantage to most warships. Yet the minuscule radio crew of HMS *Volunteer,* which did possess such equipment as it shepherded convoy HX 229 across the Atlantic, had to choose whether to spend their time detecting enemy movements or receiving anxious queries from the Admiralty—they could not do both. And almost all the small escorts were manned by scratch crews and naval reserve officers (Royal Navy Reserve and Royal Navy Volunteer Reserve), many of them a year or less away from civilian life or university—naturally, the professional naval officers were kept for the fleet destroyers, the cruiser squadrons, the Home Fleet. Britain was not playing its strongest cards, and Canada was struggling, with extreme effort, to create a very large escort navy out of virtually nothing. Its time would come, but not in the spring of 1943.

Other aspects appear less central to our story. The weather, for example, seems to have been sublimely neutral. A distinguished Canadian scholar of the convoy battles, Marc Milner, believes that the defenders had a stronger hand when winter gave way to spring and summer, with improved weather, calmer seas, and many more hours of light each day.[18] But the record itself suggests that if a convoy battle was fought in terribly rough or icy seas, then all the combatants—submarines, escorts, aircraft—were so battered by the storms that they naturally had less time and energy to devote to finding and sinking the enemy. Calmer waters and longer daylight hours did indeed offer warships and, especially, Allied aircraft more opportunity to detect U-boats on the surface, but it also provided submarines with better conditions for spotting a merchant ship's giveaway smoke. Thus, improved weather conditions gave all of the operational elements—visibility, detection, maneuverability, firepower—a better chance of being exploited by either side.

The disadvantages under which the Allied convoys were laboring in early March 1943 were, then, pretty overwhelming: inadequate naval protection, poor intelligence, nonexistent or minimal air cover (and no cover at night), together with a lot of poor, faulty, and outdated equipment. In these circumstances, it seems surprising that the shipping losses were not even greater. After all, HX 229 and SC 122 were scarcely in a better position as they steamed through the Atlantic air gap than the tragic and infamous Arctic convoy PQ 17, which had lost no fewer than twenty-three ships to German submarines and aircraft when it was forced to scatter while only halfway to north Russia in July 1942.[19]

The exceptionally pounding oceanic turbulence of this early spring of 1943—one elderly ore carrier simply snapped in half during the storms—had brought the pugilists' match to a brief close. Doenitz used this opportunity to give his boats a breather and bring them back to their French bases, while sending others to the South Atlantic. The British used it to rethink and regroup, even as Admiralty planners struggled to ensure protection for ever more distant convoys, such as those carrying troops and supplies across the Indian Ocean.

But by the beginning of April, as the ice floes diminished, the win-or-lose battle for control of the Atlantic was approaching its climax. By this time, Doenitz possessed a staggering number of operational submarines (around 240, with another 185 in training or refit), so he was capable of sending as many as forty or more U-boats against any particular convoy. The Allied navies and air forces had also rebuilt their fighting power, but perhaps the most noticeable change was in operational policy, in adopting a more aggressive stance, as proposed by the leader of Western Approaches Command, Max Horton, who swiftly gained the backing of his superiors at the Admiralty. The conclusion drawn from the grim experience of the SC 122 and HX 229 convoys being attacked on all sides, as the First Sea Lord, Sir Dudley Pound, put it to the cabinet committee on antisubmarine warfare in late March, was that "we can no longer rely on evading the U-boat packs and, hence, we shall have to fight the convoys through them."[20] In retrospect, this may be the single most important statement made regarding the Battle of the Atlantic. Without perhaps even the First Sea Lord himself fully appreciating it, the decision that the convoys were going to be defended much more vigorously, in a recognizable life-and-death struggle for control of the sea-lanes, gave the Allied side a much clearer focus than before.*

Newer Elements Enter the Fight

Admiral Pound's statement was not just one of those romantic and archaic calls to fortitude in dire times. There certainly would be fortitude in the months to come, but now the convoys would be protected by defending forces that steadily enjoyed vastly improved tools of war, and many more of them, than they had possessed previously. These new resources worked to the Allies' advantage and blunted Doenitz's strat-

---

*This did not mean that the convoys (as opposed to the independent hunter-killer groups) would deliberately seek out the wolf packs. If Admiralty routing gave them a journey free of attack, nothing could be better. But if the U-boats moved in on a convoy, they would be vigorously counterattacked, with improved weaponry.

egy, despite the great increase in the number of U-boats at his disposal. This chapter's narrative is therefore quite different from that concerning the struggle for command of the air over Germany (chapter 2), where the introduction of a single weapons system was manifestly seen to turn the tide, or that concerning the war in the Pacific (chapter 5), where a remarkable succession of breakthroughs—U.S. Marine Corps amphibious warfare, U.S. Navy fast carrier groups, Seabees construction teams, and B-29 bombers—gave America the upper hand. And although the change in fortunes between the German and Allied navies in the Battle of the Atlantic occurred much earlier in the war than did the shift in those campaigns, the Atlantic story is far messier. The improvements did not arrive according to a grand incremental plan from Max Horton's office; rather, they entered the Royal Navy's tool kit episodically, and some of these newer systems took months before they properly fitted in the whole. Yet the commander of a U-boat that had been sent south in late March 1943 to wreak havoc off Freetown would have been completely disconcerted by what he saw when he arrived back at his base in Brest in July.

That U-boat commander would have learned of a large and growing list of Allied improvements. Aerial support from RAF and Royal Canadian Air Force (RCAF) Coastal Command squadrons was rising fast, if unevenly, as additional squadrons arrived. New escort carriers were beginning to appear, some of them joined by recently formed groups of the newer corvettes and frigates. In addition, a couple of support groups consisting of very fast destroyers had been released from the Grand Fleet, due to the total inactivity of the German heavy warships based in Norway. Newer or vastly improved detection *and* killing equipment was at last making its way to the escort vessels. More and more HF-DF masts appeared even on small Allied escorts, and a fabulous 10-centimeter radar had also arrived. Wellington and Catalina bombers with the powerful spotlights known as Leigh Lights made the Bay of Biscay dangerous for U-boats on the surface. The Hedgehog grenades, together with much more powerful and better-set depth charges and aerial homing torpedoes, were all arriving on the scene. And the intelligence battle between B-Dienst and Bletchley Park's code breakers was turning in favor of the Allies. In sum, while the U-boat commander and his crew were enjoying easy kills to the south, things had gone badly wrong in the North Atlantic.

Interestingly, the poor weather in the north had obscured for a while the many improvements on the Allied side. The record storms of early 1943 continued unabated, so the convoys that did sail in late March suffered enormous physical damage; yet the two main ones, HX 230 and SC 123, got through with only one vessel sunk to the U-boats. The same was true of the early April convoys, HX 233 and ONS 3 and 4. The only convoy that met with serious attack at this time was HX 231, whose escorts fought off a whole group of U-boats on April 5 and 6, inflicted much damage, and brought the vast bulk of its cargoes to port. No fewer than twenty-two of the sixty-one vessels were oil tankers, while many of the others were carrying what might be termed "pre-D-Day" supplies—trucks, tanks, aircraft, landing craft, and vast amounts of ammunition.[21] This, of course, was key. Just getting that one huge group of oil tankers alone to Britain staved off the island's resource crisis of early 1943.

And there was to be even better news ahead. The powerful winds and great long rollers of the North Atlantic never abated, but the sound of conflict now did. Amazingly, only fifteen merchantmen were lost in those waters between June and mid-September 1943, and only one of them was in a convoy. As the Allies girded themselves for further advances in the Mediterranean and for a really massive buildup of men and munitions in Britain in preparation for the future invasion of France, and as the demands of the Pacific and Southeast Asia campaigns grew ever greater, their shipping crisis actually intensified. But that crisis was essentially one of supply and demand, eventually solved by further stupendous outputs of American industry, and no longer about the hemorrhaging of ships from the convoy routes between New York/Halifax and the Clyde/Merseyside. To the tens of thousands of crew members of the merchant vessels who for the first time steamed those rough seas without a single attack, this must have seemed incredible, inexplicable, even a bit eerie. More U-boats were being sunk than merchantmen.

One graph captures this dramatic change of fortunes in the Battle of the Atlantic during the months of 1943.

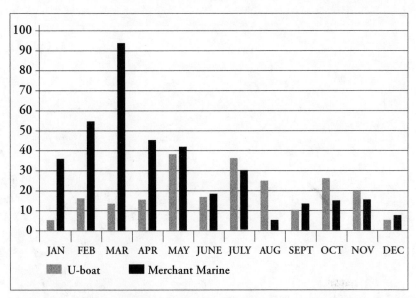

U-boat Merchant Marine

U-boat vs. Merchant Ship Losses in the North Atlantic, 1943
The dramatic rise in U-boat losses and decline in Allied merchant ship losses during the critical months of 1943 are well captured here.

Before we understand the swiftest of the changes of fortune in the five major campaigns of the Second World War, it is appropriate to study the critical convoy battles of May 1943, especially the first, since it was the most important of all. It focused, unusually, around the voyage of one slow, *westward*-bound convoy, ONS 5, which sailed from its gathering place near the Clyde for North American ports between April 23 and May 11. Here was the reprise to the saga of SC 122 and HX 229. This time the Allies won—not easily, but quite decisively. The map on page 44 captures the overall situation.

Forty-three merchant ships, emerging from five separate British ports, had gathered off the Mull of Kintyre with their escorts in late April and then set off on a great-circle route to the New World. They were not holding much cargo, but the point was that if they did not get back to America's eastern ports, there would be fewer and fewer Allied vessels left to carry oil, ores, trucks, grain, and aircraft parts to the British Isles in the future. And it was emphatically an *Allied* convoy: in addition to the twenty-eight vessels flying the Red Duster, there were five ships from the United States, three from Norway, two from the

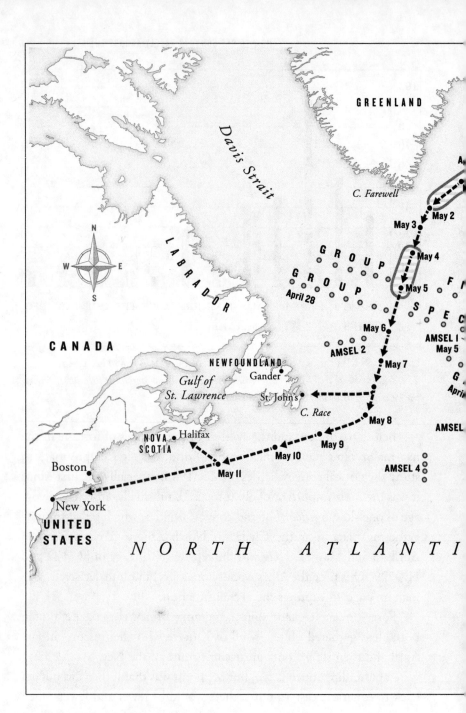

GREENLAND

Davis Strait

C. Farewell

May 3

May 2

May 4

May 5

GROUP

GROUP

April 28

May 6

AMSEL 2

AMSEL 1

May 5

May 7

LABRADOR

CANADA

NEWFOUNDLAND

Gander

*Gulf of
St. Lawrence*

St. John's

May 8

C. Race

Halifax

NOVA
SCOTIA

May 9

May 10

AMSEL 4

Boston

May 11

New York

UNITED
STATES

N O R T H   A T L A N T I

The Fiercest Convoy Battle?

◄┄┄┄┄ ONS 5 route   ▭ Air cover   ○ ○ ○ ○ U-boat wolf packs

THE FIERCEST CONVOY BATTLE?

A graphic illustration of how Allied convoys were sent on the great-circle route, yet still encountered the lines of Doenitz's wolf packs. This convoy is ONS 5, which sailed from the United Kingdom on April 21, 1943, and had to fight its way through successive lines of German U-boats before arriving in Canadian and American ports almost three weeks later.

Netherlands, two from Greece, and one each from Yugoslavia, Panama, and Denmark. Escort Group B7 consisted of two destroyers and a frigate, two rescue trawlers, and four invaluable corvettes named, deliciously, *Sunflower, Snowflake, Pink,* and *Loosestrife.* In charge of the escort was Commander Peter Gretton, whose consuming passion was the sinking of U-boats; he had recently returned from the Mediterranean, where he had received the Distinguished Service Order (DSO) for ramming and demolishing an Italian submarine. Gretton had sailed and fought in the early North Sea convoys, in the Second Battle of Narvik, in the Atlantic and Mediterranean, and with the North African landings. He had just caught his breath after fighting through HX 231, and after this present convoy would soon be out to fight with others. Nelson's frigate captains would have recognized him immediately.

The convoy headed northwestward toward its welcome "cocoon" of Allied air cover from Iceland, which was available to it between April 25 and 27. After that it skirted north of the U-boat group Star toward Cape Farewell and a period of Greenland-based air cover, and then it encountered horrendous weather. So far there had been no sight of enemy submarines, and the only casualties—apart from weather damage to many ships—were the withdrawal of some of the destroyers, including Gretton's, due to a shortage of fuel (refueling at sea during these storms was impossible). It is an indication of how bad the weather had become that the whole convoy was forced to heave to for a while and that the five destroyers of the reinforcing 3rd Support Group took more than a day to find out where it was. Ahead, Doenitz's wolf packs had been instructed to play a waiting game. By the night of May 4–5, two long lines of them (twenty-one U-boats, in the groups Fink and Specht) were waiting as the achingly slow convoy steamed toward Newfoundland. One of the submarines had been attacked and sunk by RCAF flying boats during daylight hours, but at nightfall the U-boats' time came.

What followed was two nights and one day of extraordinary fighting at sea. Attack after attack was launched by the U-boats at the now weakened escort screen; beset by so many predators, the destroyers, frigates, and corvettes could do nothing but race toward a U-boat, drive it away, drop some depth charges, then race back to the convoy. Four ships in the convoy and one straggler were sunk that night. Twenty-six merchantmen were left together, and as dawn came they were once again in an aerial gap between Greenland and Newfoundland. Another four merchantmen were sunk during the day. The second and even larger line of U-boats was waiting its turn. The only bright spot was provided by the little corvette HMS *Pink* and its intrepid lieutenant, Robert Atkinson, who had just returned to the service after eight months of battling severe seasickness. Detached by the escort commander to pick up stragglers, Atkinson soon found himself with a mini-convoy all his own, a mixed bag of three British, one Norwegian, one Greek, and one American merchantmen.[22] Short on fuel, *Pink* steamed on only one boiler, shut down one dynamo, and rationed water—until Atkinson's group was repeatedly attacked by an equally determined U 192 under Oberleutnant Happe and the corvette had to go to full power again. For hours the two vessels fought, losing and regaining contact on numerous occasions, with Atkinson almost running out of depth charges and his new Hedgehog grenades simply not working. After *Pink*'s last depth-charge volley, Atkinson decided that his ship must rejoin the straggler convoy, now at least 10 miles ahead. As the corvette turned, its crew heard a vast explosion in the water behind: U 192 had blown up. It was Happe's first Atlantic command, and his last.

As HMS *Pink* rejoined its mini-convoy, the American merchantman *West Madaket* was fatally hit by a torpedo and started to break up. After driving away any possible further U-boats, the corvette's next task was to pick up survivors—she now had sixty-one extra people on board. By this time Western Approaches Command had actually recognized Atkinson's group as a separate subconvoy and reinforced it with HMS *Sennen,* a valuable former American Coast Guard cutter that drove off further U-boat probings until the Allied ships were rescued by dense fog. On May 9 *Pink* finally escorted her charges into St. John's Harbor, Newfoundland, and Atkinson sat down to write his report.

The greater bulk of Convoy ONS 5 spent the night of May 5–6 in an epic battle, with the U-boats attacking no fewer than twenty-four times and the escorts giving no ground. HMS *Loosestrife* found itself fending off submarine probes throughout the night along the convoy's starboard flank. At two-thirty in the morning it encountered U 638, which from only 500 yards away shot off torpedoes down the warship's side before it submerged. As the *Loosestrife* raced across the submarine's wake, it dropped a salvo of ten depth charges. The explosion was so near and so violent that the men in the corvette's engine room thought their own stern had been blown off. But it hadn't. The greatest cheers came from the seventy-one merchant sailors on board the *Loosestrife* who had just been rescued when their own vessels went down. As Ronald Seth slyly notes, "For the victims of the U-boats, the hunting and killing of one of the enemy had a special relish . . . and if moral encouragement played any part in the sinking of the U-638, then the merchant seamen aboard could claim to have had a hand in it."[23]

A day and a half earlier, Horton had played his next trump card and ordered the 1st Escort Group to join the convoy. Its lead warship, the sloop HMS *Pelican,* soon pounced on U 438, but even before then the destroyer HMS *Oribi,* one of the last of the 3rd Escort Group, rammed and sank U 531. At virtually the same time the destroyer HMS *Vidette* attacked U 125 with her Hedgehog; this time the forward-launched grenades worked perfectly and the submarine was lost. This was deeply satisfying to the escort's crew; like the *Pink* and the *Loosestrife,* the *Vidette* had been in close escort of the convoy from one side of the Atlantic to the other, suffering all the strains, storms, and hazards—and, now, successes. After that dramatic night, the next days saw the sky filled with patrolling aircraft, the local Canadian escorts coming in, the U-boats pulling back, and the damaged and exhausted warships sent ahead to port. The *Oribi* limped into St. John's with its nose battered by the ramming action; the indomitable *Pink* had steamed in with a mere 18 tons of fuel in her tanks. They had won this convoy battle.

In the final tally, thirty-nine U-boats had managed to find and attack ONS 5, and together they sank twelve merchantmen. But the Allies had sunk seven U-boats, two more were lost due to a collision in the fog, and another five were badly damaged in action. What was more, the submarines were not just being sunk in close-battle encounters (the U 638 had little chance when it found itself only 500 yards from the charging HMS *Loosestrife*) but being picked off on the surface by American, British, and Canadian air patrols as much as 50 miles from the convoy. This bruising shambles of a contest marked the turn of the tide, although the grinding nature of the convoy war made it difficult to see that at the time.

In any case, if Doenitz was disturbed by these losses, he wasn't showing it. By the middle of May, reinforced by a stream of new boats, his wolf packs had assembled in the North Atlantic again. Unfortunately for the Germans, they were pushed aside by the ranks of escorting warships and aircraft attending Convoy SC 129. Then, renewing their attacks, they sought to destroy Convoy SC 130, but their opponents were far too powerful. The convoy (thirty-seven merchantmen, with eight close escorts and a nearby support group) left Halifax for the Clyde under the command of Peter Gretton, who was furious at himself for having had to pull out to refuel during the height of the ONS 5 battles. It will be noted that among the escorts was not only his own destroyer, HMS *Duncan,* but also the cutter *Sennen,* which had come to the support of the *Pink's* mini-convoy, the destroyer *Vidette,* and the corvettes *Sunflower, Snowflake, Pink,* and *Loosestrife*—the north Atlantic's equivalent of the Magnificent Seven.[24] And by now there was continuous daytime aerial support.

Thus, almost any of the thirty-three U-boats that surfaced was swiftly attacked by Liberator and Hudson bombers, which dispatched three submarines to the bottom, including the U 954—among its crew members was Doenitz's own son, Peter. HMS *Duncan* helped to destroy a fourth attacker, and *Sennen* and HMS *Jed* the fifth. Amazingly, not a single merchantman was sunk; their crews could hear distant noises and observe great explosions on the horizon, but that was the closest the U-boats got. Despite this terrible blow, Doenitz urged his wolf packs to regroup in time for the next big convoy from Halifax, HX 239, but his instructions included the following amazing sentence, swiftly translated at Bletchley Park, which had now regained its capacity to break the Shark codes: "If there is anyone who thinks that combating convoys is no longer possible, he is a weakling and no true U-boat captain."[25]

It was around this time that Horton at Western Approaches Command and senior staff at the Admiralty itself began to wonder if the German U-boat crews were starting to lose their nerve and resolution. Perhaps so (it will be discussed shortly), but one is bound to ask what even a determined submarine captain could do by this stage of the fight. For, in addition to the more distant Coastal Command aerial patrols, Convoy HX 239 was accompanied all the way by two of the new escort carriers, USS *Bogue* and HMS *Archer,* which sank two out of the three U-boats that perished. The convoy arrived in the United Kingdom on May 25, the same date as the slow convoy SC 130, which had enjoyed aerial protection both day and night. The very long-range aircraft had made twenty-eight sightings of U-boats, attacked ten of them, and sunk two. The merchantmen were unscathed. It was an extraordinary reversal of fortunes since HX 229 and SC 122.

At the end of May 1943, Doenitz, ever the sober realist, assessed the results over the past month—a catastrophic total of forty-one U-boats had been lost during those four weeks—and concluded that the North Atlantic had become too hostile an environment for his submariners. The boats would therefore be relocated to the U.S.-Gibraltar routes and farther afield, off the Brazilian and West African coasts or in the Caribbean. As they slipped in and out of their French bases, they would tightly hug the coast of northern Spain and thus avoid the western approaches. The grand admiral was certainly not beaten, but he needed time to think, and time to get newer U-boats, with better equipment, into the hands of his captains. While he did not know quite how it had been done, he could see that Allied forces had made immense leaps forward both in the art of detection and in their capacity for destruction.

What Caused This Change of Fortunes?

Another twelve months would pass, almost exactly, before many of the same Allied aircraft squadrons and escort vessels would be patrolling off the beaches of Normandy, so there was still a way to go. But this is as good a place as any to begin a more detailed analysis of the steady Allied ascendancy in the Atlantic. The first and probably the greatest factor has to be aircraft. The Second World War was the first in the entire history of human conflict in which sea power was decisively affected by airpower. Without the latter, one could not reign supreme in the former.

The best testimony to this fact comes from Doenitz. While he was personally a devoted, unpleasant Nazi, the daily War Book notes from his office were remarkably candid and soberly analytical, whichever way the Battle of the Atlantic was going. His confidential reports to Hitler about the increasing difficulties the German submarines faced were blunt and honest. Thus, despite his desperate message to his U-boat commanders (mentioned above), he was already writing a dispassionate assessment of the reversal of fortunes, just as convoys HX 239 and SC 130 were entering home waters. In his entry of May 24, 1943, he noted that "the enemy Air Force therefore played a decisive part. . . . This can be attributed to the increased use of land-based aircraft and aircraft carriers combined with the advantage of radar location."[26] And, years after the war, Doenitz wrote in his memoirs that the chief problem had been that "Germany was waging war at sea without an air arm."[27] That was like boxing with only one fist.

The growth of Allied airpower over the Atlantic took two forms. The first was the advent of VLR aircraft operating from shore-based stations, chiefly the extraordinarily robust American-built B-24 Liberator bombers, but also B-17 Flying Fortresses and the elegant Catalina flying boats, and the tough British-built Wellingtons and Sunderlands. But it was the long-range Liberators that first made the difference. The second was the appearance of purpose-built escort carriers, modest in speed and striking power compared with the *Essex-* and *Illustrious*-class fleet carriers (see chapter 5) but extremely suitable for accompanying an Atlantic convoy: their aircraft could patrol the seas for miles around and, like the VLR bombers, carried a whole armory of weapons specially designed to destroy U-boats. Rearmed and refueled on the carriers overnight, they would be ready at the crack of dawn to fly off and suppress, possibly sink, the enemy submarines. As their name implies, these carriers escorted the convoys from Halifax to Liverpool.

The whole phenomenon of "airpower at sea" became so ubiquitous during the Second World War that it can easily be taken for granted, but it is one of the most important and novel aspects of the entire conflict. Doenitz must have known it from the beginning, and it is hard not to feel some sympathy for the hand he had been dealt when he became commander in chief of the German navy at the beginning of 1943: he possessed a few vastly expensive, top-heavy surface warships (but no aircraft carriers), all badly constrained by geography, and a U-boat force (lacking much assistance from the Luftwaffe until too late) that had not only to contest the enemy's warships but also to handle aircraft that came in fast and deadly.

It is impossible to overstress what this meant for the North Atlantic struggle. Doenitz's submarine crews found themselves confronted by a whole variety of Allied countermeasures as the spring of 1943 gave way to the summer and autumn; the enemy possessed vastly improved systems of detection and destruction against which even the most modern of the U-boats had less and less chance of success. Above it all, though, there was the factor of continuous air cover for the convoys, and the increasing chance of a U-boat being attacked even as it was in transit, on the surface, perhaps hundreds of miles from a convoy. Aircraft always had the advantage of speed and suddenness over vessels on the surface of the sea. Now they had supplemented this with the newer weapons' accuracy and lethality. Above all, they now had added range. By mid-1943 a VLR Liberator could be assisting a mid-Atlantic convoy 1,200 miles from its home air base, a phenomenal improvement. Even much earlier, in late May 1941, the German battleship *Bismarck* was spotted heading for Brest by a Coastal Command Catalina based in Iceland, a thousand miles to the north. The *Bismarck*'s crippling by torpedo aircraft and sinking four days later confirmed Doenitz's prejudices against heavy surface ships, but the larger lesson of 1941 was this: without airpower, control of the seas was precarious, indeed impossible, even for submarines.*

No senior commander appears to have realized this before the Second World War began. Both the British and German navies were shortchanged by the existence of independent air services that had designed themselves for essentially land-bound purposes: the Luftwaffe in its focus upon tactical air support for the army, medium-range bombing, and (increasingly) the aerial defense of the Third Reich; and the RAF, which was dominated by Bomber Command's obsession with destroying Germany's industrial base and sapping the enemy's public morale, such that only after 1935 did Fighter Command begin to gain more resources for short-range defense of the homeland. The Third Reich's allocation of aircraft resources was at least understandable, if one assumed (as the German generals and admirals did) that Hitler's conquests would be on land and that the Fuehrer's diplomacy would keep

---

*In 1941 airpower was everywhere—the *Bismarck* chase, Crete, the Malta convoys, Pearl Harbor, the sinking of the *Prince of Wales* and the *Repulse*.

the British Empire neutral. The Royal Air Force's preferences were disastrous for an island nation so dependent for its survival upon long-range oceanic commerce. In 1939, not only was the Fleet Air Arm limited in that it could not specify aircraft suitable for carriers, as the American and Japanese navies could, but RAF Coastal Command—charged to patrol the maritime routes from land bases—was pretty much a Cinderella service compared with the aerial resources being lavished elsewhere.

By contrast, at the middle level, at least on the British side, there was extraordinary initiative as the importance of airpower at sea slowly grew. The brilliant and aggressive Peter Gretton writes in his memoirs about Coastal Command officers attending naval exercises, even a full ocean passage from the Clyde to Newfoundland, and of Royal Navy captains being taken aloft—he claims to have flown in every patrol aircraft type available between 1940 and 1943. But how much did this valuable experience count for if the aircraft squadrons were not there to be deployed? To Air Marshal Harris, appointed to be C in C of Bomber Command in June 1942 with the specific task of increasing the aerial campaign against German industry, diverting his precious long-range bomber squadrons to Coastal Command (i.e., to the Battle of the Atlantic) was, as he put it many times, "picking at the fringes of enemy power . . . looking for needles in a haystack."[28] Only immense pressures from the Chiefs of Staff compelled some reallocation; in that respect, General George Marshall and even King were a lot more appreciative than Harris of the dire need to get VLR aircraft into the Atlantic air gap by spring 1943. Doenitz had no such luck, at least not until the Fuehrer ordered additional Luftwaffe support for the Bay of Biscay battles later that year.

Closing the air gap did not happen because some great person decreed it. There was the team of chiefly Canadian air engineers who in early 1943 pulled one bomb bay from a B-24 Liberator, replaced it with extra fuel tanks, and at last created an aircraft that could reach the transatlantic gap. There were the scientists who designed the much more effective fuses for Allied depth charges, so they exploded more predictably. There was the combination of U.S. scientists and armaments manufacturers who produced the air-launched acoustic homing torpedo—nicknamed "Fido"—specifically to search for and destroy submerged enemy submarines.[29] There were the groups behind the increasing effectiveness of the HF-DF detectors. There were the designers of the ungainly but wonderful *Bogue*-class escort carriers, and the management teams at the Tacoma yards that accelerated their launchings at a critical time. Finally, there were the staffs of the formidable training schools in Scotland, Northern Ireland, and Newfoundland, who taught thousands of raw, civilian novices how to become professional aircrews and assume such grave responsibilities in the midst of a ferocious Atlantic battle.

In the early stages of the war, it was a rare event for a land-based aircraft to sink a U-boat, and there were no escort carriers at hand. From 1943 onward, aerial sinkings of German and Italian submarines rose spectacularly and steadily overtook those credited to Allied warships. Between June and August 1943, when the U-boats abandoned the North Atlantic and fanned out to prowl the seas off Zanzibar and Montevideo, they were followed by the Liberators and Catalinas. Seventy-nine submarines were sunk by hostile action in distant waters in late 1943, fifty-eight of them by Allied aircraft. Even in the middle of the Indian Ocean, there was no escape. During 1943 as a whole, 199 U-boats were sunk altogether, 140 of them by Allied aircraft, a proportion that would continue for much of the rest of the war.[30]

This leads us, then, to the destruction aspect. From the German perspective, only one weapons system mattered—the torpedo. It was and may still be the most ubiquitous and destructive naval weapon in modern times. Torpedoes could be fired by small craft, by large craft, by all sorts of aircraft. During the epic *Bismarck* chase, the battleship was crippled by a torpedo dropped by a Swordfish biplane, attacked all night by British destroyers launching torpedoes, and eventually finished off by the torpedoes of the heavy cruiser HMS *Dorsetshire*. Above all, this was the weapon of the submarine. Every modern torpedo was usually powerful enough to blow a hole in the side of an enemy vessel. Normally, this projectile had speed enough to prevent its being deflected or stopped during its drive. Once a U-boat got close enough to a convoy to fire off a salvo, its destructive possibilities—as we have seen—were enormous. Thus the key task for convoy escorts, whether surface or aerial, was to prevent enemy submarines from launching their torpedoes at the flock of merchantmen in the first place. The later introduction of acoustic or homing torpedoes, or of much more sophis-ticated U-boats near the end of the war, did not alter that basic fact.

The chief Allied counterweapon, the depth charge, was a less reli-able tool of war, at least in its early stages. To begin with, it was not a contact explosive. The theory, which was quite plausible, was that if these large drum-shaped devices could be dropped roughly in the area where a submarine had been detected, their explosive force (magnified by being underwater) would rip apart a U-boat's joints. It was an attrac-tive weapon because a depth charge launcher could be mounted on the rear deck of even small craft, such as corvettes, cutters, and sloops, as well as the larger destroyers and frigates. It was also of course the chief weapon of all of the antisubmarine aerial patrols (until later in the fight, when they could also drop acoustic torpedoes). Its major disadvantages were swiftly recognized as the conflict unfolded: a warship had to steam over the spot where a submarine had been detected, and that might give an alert U-boat commander time to dive deep (and it took the Admi-ralty a long time to appreciate just how fast and deep a submarine could dive); the proximity fuses needed to be set at the last moment, accord-ing to the U-boat's estimated depth; and the underwater explosions distorted the asdic readings. Nonetheless, to all submariners the depth charge was the most terrible weapon; waiting for the next crashing noise—and the hunt could go on for days—was an unbelievably stress-ful experience. Overall, and despite the fact that on average hundreds of depth charges were needed to sink a submerged U-boat, this was the basic destroyer of submarines because it was used throughout the war, by aircraft as well as warships, and because its efficiency was improved over time.[31]

The Hedgehog was something else, so simple in concept that one wonders why it was not produced during the First World War or much earlier in this war. This weapon was a multiheaded grenade launcher that fired forward of the warship; usually the front 4-inch gun turret was taken out and this dreadful bombard installed in its place. It had several advantages over the depth charge, and Allied escort command-ers welcomed its widespread introduction in 1943 with great enthusi-asm, even if there were many early teething problems. It really reduced the time it took to fire at the foe. Unlike the proximity-fused depth charge, it exploded on contact—either one hit the submarine or one didn't—and, with a fusillade of twenty-four grenades raining down on the desperate U-boat, the chances of a hit were rather good. Moreover, it did not distort sonar readings. After the shot was delivered, its crew could observe with pride the rows of empty spigots, looking very like the bristling spines of a hedgehog.[32]

The incubator here was a small Admiralty unit called the Department of Miscellaneous Weapons Development (DMWD), fondly known as the "Wheezers and Dodgers." Its personnel, chiefly scientists, naval officers, and retired military men, was a most eccentric group of characters but also possessed a seriousness of purpose. By the summer of 1940 Britain's strategic position was quite desperate, its resources low, its armed services battered. Under Churchill's inspiration the nation planned to fight on, by whatever means possible. It was not surprising that a society brought up on H. G. Wells and Jules Verne novels, the *Boys' Own Weekly,* and *Amateur Mechanics* should now produce a vast number of citizen-based concoctions intended to help beat Hitler. Most were truly farcical. Still, some of them might, conceivably and with much modification, turn out to be useful after all. Scrutinizing all such weird schemes, and adding its own, was the DMWD's job. Those efforts fed into Britain's war machine because a scientific-technological system existed to turn new ideas into reality.[33]

Among the DMWD's staff was Lieutenant Colonel Stewart Blacker, by 1942 in his fifties but someone who had been interested in blowing things up since he was a schoolboy in Bedford. Blacker's earliest success, when still in his early teens, was his attempt to emulate on a modest scale the mortars the Japanese army was using against Russian defenses during the 1904–6 war. Assembling a crude barrel (a circular down-pipe) and "borrowing" some black powder, Blacker managed to send his projectile—a croquet ball—into the headmaster's greenhouse 300 yards away. Thus was his career born. He served in the Royal Artillery during the First World War. By 1940 he had drafted the design for an electrically activated spigot to propel a grenade from a mortar. Further work by the DMWD expanded the weapon into a ring of multiple mortars. There then occurred all sorts of bureaucratic obstacles, plus the creation elsewhere of a rival but much less successful forward-firing mortar. By a stroke of good fortune, Churchill himself witnessed an early test during a visit to an experimental weapons base near his country house, Chertwell, and breathed life into the scheme.* Delays still plagued the project, as did the lack of good training for this new and unorthodox weapon; in 1943 a large number of frightened crews had to be hauled back to Tobermory or Loch Fyne for "retraining." Thus it was only in mid-1943 that "Blacker's bombard" really came into its own.

By the end of the war, Hedgehogs had destroyed close to fifty enemy U-boats; that was worth a greenhouse or two. Its more sophisticated replacements, the Squid and Limbo weapons systems (whose projectiles could go deeper and actually search for the submarine), added a dozen or more to that total. Unsurprisingly, Squids are still used, in vastly improved form, in today's navies.

The third improvement, on the Allied side, came in the form of detection of the enemy, especially through radar. At its simplest, detection meant a U-boat spotting an oncoming convoy (or its smoke), or an aircraft or ship catching sight of a distant U-boat. But, once seen, a U-boat could of course submerge out of sight. And a German submarine could, through newer acoustic equipment, hear the welcome sound of the screw propellers of an oncoming convoy, or the more alarming sound of a fast-approaching corvette. To the U-boat's opponents also, sonar remained the key acoustic instrument for locating another man-made device under the waters. Yet water temperatures could differ vastly from place to place, creating acoustic barriers, and grinding icebergs nearby were a nightmare to detection engineers. Above all, Doenitz's recognition—arising from his own experiences as a First World War submariner—that U-boats could neutralize Allied sonar simply by attacking on the surface gave the Germans a great advantage. A pack of eight or ten or twenty submarines, creeping on the surface toward a convoy in the moonlight, or a single U-boat, picking off American freighters as they steamed along the floodlit Florida coast, simply dumfounded *all* underwater detection systems. Sonar could work well, but only when the conditions were right.

Doenitz's altered strategy meant that the U-boats had to be identi-

---

* So the story goes. Or possibly not—the rumor is that the two Royal Navy lieutenants attached to this experiment were extraordinarily handsome, and they persuaded Churchill's youngest daughter, Sarah, who was also watching that day, to get her father to stay on and see the Hedgehog in action.

fied at the surface, and before they launched their assaults. The convoy escorts had to know where the attackers were, how far away they were, how many of them there were, and what their line of attack was. If all that was known, the defense of the merchantmen could be undertaken. The submarines could be countered and, with the right weaponry, driven off, perhaps sunk.

It was not really until 1943 that the Allied navies possessed increasingly effective surface-detecting systems, the first of which was the tactical HF-DF radio-beam identifying mechanism mentioned earlier, but there was also the Admiralty's invaluable long-range tracking system, which allowed Doenitz's own messages (if decrypted in time) to be read. This type of direction finder had first been set up early in the war on the east coast of England, to locate German radio senders on the other side of the North Sea. It was not too difficult to reduce the size of the apparatus and move it onto individual warships by mid-1943. "Huff-Duff," as the Allied sailors fondly called it, was relatively simple and reliable, and it worked well at the ranges that mattered. It could pick up a U-boat's radio signals close by, and thus bring an escort to the threatened flank for a counterattack, but it could also detect the submarine's radio traffic as much as 15 miles away, thereby allowing the redirection of the convoy and/or the summoning of additional Allied naval and aerial support by the Admiralty, which was reading the long-range enemy messages. Even now, it seems rather remarkable that Doenitz allowed his U-boat commanders to chatter so much, although he himself was guilty of the same sin. While the actual messages might not be understood by the Allies until Bletchley Park decrypted them, the location of the transmitting submarine near a convoy was easily identified. When an Admiralty plotting team picked up reports from its escort commanders that, say, eight U-boats were forming into a wolf pack, it could send an alert to British forces in the area. Rohwer, reflecting on the outcome of the critical convoy battles of March 1943, identifies HF-DF as "decisive."[34]

Centimetric radar was even more of a breakthrough, arguably the greatest. HF-DF might have identified a U-boat's radio emissions 20 miles from the convoy, but the corvette or plane dispatched in that direction still needed to locate a small target such as a conning tower, perhaps in the dark or in fog. The giant radar towers erected along the coast of southeast England to alert Fighter Command of Luftwaffe attacks during the Battle of Britain could never be replicated in the mid-Atlantic, simply because the structures were far too large. What was needed was a miniaturized version, but creating one had defied all British and American efforts for basic physical and technical reasons: there seemed to be no device that could hold the power necessary to generate the microwave pulses needed to locate objects much smaller than, say, a squadron of Junkers bombers coming across the English Channel, yet still made small enough to be put on a small escort vessel or in the nose of a long-range aircraft. There had been early air-to-surface-vessel (ASV) sets in Allied aircraft, but by 1942 the German Metox detectors provided the U-boats with early warning of them. Another breakthrough was needed, and by late spring of 1943 that problem had been solved with the steady introduction of 10-centimeter (later 9.1-centimeter) radar into Allied reconnaissance aircraft and even humble *Flower*-class corvettes; equipped with this facility, they could spot a U-boat's conning tower miles away, day or night. In calm waters, the radar set could even pick up a periscope. From the Allies' viewpoint, the additional beauty of it was that none of the German systems could detect centimetric radar working against them.[35]

Where did this centimetric radar come from? In many accounts of the war, it simply "pops up"; Liddell Hart is no worse than many others in noting, "But radar, on the new 10cm wavelength that the U-boats could not intercept, was certainly a very important factor."[36] Hitherto, all scientists' efforts to create miniaturized radar with sufficient power had failed, and Doenitz's advisors believed it was impossible, which is why German warships were limited to a primitive gunnery-direction radar, not a proper detection system. The breakthrough came in spring 1940 at Birmingham University, in the labs of Mark Oliphant (himself a student of the great physicist Ernest Rutherford), when the junior scientists John Randall and Harry Boot, working in a modest wooden building, finally put together the cavity magnetron.

This saucer-sized object possessed an amazing capacity to detect small metal objects, such as a U-boat's conning tower, and it needed a much smaller antenna for such detection. Most important of all, the device's case did not crack or melt because of the extreme energy exuded. Later in the year important tests took place at the Telecommunications Research Establishment on the Dorset coast. In midsummer the radar picked up an echo from a man cycling in the distance along the cliff, and in November it tracked the conning tower of a Royal Navy submarine steaming along the shore. Ironically, Oliphant's team had found their first clue in papers published sixty years earlier by the great German physicist and engineer Adolf Herz, who had set out the original theory for a metal casement sturdy enough to hold a machine sending out very large energy pulses. Randall had studied radio physics in Germany during the 1930s and had read Herz's articles during that time. Back in Birmingham, he and another young scholar simply picked up the raw parts from a scrap metal dealer and assembled the device.

Almost inevitably, development of this novel gadget ran into a few problems: low budgets, inadequate research facilities, and an understandable concentration of most of Britain's scientific efforts at finding better ways of detecting German air attacks on the home islands. But in September 1940 (at the height of the Battle of Britain, and well before the United States formally entered the war) the Tizard Mission arrived in the United States to discuss scientific cooperation. This mission brought with it a prototype cavity magnetron, among many other devices, and handed it to the astonished Americans, who quickly recognized that this far surpassed all their own approaches to the miniature-radar problem. Production and test improvements went into full gear, both at Bell Labs and at the newly created Radiation Laboratory (Rad Lab) at the Massachusetts Institute of Technology. Even so, there were all sorts of delays—where could they fit the equipment and operator in a Liberator? Where could they install the antennae?—so it was not until the crisis months of March and April 1943 that squadrons of fully equipped aircraft began to join the Allied forces in the Battle of the Atlantic.

Soon everyone was clamoring for centimetric radar—for the escorts, for the carrier aircraft, for gunnery control on the battleships. The destruction of the German battle cruiser *Scharnhorst* off the North Cape on Boxing Day 1943, when the vessel was first shadowed by the centimetric radar of British cruisers and then crushed by the radar-controlled gunnery of the battleship HMS *Duke of York,* was an apt demonstration of the value of a machine that initially had been put together in a Birmingham shed. By the close of the war, American industry had produced more than a million cavity magnetrons, and in his *Scientists Against Time* (1946) James Baxter called them "the most valuable cargo ever brought to our shores" and "the single most important item in reverse lease-lend."[37] As a small though nice bonus, the ships using it could pick out life rafts and lifeboats in the darkest night and foggiest day. Many Allied and Axis sailors were to be rescued this way.

To this list of crucial detection instruments that arrived in 1943 might be added the creation of an idiosyncratic former Great War pilot, Humphrey de Verd Leigh, who conceived of patrol aircraft carrying powerful searchlights (named, in his honor, Leigh Lights) that would catch in their stunning glare and paralyze U-boats recharging their batteries at night. Leigh had invented, tested, and paid for this light on his own, solving the technical problems (including flying with the prototypes), but then had a hard time getting Air Ministry support, which delayed at least by a year its general use in Allied aircraft. No doubt U-boat commanders crossing the Bay of Biscay at night wished it had been delayed forever.[38]

Although all were developed separately, when they were brought together, the Leigh Light, the increasingly improved variants of HF-DF, the cavity magnetron, and the airborne homing torpedo became interacting pieces in a single platform such as a Sunderland bomber. This much diminished a U-boat's chances of escaping detection, attack, and destruction.

By comparison, the contest between the code breakers at Bletchley Park and their archrivals in the B-Dienst seems a less decisive factor in the Battle of the Atlantic once a detailed analysis of the key convoy battles of 1943 has been undertaken. It was certainly far less important than most of the popular literature about the code breakers suggests. From the first revelations of around 1970 to the present day, the reading public has been fascinated by the idea that before the war the German armed forces possessed a super-clever mechanical gadget that could turn all messages and reports into an undecipherable mishmash of symbols that could be understood only if the recipient had a similar machine to decrypt these documents. They were even more fascinated by learning that, at a top-secret location, the British, aided first by the Poles and later by an American contingent, had machines that could

read those messages—until the Germans made transmission more difficult, frustrating the Bletchley code breakers for months until they
cracked the newer system. Equally exciting was that Doenitz's own
B-Dienst was doing the same thing—reading Admiralty codes. And in
the Pacific the U.S. code-breaking services were reading Japanese military and diplomatic messages through their Magic and Purple systems.
Here, surely, was a new way of understanding the outcome of the war,
an ultimate explanation of why the Allies won.

Not quite. We should not underestimate the extraordinarily important role of the intelligence war that was fought between the Axis powers and the Grand Alliance, but we must recognize that, operationally,
Ultra could do only so much, and so we should not overestimate it,
either. Depending on each side's skills in concealing their own messages
and reading those of the enemy, code breaking offered vital information that permitted the pre-positioning of forces for an impending
battle. Both the Germans and the Allies benefited greatly from being
able to read each other's messages. Even if a decryption system worked
imperfectly, a one-eyed man has an inestimable advantage if his opponent is blind. In 1942, it is true, the British lost their capacity to read
Doenitz's messages (or at least they took much longer to decrypt),
whereas B-Dienst was providing the grand admiral with virtually immediate information about the Allied convoys; in the grim months of
late 1942 and early 1943, that meant a lot.

But what if the convoys were better protected *and* the British could
see as well as the Germans? Then things were greatly different.
Consider, for example, the situation during mid-May 1943 in which
each side was rather smartly reading the other's traffic and sending the
new intelligence to its forces at sea. By the time of the great battles
(May 7–14) around convoys HX 237 and SC 129, the intelligence war
had reached a remarkable level of sophistication, of point and counter-
point, so that neither side had the lead for very long. The same was true
a week later with regard to the fighting that surrounded convoy HX
239, coming from Halifax to the Clyde. First, B-Dienst detected the
location and course of this large group of merchantmen. Shortly afterward, the British cryptographers read Doenitz's Shark instructions for a
wolf pack ambush, so Western Approaches Command ordered the con-

voys to alter course. But B-Dienst then read those signals and redirected the U-boat lines—no fewer than twenty-two submarines—to attacking positions. Unfortunately for the U-boats, however, at that point they met the massed forces of Allied air- and sea power over the Atlantic. The new escort carrier USS *Bogue* destroyed U 569, while HMS *Archer*'s aircraft, now fitted with rockets, sank U 752. No merchant ships were lost.[39] The plain fact was that Ultra assisted in but could never win the hard-fought convoy battles. When the British decision was made to fight the convoys through the U-boat lines, victory went to the side with the smartest and most powerful weaponry, not the one with the better decrypts.

We can now understand how these great advances in the Allied systems of detection and destruction interacted with the two other factors mentioned above: numbers and morale. When, in mid-March 1943, HMS *Volunteer* was the only escorting warship with convoy HX 229 to possess HF-DF, and when its radio operators had time either for detection of U-boats or for receipt of Admiralty orders but not both, its potential was limited. When, about six months later, all escorts were fitted with HF-DF, the Allies gained a huge advantage, as they did also with the coming of short-range radio between warships. Similarly, when there were just a few hard-stretched aircraft operating for only several hours each in the mid-Atlantic, the U-boats benefited enormously. When whole squadrons of very long-range aircraft were operating out of bases in the Shetlands, Northern Ireland, Iceland, Greenland, and Newfoundland (and, after mid-1943, the Azores), and when the Bay of Biscay could be patrolled all through the night by aircraft equipped with centimetric radar, Leigh Lights, depth charges, acoustic torpedoes, even rockets, Doenitz's submarines knew no rest. A single escort carrier such as HMS *Archer* was a welcome sight to a convoy but could not affect the tactical balance very much. With a dozen escort carriers at sea, by late 1943, the whole situation had changed.

It was small wonder, then, that the Ultra decrypts detected a sharp decline in the morale of the U-boat crews after the May 1943 battles; small wonder that Doenitz's urgings sounded more desperate. While he possessed more submarines than at any previous time, fewer now pressed home their attacks, and more reported damage and returned to base. Many of their crews were taken into Allied captivity as their stricken boats slowly sank beneath the waves. Captain Herbert Werner, a U-boat commander himself throughout this long war, later wrote a moving memoir; his title, *Iron Coffins,* says it all.[40]

A Few Final Blows

The overwhelming importance of airpower received further confirmation during the next two phases of this history: the Allies' aerial offensives in the Bay of Biscay and the failed U-boat attacks on the U.S. convoys to North Africa. These two battlefields were separate geographically, but the struggles unfolded in roughly the same months of 1943 and were in many ways coeval, for if the submarines tiptoeing westward past northern Spain could be destroyed en route, there would be fewer of them surviving to attack the U.S.-Gibraltar convoys or to range farther afield. And if the American escorts on the North African route could batter Doenitz's wolf packs as savagely as the predominantly British and Canadian fleets had done earlier in the North Atlantic, there would be fewer submarines returning to French bases in any event.

U-boats could cross the Bay of Biscay either on the surface or submerged, and by day or at night, and it was from this that their tactical-technical problem emerged. If they traveled submerged, their speed was much slower, their batteries drained faster (which certainly didn't help crew morale), and they could still be picked up by the sonar of an enemy warship. On the other hand, if they traveled on the surface, they could reach the wide Atlantic waters faster, though with a much greater possibility of detection by enemy radar, and then they would be at serious risk of aerial as well as surface-vessel attack just when they were most vulnerable. Additionally, by the second half of 1943 Bletchley Park was much more successful at reading German naval codes, and more swiftly, than B-Dienst was in adjusting to the improved Allied ciphers.

These problems were compounded by the fact that Allied aircraft and surface escorts had been transformed by the many technological innovations described above into horrifying submarine-killing machines. It was not just that the Allies were now deploying the new support groups of powerful sloops and fleet destroyers and the purpose-built escort carriers. It was also the case that the standard, rather plodding workhorses in the Allies' armory, such as the *Flower*-class corvettes and the medium-range aircraft such as the Wellingtons, the Hudsons, and the Sunderlands, were being steadily improved. In 1940–41 the corvettes were underarmed, underequipped, and horrible sailors against the giant North Atlantic waves (one American observer thought their crews should get submariner's pay because the vessels seemed to be more than half underwater much of the time). After 1943, those vessels, and the newer types of corvettes and frigates that gradually replaced them and were especially designed for the Atlantic, had much greater horsepower, better detection and destruction equipment, a raised bow, and much more space for the commander's many functions. On larger vessels, one of the forward guns was now replaced by a Hedgehog; the depth charges were much more powerful and had improved timing devices; and the superstructure sprouted all sorts of antennae, not just the radio mast but the HF-DF and 10-centimeter equipment as well.

The same metamorphosis was true of a Wellington, a sturdy but basic twin-engine bomber originally designed by Barnes Wallis for raids into the western parts of Germany. By 1943 it had, under RAF Coastal Command, become an amazing acoustic machine, spouting aerials along its spine, out of its side, and out of its nose. Its armaments could include forward-firing cannon, rockets, more efficient depth charges, even the new acoustic torpedoes. At night, when its radar brought it close enough to a U-boat on the surface, its massive Leigh Lights would go on, and then the mayhem against the submarines would begin.

Doenitz was nothing if not a fighter, and the German shipyards were producing more U-boats than ever in 1943, which could still reach the Atlantic via roundabout routes through the Greenland-Iceland-Faroes gaps. Recognizing that Allied airpower was now key, Doenitz also persuaded Berlin in autumn 1943 to give his U-boats much more support in the Bay of Biscay than before. In fact, so seriously did the German air attacks upon Coastal Command's antisubmarine squadrons become at that time that the RAF was compelled in turn to allocate Beaufighter and Mosquito night fighters to protect them. Moreover, a number of U-boats were equipped with far heavier antiaircraft guns, including four-barrel flak guns, for Doenitz was also encouraging his captains to fight it out on the surface if they could. Finally, the Luftwaffe itself was also being equipped with acoustic torpedoes, as were the newer submarines; in late August German aircraft made their first attacks with glider bombs on two of the escort groups.[41]

Thus, apart from the reduction of attacks in the North Atlantic after June 1943, the fighting over the convoys was not getting any less hectic, nor the wolf packs any less dangerous. Individual actions were as wild as ever. One Sunderland, flying back home across the Bay of Biscay, was attacked by no fewer than eight Junkers Ju 88s, shot down three of them, and got back to land on the pebbles of Chesil Beach on three engines; the German nickname for the Sunderland was "the flying porcupine." A little later, a furious surface-to-air battle occurred between five U-boats traveling on the surface across the bay together and four Polish-flown Mosquitos. There were British, American, Australian, Polish, Canadian, and Czech squadrons in this phase of the fighting.[42] Some of the planes crashed onto the U-boats, for the shoot-outs between two determined foes distinctly increased that possibility. Occasionally the survivors of any combination of merchantmen, escorts, aircraft, and U-boats might be in their lifeboats and rafts in the same waters and picked up by the same rescue ship. Ramming was common, and among some destroyer captains it seems almost to have been the preferred form of attack—they made sure they crushed the enemy vessel, and they probably also got three weeks of home leave for repairs as well.

But the speed and ferocity of these episodes could not disguise the fact that "the Allied navies and air forces . . . at last enjoyed adequate strength in the right kinds of ships and aircraft, were trained to high efficiency and equipped with a comprehensive armory for finding and destroying U-boats."[43] Levels of training had been constantly racheted up, and in any case many of the Allied naval groups and air squadrons had by now considerable experience, which played to their advantage time and again. When a large battle developed in mid-October around convoys ON 206 and ONS 20, the merchantmen and close escorts were supported by no fewer than three highly competent Liberator squadrons (nos. 59, 86, and 120) and by a support group under Gretton, including, once again, *Duncan, Vidette, Sunflower, Pink,* and *Loose-strife.* One merchantman was lost, and five U-boats went to the bottom.

Operating farther afield from close-escort support, Captain Johnny Walker and his 2nd Escort Group seem to have been given virtually a free hand to pursue his all-consuming passion: sinking U-boats, which they would sometimes stalk for days. By this time Walker had perfected his "creeping attack," in which only one of his group kept its asdic sender/receiver on and then stayed at a fair distance from a submerged U-boat, whose captain was unaware that the other vessels, with their asdic switched off, were quietly coming right overhead, directed by radio from the control boat. On other occasions, Walker would have his sloops—HMS *Starling, Wild Goose, Cygnet, Wren, Woodpecker,* and *Kite*—advance in line abreast and drop their charges simultaneously, like an artillery barrage on land. Of the two schools of strategic thought about convoy warfare, Walker was definitely *not* of the just-get-the-merchantmen-home-safely persuasion. His operating instructions to the 2nd Escort Group, preserved after the war by the Captain Walker's Old Boys Association in their home base of Liverpool, reads: "Our object is to kill, and all officers must fully develop the spirit of vicious offensive. No matter how many convoys we may shepherd through in safety, we shall have failed unless we slaughter U boats."[44] Walker was as good as his word. His little group sank no fewer than twenty U-boats from 1943 onward, and he himself gained an astonishing four DSOs before his premature death from exhaustion in 1944.

Walker's cat-and-mouse tactics were part of a larger Admiralty plan. It will be recalled how, during the convoy battles of early 1943, escort vessels such as *Vidette* and *Pink* had to keep giving up the pursuit of a submarine because of the dire need to return to guard their flocks. In the critical months of March through June 1943, the Admiralty's Operational Research Department made a detailed statistical analysis of all the encounters between U-boats and Allied escorts over the previous two years.[45] It found that if a warship went after a fast-submerging submarine for only a few attacks, the chances of the foe's getting away were good. If the surface craft could remain in pursuit, perhaps losing but then regaining sonar contact an hour or so later, and especially if it could stay around to drop depth charges in six or more attacks over more than ninety minutes, then U-boat losses markedly increased. Here, then, was the statistical case for the creation of the support groups, at least if they remained in the general area of the convoys and did not go out hunting for a needle in a haystack. Horton recognized this all along, and it was only the overall shortage of surface warships that had prevented groups such as Walker's being formed much earlier. But those Operational Research figures did pinpoint Doenitz's dilemma in a stark way. His boats could stop the flow of Allied supplies to Europe only if they were willing to close with the convoys, but in so doing after July 1943, they were increasingly likely to be detected by aircraft and surface support groups, and hunted down to the end.

If it was difficult, almost impossible, for the U-boats to deal with Walker, it was hardly less dangerous for them to go against the U.S. escort carrier groups on the Gibraltar and North Africa routes. Although the U.S. Navy had come into anti-U-boat warfare in a woefully unprepared state, it had learned fast (and despite the far larger calls upon its attention and resources from the Pacific War). By 1943 a stream of new, small, but powerful escort carriers was pouring out of the Tacoma, Washington, shipyards, their crews and aircrews going into intensive training, steaming through the Panama Canal, then forming the core of the escorts for the enormous numbers of American troops, munitions, and other supplies heading toward the Mediterranean theater as the invasions of Sicily and Italy developed. The first three of the *Bogue* class—USS *Bogue* itself, USS *Core,* and USS *Card*— had particularly aggressive and competent aircrews; in a sort of variant of Walker's creeping attacks, two of their planes would strafe a U-boat on the surface, obscuring the fact that a third aircraft was approaching with bombs or homing torpedoes. Roskill's official naval history records the USS *Bogue*'s aircraft sinking six U-boats on these 1943 Gibraltar runs, and the total kills achieved by the USS *Card*'s aircraft and close escorts was even higher: at least ten U-boats, perhaps an eleventh.[46]

Overall, the escort carriers destroyed many fewer enemy submarines during the war than did land-based aircraft, yet their role in driving away U-boats from the North Atlantic convoys in the critical months of spring and early summer 1943 was great, and their protection of the flow of American troops and goods to North Africa and the

Mediterranean was absolutely invaluable. Later in 1943, some escort carriers were detached to ambush U-boats meeting at refueling points in the Central Atlantic. Although British naval intelligence agonized about the risk of their Ultra source being detected—how many coincidences could there be of U.S. planes arriving in the bare oceans just when a U-boat was being refueled by the "milk cow" vessel?—the attacks certainly destroyed a disproportionately large share of refueling submarines, leaving the regular boats perilously low on supplies and forced to return to base early.

From these dire trends the German submarine fleet could not recover. Though it resumed its attacks on Allied convoys in the first five months of 1944, including a renewed series of onslaughts upon the North Atlantic routes against the awful odds, it could never again achieve the success rate of its glory years of 1942 and early 1943. Its failure was evident, in spectacular form, on D-Day itself. There, as we shall see in chapter 4, the balance of forces—in the air, at sea, even on some of the contested beaches—was totally disproportionate. Not only were the German high command's few Luftwaffe squadrons promptly destroyed, but the same was true of the Kriegsmarine's "iron coffins"; Werner recounts how he and a group of highly experienced U-boat commanders, amazed at being held back in early June, were then ordered to attack the D-Day landings, including ramming an enemy ship like a kamikaze. Nothing could have been more remote from a submarine's true vocation. Five of the eight U-boats so ordered were destroyed in such vain attacks; the other three limped back, damaged and ashamed.[47] Allied control of the air and the waters was complete, and between June and August 1944 Doenitz's submarines sank only five vessels during the greatest amphibious invasion of all time (one of those sunk, alas, was HMS *Pink*).

But the failure of the U-boats to destroy, or even dent, Operation Overlord was not just a mark of an imbalance of power at the local, tactical level. It was also evidence that, whatever their earlier remarkable and sustained successes, the German submarines had failed to stop the gigantic surge of troops, munitions, fuel, food, and all the other vital goods being brought from the New World to the Old. The submarines had done their best—U-boat "aces" such as Kinzel were extraordinary fighters—but they had been beaten by a well-organized, multiple-armed counterattack in the middle of 1943. Because of the new Allied strengths, the convoys were getting through, increasingly unscathed; because they were unscathed, the island fortress of Great Britain was turned into an enormous launchpad for the invasion of Europe; and because that invasion took place without severe (or even much) disruption and the Allied armies could push ahead and into the Third Reich, the days of the U-boats were numbered. Their home bases, whether in western France or northern Germany, would be captured from the landward side.

It is true that the German military-industrial complex was, in 1944 and early 1945, still pouring out dozens and dozens of newer, larger U-boats. Moreover, German scientists were working on some remarkably advanced types of U-boat (Hellmuth Walter's brilliant designs, and the Type XXI "elektro-Booten"), which concerned the Admiralty all through the last year of the war in Europe. Doenitz's U-boat fleet was still receiving and sending new boats out as late as April 1945, despite the more intensive Allied bombing of their construction yards and railway supply routes. The German submarines fought to the very end, impressively and ferociously. Since they made little impression upon the ceaseless flow of men and munitions across the Atlantic—and by the autumn of 1944 the new U.S. divisions could sail directly to France—their story fades into the background in most histories of the Second World War. Yet the period from January to April 1945 witnessed some of the fiercest duels ever between the U-boats and their long-lasting foes, RAF Coastal Command, the Royal Navy, and their Canadian equivalents.

But those duels were no longer fought for control of the Atlantic routes; they were contests that took place chiefly across the North Sea and in British and Canadian coastal waters. These were not easy contests, for the U-boats were now much harder to detect when in inshore waters, possessed far more sophisticated weaponry, and often steamed on the surface in small groups for protection. At one stage, squadrons of RAF tank-busting Typhoons supporting the military campaign in the Netherlands and northwest Germany were actually diverted to reinforce Coastal Command's Beaufighters and Mosquitos against U-boats fighting it out on the surface, near the Dogger Bank and Jutland. But dogfights are not convoy battles, and in the last five months of the war only forty-six Allied merchant ships were lost worldwide (one-seventh the rate of 1942); as usual, most of those losses were of ships sailing independently rather than in convoy. By contrast, another 151 submarines were destroyed. During the war as a whole, the casualty rate among the U-boat crews was a staggering 63 percent, or 76 percent if captured sailors are included. No other major service in the struggle came close to these terrible rates of loss.[48]

The larger point was not that the German submarine forces were too few, by any measure, but that they had reemerged too late to win the critical campaign of the Central Atlantic. The striking power of their enemies kept increasing and, besieged in those closing months from both west and east, the U-boats had, like the Third Reich itself, nowhere to go. The last act for this proud service was to surrender. When British troops reached U-boat harbors such as Flensburg in May 1945, they found rows and rows of submarines either tied up alongside the jetties or partly sticking out of the water, sunk by Allied bombing or by their own crews. Many crews deliberately scuttled their submarines offshore rather than obey the order to turn themselves in. Across the oceans, German vessels steamed into Allied ports flying the black surrender flag, or into neutral harbors to be interred. It is ironic that the final order to abandon the fight and surrender peacefully, drafted by the British Admiralty on May 8, 1945, came from the man who had been appointed by Hitler as his successor: Grand Admiral Karl Doenitz.

Final Thoughts

Is the story of the Battle of the Atlantic yet another, and very famous, example of victory simply by brute force? Many historians, almost mesmerized by the staggering output of new merchant ships from the American yards after 1942 onward—overall, Anglo-American shipyards laid down 42.5 million tons of vessels between 1939 and 1945—believe that it was.[49] This account seeks to place a caution against that assumption.

Clearly, the material advantages were in the Allies' hands by the summer of 1943 onward, although the continued capacity of Doenitz's submarines to sink merchantmen stayed challenging, and Horton and Slessor felt that they could never relax against this most formidable enemy, even when the odds were becoming more favorable to them. Churchill was right when he talked about "the *proper* application of overwhelming force." Sheer numbers meant a lot, but mass could not be turned into victory without two other vital ingredients: organization and quality. Without them, the strategic directives given by the men at the top were as nothing. (After all, Mussolini issued grand strategic orders frequently, but to what effect?) Nor could the men at the bottom of the chain of command be expected to win even a single encounter with a U-boat, not to mention a weeklong convoy battle, if they did not have the right tools for detection and destruction and if they were not properly organized so that those tools could be used effectively.

Time and again in this particular story we see how the "proper application" of resources led to endeavors that gave the frontline forces the instruments for winning, and how those victories in the Atlantic and more distant seas steadily tilted the overall campaign balance in favor of the Allies and thus toward the fulfillment of the January 1943 grand strategy. Time and again, too, we can identify where the newer applications became turning points: where a certain idea was turned into reality, which people and/or organizations were responsible, and how their breakthroughs directly affected the field of battle.

Among those many advances, four were described above: the coming of very long-range and heavily equipped bombers that could stay with the convoys all the way; centimetric radar, and the successive roles of the original Birmingham University team, the Tizard Mission, and thereafter Bell Labs and the MIT Rad Lab; the Hedgehog mortar, from a quirky schoolboy's dream, via the Admiralty's unusual DMWD, to its use at sea by Allied escorts; and the creation of the hard-hitting convoy support groups, which under commanders such as Gretton and Walker married the newer weaponry with the more effective stalking tactics they had developed. Not all of these participants were engineers per se, but all helped to engineer the Allied victory.

Thus was the first operational objective of the Casablanca statement achieved, although no one on the Western planning staffs was counting their winnings even at the end of 1943. The strategic air campaign against Germany had failed throughout that entire year. Plans for an invasion of France were repeatedly postponed as the difficulties and complexities of such a vast operation became more daunting, especially to the British chiefs. The Mediterranean campaign was going far too slowly. The mid-Pacific campaign was still in its early stages, and all operations through the Southwest Pacific or Southeast Asia were painfully slow. Only in the grinding maw that was the Battle of Kursk in the summer of 1943 was there another indicator that the tide was turning against Germany. In the great expanses of the Atlantic, and in the limitless fields of western Russia and the Ukraine, there were the early hints of Goetterdaemmerung, the signs that the Second World War was swinging, after years of brutal conflict, in favor of the Grand Alliance.

CHAPTER TWO

HOW TO WIN COMMAND OF THE AIR

For I dipt into the future, far as human eye could see,
Saw the Vision of the world, and all the wonder that would be;
Saw the heavens fill with commerce, argosies of magic sails,
Pilots of the purple twilight, dropping down with costly bales;
Heard the heavens fill with shouting, and there rain'd a ghastly dew
From the nations' airy navies grappling in the central blue.

—ALFRED LORD TENNYSON, "LOCKSLEY HALL"

In the early morning of October 14, 1943, an armada of 291 American B-17 bombers (Flying Fortresses) took off from airfields in the eastern flatlands of England and began their long journey to drop bombs on the cities of Schweinfurt and Regensburg, home respectively to the manufacturers of such critical German wartime items as ball bearings and Messerschmitt fighters. Viewers who watched their takeoff and rise into the skies were witness to what was probably the twentieth century's greatest demonstration of military power to date (not even the A-bombs would compare in terms of the number of lives destroyed). They were also witness to what would be a mission of failure and much unnecessary loss. This was not the first time those squadrons had attacked Schweinfurt (an earlier raid had taken place on August 17), but it was going to be the last for quite a while.

Among the almost three thousand fliers that day was Elmer Bendiger, a former journalist turned navigator. Thirty-seven years later, and after visiting a poppy field in England that once had bordered his USAAF base, he published his account of that day.[1]

Bendiger, a professional writer, had a gift for words ("The cloud cover stopped close to the English shore, and the Channel sparkled. White froth curled over the blue waves"), so the reader has to make sure he keeps close to the hard narrative amid the magical language. The P-47 Thunderbolt escort fighters were "like shining angels," but after Aachen they waggled their wings and flew home, because of their limited range. Only then—of course—did Adolf Galland, the head of the German fighter arm and like Doenitz an utter professional, order in the waves and waves of attacking aircraft; once each squadron of Focke-Wulfs and Messerschmitts had finished its run, it flew to the nearest airfield, refueled and rearmed, and came up again. Even if some B-17s (amazingly) got through to bomb the intended targets, the real issue of the Schweinfurt-Regensburg raids became less what happened to German ball-bearing production than the question of how many Flying Fortresses could be shot down.

As it turned out, twenty-nine Fortresses were lost that day before they reached the targets, and another thirty-one on the way home: sixty aircraft and 600 fighting men, a loss of well over 20 percent in one mission. Coincidentally, the August raid on Schweinfurt also had led to the destruction of exactly sixty B-17s. The raids on the following days simply deepened the losses. "In six days of warfare," notes Bendiger, "we had lost 1,480 crewmen over Europe." Only three or four of his squadron of eighteen planes got back to Kimbolton that night; others limped in during the next day, and "six of our planes were burned-out hulks somewhere in Europe." A little later he adds, "We airmen of '43 had demonstrated in our own flesh and blood the fallacy that a formation of heavily armed bombers, alone and unescorted, could triumph over any swarm of fighters. . . . It may now be asked whether in fact we had vindicated Billy Mitchell"—a tart comment on the well-known American theorist of airpower.[2]

In the autumn of 1943, the Luftwaffe was clearly winning against the attacking armadas of American aircraft; exactly the same setbacks were occurring in the parallel and even longer-lasting strategic bombing campaign carried out by RAF Bomber Command across western Europe every night.[3] Even as late as March 30, 1944 (fewer than ten weeks before D-Day and fourteen *months* after Casablanca), 795 RAF bombers were sent to attack Nuremberg. While the British squadrons steadily lost direction and coherence—some of them, ironically, bombed Schweinfurt, 55 miles away from the planned target—the German night fighter counteroffensive was superbly orchestrated, slashing away at the Lancasters and Halifaxes. The veteran Pathfinder leader Wing Commander Pat Daniels (on his seventy-seventh mission) later recorded: "So many aircraft were seen to blow up around us that I instructed my crew to stop recording aircraft going down for my navigator to log them because I wanted them to spend their time looking out for fighters."[4]

The result was that a colossal total of ninety-five RAF bombers failed to return, another dozen were scrapped when they got back to England, and fifty-nine were badly damaged. But the loss of the machines paled in importance beside the slaughter of the trained crews, whose attrition rate was higher than that among British junior officers during the 1916 Battle of the Somme. Similarly, "a young American . . . had a better chance of surviving by joining the marines and fighting in the Pacific than by flying with the Eighth Air Force in 1943."[5] As the later American and British official histories admit, by early 1944 things were actually going backward. Most of the nonofficial histories are much more scathing. Quite the grandest of the Anglo-American stratagems for winning the war had collapsed.[6]

The Theory and Origins of Strategic Bombing

Well before the Wright brothers had stunned the world with their flying achievements of 1903, mankind had speculated about the coming of extraordinary machines that would not only transport humans and commerce through the air but also be capable of inflicting terrible damage and panic upon an enemy. Tennyson was not the first to foresee air warfare—Leonardo da Vinci dabbled with some heavier-than-air machines in his military sketches—and the massive technological advances of the nineteenth century only fueled these futuristic speculations.[7] Alfred Gollin records that within only a few years of the historic flights at Kitty Hawk, the war ministries of Europe and the United States were planning for aerial forces and knocking at the Wright brothers' door. And during a public debate in April 1904 about Britain's global strategic fate, the prescient imperialist Leo Amery had argued that in the not-too-distant future airpower would join (if not overtake) power on land and sea as an instrument of national policy.[8]

It is important to realize that even in its early years airpower wore many faces and could be thought of in many different ways.[9] The period 1880–1914 witnessed a staggering number of new inventions or vast improvements in earlier, eccentric prototypes—the wireless, the torpedo, the submarine, destroyers, the machine gun, high explosives, calculating machines, the automobile, and the aircraft itself. It is therefore understandable that some commentators thought of the airplane as just another addition to the armory of weapons for combating the enemy's soldiers and sailors (much as, perhaps apocryphally, President Truman thought that the atomic device was simply another big bomb). Aircraft, in this conception, were a sort of aerial cavalry, useful chiefly for reconnaissance or gun spotting—pre-1914 navies, grappling with the problems of modern fire control, were particularly interested in this aspect. Pilots could perhaps machine-gun the enemy's troops or toss a few grenades onto their heads, but the skimpy aircraft could not carry any heavy weaponry. Those views would change as planes became stronger and their engines much more powerful, and as airships revealed their vulnerabilities, but the basic point remained: these newfangled instruments were chiefly to be used for observation of, or direct attack upon, the enemy's forces.

Still, there was the further possibility that, because planes could fly over the enemy's lines, they might also be used to launch attacks in the rear, against supply lines, wagon trains, rail lines, bridges, or even army headquarters. In an era when armies in battle used up an unprecedentedly large amount of munitions and other supplies every day, wrecking a road or destroying a depot 20 miles behind the front lines might really influence the fighting, and very quickly. One might not be very accurate in such attacks, but one certainly could be very disruptive, forcing troops and wagons off the roads, possibly compelling them to march only at night. One might, of course, also provoke the other side to deploy aircraft to shoot down one's own aircraft. Both of these options were soon to be described as *tactical* air warfare, and they called for their own, appropriate weapons systems. To be sure, any fighter plane could swoop down to strafe a column of enemy soldiers on the march, but its true function was to shoot opposing fighters out of the skies; if one wanted to have a serious tool for ground attacks, far better to construct more heavily armed and heavily armored aircraft, equipped with rockets and bombs, to roam low across and behind the battlefield in a campaign of disruption.

All this was pure theory in 1918, and scarcely in sight throughout the 1920s; by the advent of the Second World War, it was beginning to become true, and would be manifested in the Luftwaffe's Ju 87 Stuka dive-bombers, the RAF's Typhoons, and the USAAF's Thunderbolts. Yet since an aircraft's range theoretically knew no limits (depending upon sufficient fuel capacity), early airpower enthusiasts also envisaged deploying them much farther afield and in a very different capacity, against the sources of the enemy's fighting capacities: the industrial plants and shipyards, hundreds of miles behind the front lines, that were pouring out guns, ammunition, trucks, ball bearings, sheet steel, warships, and so on. These would be aerial attacks upon the enemy's homeland, perhaps a great distance from the fields of battle or the naval struggles, and called for other, different types of aircraft, ones that would be larger, heavier, multiengined, and capable of carrying bombs over long distances. Since such bombers would be slower than the nimble fighters, they would also carry, in addition to a navigator and a bombardier, several gunners to beat off aerial attacks—thus vastly increasing the demand for aircrews.

The military implications here were revolutionary. To airpower theorists, the newer long-range bombers should no longer play a supporting role over some future putative Western Front or in support of fleets in the North Sea. They were an independent third force, and had to be organized as such. Their mission was no longer tactical and short-term; it was *strategic,* because it independently struck at the enemy's capacity to fight, and could bring war to the home front. This was the revolutionary part, the revolutionary appeal: longer-range bombing would cut the enemy's arteries, starve his armies of supplies, and shorten the war. It was, its advocates insisted, a far better investment than the slow slaughter of trench warfare and grinding maritime blockades; it could, they said, avoid the losses of hundreds of thousands of soldiers' lives in land campaigning. This was not just a new offensive instrument, then, but a fighting arm that one day would relegate armies and navies to a secondary role—hence the resistance of traditional generals and admirals to independent air forces from the early 1920s on, and their profound mistrust of their advocates.

There was one further leap, which would be the most controversial of all: why not use airpower to crack the morale of the enemy's entire population, that is to say, deliberately to destroy their will to support the fight by ensuring that, like their fighting troops, they too were going to be directly hurt by their support of the war? It was a natural further step. Having taken a step away from immediate fighting in the field, why not go after the munitions workers who produced the enemy's weapons—and then those who supported those workers? Bombing a bakery was in this sense as logical as bombing a power station, just as destroying the power station was as logical as bombing a railway line pointing toward the front. But this idea of area bombing was not something that simply emerged from the furnace of war around 1940 or 1942. The vision of monstrous aerial raids that would cause widespread civilian panic and loss of life was common to all the futuristic war literature over the previous century—the most famous being H. G. Wells's 1908 bestseller *The War in the Air,* with its lurid description of New York City being consumed in flames. And had not Napoleon himself declared that the factor of morale in warfare was many times the physical element?

As we shall see, both of these strategic aims—bombing to destroy the enemy's military output and bombing to destroy his morale—tended to merge as the Second World War unfolded, for a number of very different reasons. In the first place, since the enemy's shipyards, steelworks, arms plants, and railway junctions were almost always located in big cities, the workers and their families traditionally lived right next door.* Thus there was bound to be what, in later euphemistic military jargon, was termed "collateral damage" to nonmilitary targets. Second, there was the awkward fact that the pinpoint bombing of a defined military target such as a tank factory was not, and would never

---

*The terraced houses next to the great warship-building firm of Swan Hunter in Wallsend-on-Tyne, where I grew up, were so close to the yard that the nearest ones were overshadowed by the profiles of giant battleships such as HMS *Nelson* and HMS *Anson* as they were being assembled. How could those houses not be hit during an air raid upon the yard itself?

be, "pinpoint," unless it was a very rare event such as the low-altitude attacks of, say, Mosquito special squadrons on a particular object. The accuracy of higher-altitude and larger-scale bombing over the Brest shipyards or the Duisburg steelyards was badly affected by almost constant cloud cover and high winds for most of the year, making the bombsights inadequate. Somewhere down there was the enemy factory, but the flak was getting heavy and the enemy fighters were approaching, so the aircrews dropped the bombs and got back to base. Many of the remarkably candid memoirs of American and British fliers during the strategic bombing campaign admit that they just wanted to get rid of their dead weight of 4,000 pounds of bombs and escape home.

But these points relate to the inaccuracy of strategic bombing, not to its intention. An aerial attack against specific enemy military targets, even if it went badly astray on any particular night, was still an assault upon the other side's total fighting capacity; it was clearly within the long-established traditions of the "rules of war." But a dedicated air campaign to weaken the enemy's resolve by deliberately plastering his major cities was something else. It disregarded the West's long-established principles of proportionality. Now, the purpose of this chapter is not to analyze the *ethics* of the Second World War's strategic bombing campaigns but to examine their *efficacy*, and especially why the Anglo-American bombing in particular failed to live up to its forecasted performance until 1944, when it finally gained command of the air over western Europe. Still, it will be important to keep these distinctions in mind as the story unfolds: tactical air fighting was clearly very different from strategic bombing, but the nature and meaning of discriminate versus indiscriminate aerial warfare introduces an even greater distinction in most people's eyes. This is why so many books on this topic have been written with such passion.[10]

There is one final, intriguing point to notice about airpower theory: it lacked historical experience. When Clausewitz wrote *On War*, he distilled the lessons of hundreds of years of European land warfare that had culminated in the gigantic struggles of Napoleonic war. And when Mahan wrote *The Influence of Sea Power upon History*, his grand-theoretical arguments were followed by a detailed account of naval warfare in a specific time frame that was also part of the book's title: 1660–1783. Airpower theorists could only look forward: there was no past, merely hints of the potentially revolutionary effects upon war of humankind's newly found capacity to fly. Their visions, no less than Tennyson's, were wildly conjectural. They were hypotheses. And, the theorists felt, they needed testing.

From Folkestone (1917) to Dunkirk (1940)

Strategic bombing was forged in the furnace of the First World War. By 1914 most of the great powers possessed some military aircraft, often divided into army planes and navy planes, albeit of the same primitive designs. Early wartime aircraft were chiefly used for reconnaissance and artillery spotting, and it was some time before rival squadrons began tangling with each other over the trenches. But the industrialization of war in all its forms swiftly impacted the nascent aircraft industries, which began to turn out more powerful, better-armed, and longer-range planes. The first raids against enemy soil were chiefly carried out by German zeppelins against England in a campaign between January 1915 and November 1916 that claimed dozens of lives and did much damage, but more indicative of the future was the attack upon Folkestone on May 25, 1917. In this daylight raid twenty-one twin-engine Gotha bombers hit the seaside town, killing 165 and seriously wounding another 432 civilians. A new era had begun.

The Folkestone raid, and the bombings of London that soon followed, produced panic and riots, galvanizing Lloyd George's government not just into providing barrage balloons, high-level guns, and other air defense measures but also into taking the more dramatic institutional step of creating an independent Air Ministry that would take over the separate forces of the Army's Royal Flying Corps and the Navy's Royal Naval Air Service. In the Smuts Report of August 1917, its author, the talented South African general who was at that point a key member of the Imperial War Cabinet, advanced in a single paragraph the case for strategic airpower:

> Unlike artillery an air fleet can conduct extensive operations far from, and independently of, both Army and Navy. *As far as can at present be foreseen,* there is absolutely no limit to the scale of its future, independent, war use. And the day may not be far off when aerial operations,

with their devastation of enemy lands and destruction of industrial and populous centers on a vast scale, may become the principal operations of war, to which the older forms of military and naval operations may become secondary and subordinate.[11]

Persuaded by Smuts's elegant prose and by the nationwide alarm, the government not only established the independent Royal Air Force in April 1918 but also authorized funds for the building of long-range, four-engine Vickers Vimy and Handley-Page bombers that would carry the fight to the center of Berlin.

As it turned out, the war ended before such strategic bombing could begin, but there was no doubt that the aerial events of 1917 and 1918 had given birth to a totally new dimension of human conflict. Britain went further than any other country in creating an independent air arm, in part because it wanted to deter any future attacks on its island fastness by threatening to inflict even more damage upon an enemy, in part because it tended to see its own strategic bombing as a natural aerial extension of the naval blockade that had weakened Germany, and finally because it was a relatively easy substitute for sending vast future armies back to the bloodied trenches of Europe. France's obsession, understandably, was with land security, so air forces continued as a supplement to the army—the service itself was called L'Armée de l'Air. Russia was first convulsed in civil war, then preoccupied by early Bolshevik reconstruction, then concerned almost exclusively with homeland defense. In Japan the mutual dislike between its army and navy was such that two parallel air forces were created, each to support the operations of the mother service, and neither of which called for building large, expensive bombers. The same service bipolarity occurred in the United States, although that did not stop the U.S. Army Air Corps from developing significant theories and designs for long-range strategic bombing during the interwar years. Only in Mussolini's Italy was a separate air arm developed to match the British structures, though in that case only with a mix of medium-range bombers and fighters.

Though Smuts's concept of a transcendent war-winning weapon was nowhere near developed in the years after 1919 because of service jealousies and because the newfound air services were being slashed to the bone by postwar financial retrenchments, nevertheless the *theory* of airpower's unique capacities and even greater potential gained more and more conviction. This was due, in no small measure, to three remarkable prophet-propagandists: the redoubtable General Billy Mitchell in the United States; the first head of the Royal Air Force, Air Marshal Sir Hugh Trenchard; and the Italian airman-author Giulio Douhet.[12] Each of them had begun his career as an army officer and been witness to modern, expensive land warfare before he was moved on to his country's fledgling air service. Each was attracted to the possibilities of fighting offered by this new dimension of conflict. Each faced hostile criticism from many sides and, as a consequence, tended to advance the claims of airpower much further than was practicable at the time: they were laying claim to the future, and the more they did so, the more the traditionalists tried to block them. Mitchell, indeed, was reverted to the rank of colonel in 1925 and shortly afterward court-martialed for his public attacks upon the army and navy's leadership.

None of the three was as simplistic—or as immoral—in his advocacy of bombing the enemy's heartland as later portrayals would suggest. For example, Trenchard, in his first comprehensive statement of air doctrine (1927), made a distinction between scaring enemy workers away from their factory or dockyard by frequently bombing their workplaces and "the indiscriminate bombing of a city for the sole purpose of terrorizing the civilian population," which would be "contrary to the dictates of humanity."

Notwithstanding that caveat, Trenchard's claim that the best way to defeat any future foe was to strike at his industrial heartland produced robust and withering retorts from the professional heads of the British Army and the Royal Navy, both of whom pointed out not only the dubious accuracy and thus immorality of bombing civil-industrial targets but also the lack of empirical evidence that such a strategy would bring an enemy to his knees.[13]

If any group was responsible for painting the horrifying specter of cities and civilians devastated by future aerial attack, it was the professional politicians of the interwar years and the popular press. The best-known and most-quoted statement of all came from the Tory leader Stanley Baldwin in November 1932:

I think it well for the man in the street to realize that there is no power on earth that can protect him from being bombed. Whatever people may tell him, the bomber will always get through. . . . [T]he only defence is in offence, which means you have to kill more women and children more quickly than the enemy if you want to save your-selves.

Grim words; and the fact that Baldwin was generally regarded as a mild-mannered and soothing figure made this opinion seem even more disturbing to public opinion. "Fear of the bomber" grew and grew, and as the 1930s progressed, news from the fighting in China, Spain, and Ethiopia—especially the reports of poison-gas bombings—only increased the feelings of terror.[14]

Yet the fact was that powerful, long-range, independent strategic air forces had still not arrived when the Second World War broke out. For the geopolitical and political reasons mentioned above, neither the French, Soviet, nor Japanese armed service placed value or priority upon strategic air campaigning—the threats were much closer to home. The Luftwaffe was more balanced in its air doctrines and weapons systems, and by 1939 it possessed large numbers of medium-range bombers, but the building of a true strategic force (like, perhaps, German aircraft carriers) had been postponed because Hitler was obsessed with nearby European conquests. The Italian air force still dallied with Douhet's strategic boldness, at least in theory, but its situation as by far the weakest of the powers in industrial and technological terms meant that there was a massive gap between rhetoric and reality.

Thus in 1939 the only air forces that even came close to independent aerial power were the RAF's Bomber Command and its American equivalent. Both had carved out self-standing structures from skeptical or even hostile military and naval establishments to pursue that aim (the American strategic bombing enthusiasts worked in disguised form at the U.S. Army Air Corps's so-called Tactical School at Maxwell Air Force Base until Congress finally became alarmed at the rise of German and Japanese airpower). Both inherited the Mahanian, navalist assumption that economic pressure upon the foe, now applied from the air as well as by sea, would cause a crumbling from within, and also assumed that the Allies had deeper purses than the Fascist states to sustain such a conflict, especially after the United States joined the conflict openly in December 1941. Both countries also enjoyed a "moat"—a stretch of water that kept them from a foe—but, curiously, they felt that bombers (rather than short-range fighters) were the best instrument to effect a surprise attack from over the horizon. Here they were totally different from the continental European powers. How could Prussia-Germany, surrounded on all sides, not stress the primacy of land power and, with the coming of aircraft, not emphasize planes that chiefly supported its armies? Only the British and the Americans had the privilege of choice.

Ironically, however, those American and British bombing advocates still did not have the capacity to implement what their theories of strategic warfare called for. They had prototype four-engine "heavies" on their testing fields or still on the drawing boards, but their air staffs were also under increasing pressure to give priority to antisubmarine aircraft, carrier planes, and fighters for defense of the homeland. Scared of the Luftwaffe's rise, the British government and public in the late 1930s wanted the security of fighters overhead more than anything else. Thus bomber programs had no monopoly on production during those years, nor would they when war broke out.

What was more, the early stages of the air war pushed strategic bombing further to the margins. The United States military would be sitting on the sidelines for another twenty-five months, watching the conflict spread across Europe and planning for its great future demonstration of airpower, but (obviously) without any direct experience of what it was like to conduct an aerial offensive day after day. By contrast, RAF Bomber Command was in the thick of it from the beginning, although under highly constrained circumstances. The French would not let them fly from French bases out of a fear of reprisal attacks, and the Air Ministry at first thought its bombers should drop propaganda pamphlets on northern Germany rather than explosives. Even in assaults using explosives, the existing aircraft were rather slow, did not fly at very high altitudes, carried a meager amount of bombs, and had few defenses against Luftwaffe fighters in daytime encounters. As a result, by April 1940 Bomber Command had confined itself to nighttime bombing. While this reduced losses, it massively reduced accuracy as well.

In any case, all eyes were on the Luftwaffe's stunning successes in 1939–40. Working with the relatively few but massively effective Wehrmacht panzer and motorized infantry divisions, the German air force swept into Poland, not only wiping out the less modernized Polish army but also inflicting horrific, indiscriminate bombing raids upon Warsaw and other cities. All the imagined nightmares of civilian slaughter, collapsed buildings, and panicked crowds so luridly described in the interwar literature were now realized in the attacks of the loud, screaming Stukas or the steady aerial bombardment of the Dorniers and Heinkels. After seven months of "phony war," the same happened to the Netherlands, Belgium, and Norway in the late spring of 1940. What could stop the vultures from the sky, especially when they were followed up by paratroops, tank columns, and assault forces? How could, say, the city of Rotterdam avoid the devastation that had hit Warsaw? Nothing would suffice—or, at least, nothing in the slim armories of the Dutch and Polish states in 1939–40. To all observers, then, it appeared that Hitler's massive investments in the Luftwaffe had given Germany an unbeatable weapon. The astonishing defeat of France in May and June 1940, the eclipse of what had been one of western Europe's greatest military nations for the past three centuries, totally confirmed this impression. Traditional principles of warfare were obsolete.

Or were they? However stunning the Nazi successes of 1939–40, the Luftwaffe had operated within specific and unusually favorable circumstances that might not be repeated elsewhere, and certainly not against a sizable opponent. In all those early campaigns, the German air force enjoyed the all-important advantage of operating within favorable range. From airfields in East Prussia or Germany itself, the distance to Warsaw was negligible; from Rhineland bases to Holland was a hop and a skip. French military positions in Champagne or Lorraine could be assaulted from three sides within twenty minutes. Too, the Luftwaffe encountered no serious opposition in the air. There was no equal sparring partner. The large French air force of the early 1930s was sadly neglected and dilapidated after a decade of economic decay.

This meant that Hermann Goering's vast air flotillas were actually *not* equipped to fight against the only other European power that had seriously invested in airpower during the late 1930s. Perhaps the Germans were led astray by the RAF's piecemeal and limited effort in the battles over France in April and May 1940, where British ground and aerial forces had been blown away by Germany's blitzkrieg methods—even if the RAF's Hurricanes inflicted a fair amount of damage upon the Luftwaffe when covering the retreat of the Allied land forces to Dunkirk. Yet the great core of RAF Fighter Command had not been broken by the fall of France; its highly respected chief, Air Marshal Sir Hugh Dowding, despite intense political pressure, had refused to release further fighter squadrons from U.K. air bases, apprehending the challenge that was to come. This meant that in April 1940 he and his Fighter Command groups in the south and east of England had a modern air force that could take on the Luftwaffe. At last the fight in terms of aerial equipment and supporting services was equal. What was more, for the first time since the war began, the Germans would lack the range to get at the heart of their enemy.

The Battle of Britain: Lessons Learned, and Those Not Learned

The first real chance to test the theory and the full possibilities of strategic bombing came during the Battle of Britain, waged during the summer and autumn of 1940, and, though less dramatically, during the next several years of German nighttime bombing of the cities and factories of the British Isles. As suggested above, no previous aerial conflict could compare. The First World War's aerial battle had consisted chiefly of dogfights over the trenches, plus a few tentative forays at longer-range bombing. Italy's merciless aerial attacks upon the tribes of Tripolitania and Ethiopia, the Kondor Legion's devastation of towns such as Guernica during the Spanish Civil War, the Japanese air assaults on various Chinese cities in the 1930s—all of them were one-sided airborne attacks by an industrialized state upon poorer, usually defenseless populations, and the Luftwaffe's onslaught against Poland in September 1939 was not much different. The spectacular Nazi conquests of the Low Countries, Denmark, Norway, and even France itself in April through June 1940 were manifestations of *tactical* airpower, for which the bulk of the Luftwaffe had been designed and trained.

Only with the Battle of Britain did there arrive that Tennysonian vision of rival aerial navies grappling for command of the skies, day after day, month after month.[15] During the long, hot summer of 1940, over the wheat fields and orchards of Kent and Sussex, strategic theory encountered logistical and organizational reality. It was not just the emotion-charged images—of aircraft's vapor trails entangled across a clear blue sky, of St. Paul's Cathedral standing out above the flames, of Churchill visiting the bombed-out houses of Londoners—that counted. This struggle also consolidated the resolve of the British to soldier on, and had enormous effects upon foreign opinion abroad, especially in neutral America. Strategically, it was also the first time the Nazi juggernaut had been checked. Britain and its empire would fight on, in their time carrying the battle back to Hitler, in the air and at sea, in commando raids and in support of the embryonic European resistance movements, and in the Mediterranean against Mussolini's Italy, which had rashly joined the fray as France fell. Hitler himself, eagerly anticipating his eastern campaign by late 1940 and still confident in his destiny, might dismiss this as a setback. But the American special envoys and military experts sent to Britain in the summer of 1940 could return to an anxious Washington with assurances that the island state would not surrender, a message much reinforced by Churchill's great radio speeches of this period and his many personal letters to FDR.

Yet neither the human dimension of the Battle of Britain nor its larger strategic implications concern us here so much as the operational and tactical pointers it gives to understanding the successful waging of modern aerial warfare when two opponents possess large resources and good organization. It also offers us a nuanced, multilayered contribution to the eternal, more general debate about offense versus defense in war. The German strategic purpose here was clear: Hitler wanted Britain knocked out of the war, and if that could not be achieved through intimidation and some Vichy-like deal—surely unlikely after Chamberlain stepped down—it would be secured by an invasion. That in effect meant obtaining control over at least the narrow waters of the English Channel, across which the German armies would flow. Yet that operational aim depended in its turn upon gaining unqualified control of the air, not just by destroying the RAF but also by keeping at bay Britain's powerful naval forces. To the newly formed Churchill government, its own strategic logic chain was also perfectly clear: by denying the attacker command of the air, he was also denied control of the waters around the planned invasion beaches, and landings themselves could never take place. This was to be Germany's own situation from 1942 onward, when the Allies moved from defense to offense. Before launching a seaborne invasion, one needed to dominate the skies.

In terms of institutional organization, the Germans were remarkably good. The Battle of Britain was not one for which they had really planned, either in air force doctrine or pilot training or in basic infrastructural investment in airfields, fuel tanks, and the rest; certainly the Luftwaffe had inherited an array of northern French air bases, but those were of mixed utility—the Bf 109s, with their delicate undercarriages, often found themselves flying from bumpy grass fields. There were many other weaknesses—German radar was way behind that of Fighter Command, intelligence was poor, and there was far too much interference (about targeting or close escort tactics) from Goering and more distantly from Hitler. The latter had only ordered preparations for Operation Sealion, the maritime invasion of Britain, on July 16, and it was not until August 6 that the operational commanders, Albert Kesselring and Hugo Sperrle, got detailed instructions for their air strike. So a fair-minded observer can only be struck (and not for the first or last time) by how impressive German organization was, even when switched from one operational challenge to a different one. It was no simple task to fling hundreds of fighters and bombers across the Channel—on September 7 it was almost one thousand aircraft—day after day, week after week. But that's what they did.[16]

Yet the British organization was even better prepared—in its chain of command and integrated defense systems, with radar, spotter stations, air raid precautions, sector command posts, and air traffic controllers, all decentralized yet networked, and all feeding into Dowding's Fighter Command headquarters. To be sure, it *should* have been well organized, for this was the contest that British governments had been bracing themselves for all through the 1930s (even Neville Chamberlain did not stint on investments in air defense). Given the British Empire's vast obligations and limited resources in the interwar years, it is noteworthy that a relatively large share of financial and industrial capital had gone into the very careful and sophisticated defense of the

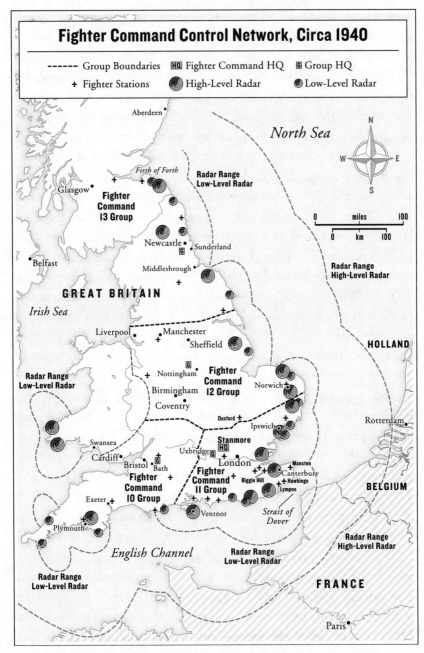

Fighter Command Control Network, Circa 1940

----- Group Boundaries   [HQ] Fighter Command HQ   [G] Group HQ

+ Fighter Stations   High-Level Radar   Low-Level Radar

FIGHTER COMMAND CONTROL NETWORK, CIRCA 1940

By this time, the British possessed the most integrated aerial defense system in the world: even this fine map of UK radar stations and RAF Fighter Command arrangements could not possibly also include the dense networks of aerial observation posts (linked via underground phonelines), antiaircraft batteries, and barrage balloons.

United Kingdom itself.[17] Thus, while Britain encountered many set-backs in the early years of the war, it certainly was prepared for enemy assaults from the air.

In this epic and complex struggle between these rival, sophisticated war-making organizations, three aspects stand out above all the others in explaining the course of the campaign: geography, targeting, and men and their weapons—the pilots and their planes.

By geography here we mean the critical roles of distance and space. The flying machine liberated human beings from the natural con-straints of land and sea—man was now, writers claimed, "as free as a bird."[18] There was, unfortunately, one massive constraint: heavier-than-air craft had very limited range, even more so when they carried hun-dreds of pounds of bombs. A frigate in Nelson's day was bound to the sea, but it could (with local provisions) stay out for months, even years, at a time. When the Second World War broke out, an aircraft's time in the sky was shockingly limited by the size of its fuel tanks and the not inconsiderable fact that it also had to fly back to base. Liddell Hart noted long after the war that the superb single-engine Messerschmitt Bf 109, based in Pas-de-Calais, could just about reach the outskirts of London, and then had to turn back. If it had already had to fight with the Hurricanes and Spitfires over the Kent bases, it had about 75 min-utes of tactical flying time and was, of course, no longer protecting German bombers.[19] Obviously, Dowding's squadrons, rising from those very bases with full tanks, had much more fighting time. And if an RAF plane was damaged, it had a fair chance of gliding down to an airfield or cornfield somewhere, and of the pilot surviving; a German plane force-landing on English soil was an aircraft lost and a crew captured.

It is precisely here, in the matter of boosting Fighter Command's range and accuracy of counterattack, that the remarkable air-tracking radar system developed by Sir Robert Watson-Watt and his associates in the 1930s made its greatest contribution.* Its value was immeasur-able. Those tall trifid-like pylons with pulsing machines, set up in a chain along the east and south coasts of England, could pick up objects

---

*By a nice coincidence, Watson-Watt was a descendant of an even more famous Scot-tish engineer-inventor, James Watt, who had created the world's first practical steam engine.

in the sky a hundred miles away and relay that information into a home defense network; at the time, there was nothing like it.[20] To Dowding's hard-pressed squadrons, it gave two vital gifts. First, it gave the traffic controllers the direction and the size of the oncoming air fleets well before they reached the English coasts, and thus allowed Fighter Command to prepare itself. Perhaps the best example here comes not from the fighting in the southeast but from a flank attack on August 14 by the Luftwaffe's No. 3 fleet (from Norway) against Newcastle, which was handily intercepted and cost the Germans fifteen bombers with no RAF losses; without radar, no such interception would have been possible.

The second gift, of time, was equally precious. If those strange tall aerials sited along the shore really could pick up, say, sixty dots rising into the skies above their bases at Abbeville, then the RAF squadrons did not need to have more than a few patrol planes in the air (and wasting fuel) at any one time. The great bulk of their squadrons and their overworked aircrews and support mechanics could rest on the ground until the order was given to scramble. Those popular photographs of blond, tousle-haired young men resting on the tarmac beside their Hurricanes and having one last cigarette are really telling us that the British fighters had the great defensive advantage of not needing to take off until the enemy's aircraft might already have depleted 30 or 40 percent of their fuel supplies. What is more, an RAF plane could slip down to any airfield and get swift refueling. The pilot of a Junkers bomber or a Messerschmitt fighter had to recross the Channel, get home to a base perhaps 30 miles from the coast, and then rise into the air again—which was, obviously, much more exhausting (as the Luftwaffe's own detailed record shows), and was not helped by Goering's refusal to give his aircrews any significant rest. Unsurprisingly, a lot of German crew casualties came at the end of a long, aching day.

Then there was the factor of targeting. The issue was a pretty simple one, but the Luftwaffe did not think about it enough (and, as we shall see, they would not be alone in this defect). As a number of historians have observed, in this the Germans had, ironically, forgotten their Clausewitz—that is, the need to identify the key object of the campaign and the best means to achieve that object. The basic aim, as noted above, was that elusive concept "command of the air." Yet by the beginning of September the Luftwaffe had failed in that aim and had lost more than eight hundred planes, with many more in need of repair. Thus it was clear that the RAF, while also losing hundreds of aircraft in these weeks, had a far greater capacity to contest the skies than the German commanders had originally expected. A continued failure to eliminate the British fighters and their airfields meant, simply, that the battle as a whole would be lost. This was the key. Fighter Command had to be destroyed.

Unfortunately for Hitler's ambitions, Goering and his air marshals made a couple of fateful decisions at a critical meeting held on September 3, 1940 (ironically, in The Hague, city of the international peace conferences). The bombers of Luftflotte No. 3, which always had a longer way to go from their Normandy bases to fight over Kent, were, at Sperrle's urgings, released for night attacks upon the many ports of the southern British coast—Southampton, Portsmouth, Devonport, Cardiff, and the rest. More important still, it was resolved that Kesselring's Luftflotte No. 2, based in Pas-de-Calais and heavily reinforced once again, would wage an unrelenting daylight and nighttime campaign against London. The result was to give the capital city, and especially its East End working classes, an ordeal by fire. The first mass attack on September 7 killed more than 300 civilians and seriously wounded another 1,300; the flames raging in the docklands and around St. Paul's Cathedral could be seen for 50 miles and were an easy beacon for Goering's night bombers. This pounding continued over the weeks to follow, as each side's air force threw more and more planes into the struggle. On some occasions the Luftwaffe found a chance to ravage London; on other days Fighter Command's counterattacks were punishing—sixty bombers were shot down on September 15 for the loss of twenty-six RAF fighters (half of whose pilots were saved by their parachutes). The struggle was new and titanic, and the world watched in awe.

The London Blitz concealed the critical fact that the Luftwaffe's decision to switch targets saved Fighter Command. Given the inherently strong defensive position held by the British, operational logic pointed to the priority of targets for the German attackers: (1) the key radar stations along the coast, (2) Fighter Command's airfields in southeast England, and (3) the numbers and morale of the Allied pilots. Amazingly—and especially given the fact that the tall detection masts were being erected on coastal hilltops as early as 1938—German air intelligence failed to appreciate the critical importance of radar to the entire British defense system even when, again and again, RAF squadrons seemed to be waiting in the right spot to meet the raiders. The utility of some early attacks upon the stations was actually questioned by Goering on August 15, and such attacks ceased thereafter. Yet without radar, the defenders would have been reliant solely upon the Observer Corps and hugely expensive aerial patrols over the Channel. Goering's September 3 decision to alter the targets also took the pressure off Fighter Command's forward airfields, which had been under unrelenting attack during August—RAF Manston in Kent had had to be abandoned—and especially off-key sector stations such as Biggin Hill and Northolt, which controlled whole fighter groups. The absence of bombing and strafing of the airfields gave the weary crews a bit more time to rest and the engineers time to repair and refuel the planes without interruption. But the failure to attack the blatantly vulnerable radar stations continually amazed the British.

Instead of concentrating upon the critical points of Britain's aerial defense system, Goering, urged on by the Fuehrer, chose to attack the largest, most sprawling residential target in the world. The damage inflicted upon the city, as well as upon England's southern and western ports, was huge and brutal. But it is doubtful that the Luftwaffe could have destroyed London even if it had had twice as many bombers—and even if it had done so, that did not guarantee a way of surmounting the rising problem of invading the entire island as summer 1940 gave way to autumn. Small wonder that the skeptical German naval and army chiefs just sat on their hands, content to let Goering prove himself or fail. Finally, attacking London from French air bases involved a lot more time over England and thus allowed the RAF a second chance to get at the weary German bombers on their return; it also allowed Dowding the chance to pull in fresh squadrons from East Anglia and the Midlands for afternoon counterattacks. "I could hardly believe the Germans would have made such a mistake," Dowding later wrote.[21]

"Fresh squadrons" meant the conjunction of two elements, planes and pilots. Each side deployed a small variety of aircraft types, the RAF squadrons flying the slower (and soon to be replaced) Defiants, most predominantly the Hawker Hurricanes, and a small but increasing number of Spitfires, while the Luftwaffe relied upon its fearsome Stuka Ju 87 dive-bomber, a mix of twin-engine fighter-bombers and medium bombers such as the Bf 110, the Dornier 17, and the Heinkel 111, plus the larger, slower Ju 52, together with the mainstay of the fighter force, the single-engine monocoque Messerschmitt Bf 109. For an entire year, all of these famous—or notorious—aircraft had terrorized Europe (British railway stations, post offices, and schools all had wall charts showing their dread silhouettes); now they were being shot out of the sky, and there were no newer, improved, longer-range bombers in the pipeline. As noted above, the Luftwaffe's exploration of heavy bombers had been diverted in the 1930s, so there was no German equivalent of the enormous Anglo-American strategic air forces involved in the later attacks upon the Third Reich. Dorniers bombing the London dock-yards were, at this stage in the war, the closest equivalent to strategic airpower, and with limited bomb capacity they were clearly inadequate to the task.

The aerial campaign against London and the southern counties thus became another rock-paper-scissors game. If the German bombers reached their targets as planned—and many of them did, as was apparent from the raids on the East End—they could inflict appalling damage on the British people, housing, factories, and ports. If the Hurricanes and Spitfires got to them first, they suffered heavy losses; even the tough Bf 110s found it hard to handle a Hurricane, let alone a Spitfire. It was no doubt particularly satisfying to members of those Anglo-French regiments that had been battered at Dunkirk, and to those civilians who had escaped the relentless dive-bombings of refugee columns across northern France, to learn that the Stukas were easy prey for much faster RAF fighters and had to be withdrawn from the fight. On the other hand, if the Bf 109s tangled with the British squadrons before the latter got to the German bombers, it was a different story, and the odds were much more even. When, in late August, Goering showed alarm at his bombers' losses by ordering the Bf 109s to stay close to the Bf 110 bombers, an enormous amount of tactical flexibility was lost. Coming under more and more criticism for failing to protect the bombers, the Messerschmitt fighter squadrons were forced into flying as many as three sorties a day to England, which was simply unsustainable.

Thus, although both sides experienced massive attrition rates in frontline aircraft, that was the lesser worry. Both Germany and Britain were moving to "total war" production and receiving fresh aircraft every month, with output from U.K. factories surprisingly some way ahead. The greater mutual concern was the attrition rate of skilled pilots, who were much harder to replace than single-engine aircraft. As Williamson Murray's detailed analysis reveals, both air forces were suffering enormous losses of fighter pilots as August 1940 unfolded into September. The RAF switched pilots from Bomber Command, borrowed from the Fleet Air Arm, cut the training time for new pilots in half, and—an enormous blessing—opened its arms to any Free French, Poles, Czechs, Norwegians, and Americans who could fly, not to mention the stream of fresh pilots from Canada, Australia, New Zealand, South Africa, and Rhodesia.* (The Luftwaffe, by contrast, received only an ill-fated Italian show of support during the Battle of Britain.) The key statistic is also supplied by Murray: at the beginning of May 1940 Goering commanded more than 1,000 "operationally ready" Bf 109 pilots, and their attrition rate that month was a mere 6.8 percent; at the beginning of September he had only 735 such pilots, and 23.1 percent were lost as the month unfolded.[22] The great RAF counterattack of Sunday, September 15, when sixty fighters from No. 12 Group in the Midlands swept southward into the German formations late in the day, shocked Luftwaffe commanders by demonstrating the depth of Britain's aerial defenses.

Overall, the strategic lesson of the Battle of Britain was very clear: against a well-defended and well-organized aerial defensive system, a force of bombers could not always "get through" in the Baldwinian meaning of the term; a few might make it, but most would suffer if flying alone or with only partial protection from escort planes. Getting protection for half of the journey was like getting hardly any protection at all, for the enemy simply moved his chief defensive lines farther back, waited, and then assaulted. A strategic bombing campaign by one in-

---

*Two Polish squadrons, 302 and 303, flew in the Battle of Britain, and with astonishing results. The 303 (Warsaw) squadron had the highest results of *any* RAF squadron in the struggle, shooting down no fewer than 110 German planes between July 10 and the end of October—a small but nice revenge for the awful destruction of Warsaw. By the end of the war, the RAF's battle order included fifteen Polish squadrons, with 19,400 personnel.

dustrialized nation against another thus depended for its outcome upon which side's fighter forces had cleared the skies of their direct opponents. Nothing else would do.

It is for this reason that we have spent much space and detail on a campaign that was waged almost two and a half years before Casablanca. Of course, the many differences between the (essentially) four months of the Battle of Britain and the unremittingly harsh fifty months (if one counts from January 1941 onward) of the British and American strategic bombing campaign against Germany were massive. But the basic operational principles of aerial warfare—that is, a true appreciation of geography, targeting, and men and planes—remained the same.

It is therefore remarkable how swiftly and completely both the American and British commanders dismissed the idea that there was a larger lesson to be learned from Fighter Command's defeat of a numerically larger Luftwaffe in those epic months of 1940. American observers concluded that the German bombers had inadequate armament, flew too low, and had poor formation discipline; this would not happen to them. Since the Americans planned to deploy "large numbers of aircraft with high speed, good defensive power, and high altitude," the USAAF would avoid all of Goering and Kesselring's problems. True, the B-17 Fortresses and B-24 Liberators were vastly superior to, say, a Heinkel 111, yet all this assumed that the Luftwaffe's fighting capacities would remain static and that flying at 24,000 feet rather than 12,000 feet would minimize losses. But what if newer German fighters flew at 30,000 feet? The conclusions of the RAF chiefs (though probably not Dowding himself) were even more egregious. The remarkably high morale and fighting spirit of the civilian populations of London, Portsmouth, and Coventry when under heavy bombing did not lead to any questioning of air force doctrine about striking at the enemy's heartland. Instead, the C in C, Sir Charles Portal, and others simply assumed that the German people would not be as tough as the British— an exact mirror of the conclusion that the Luftwaffe's prewar planners had come to regarding the weaker determination of the Western democracies compared with the iron will and national unity established by the Nazi movement.[23] If the operating assumption was that the other side would crack first—which sounds awfully like Douglas Haig,

Erich Ludendorff, and trench warfare during the First World War—
then it is no surprise that little attention was paid to really important
issues such as distance, targeting, and detection.

The Allied Bombing Offensive and Its Collapse, Late 1940 to Late 1943

On October 12, 1940, Hitler postponed Operation Sealion until the
following spring at the earliest. By January, with his mind now fixed
upon the coming attack on the USSR, he ordered a halt to most preparations for an invasion of Britain. Goering in turn instructed Luftflotte
No. 3 under Sperrle just to concentrate upon night attacks on industrial targets. There would be no more epic fighter duels over the apple
orchards of Kent.

And in fact there would rarely be any German fighters over England again. Sperrle's force was augmented by the transfer of some
bomber squadrons from Luftflotte No. 2 and became the single command for the continuation of the air war against Britain. He thus
possessed a considerable fleet of up to 750 bombers (although repair,
maintenance, and training made total operational numbers much
smaller than that). There was no letup for the inhabitants of London
and other British cities during the night attacks of the rest of the year—
even after Hitler's postponement of Operation Sealion, London was
attacked for fifty-seven nights in a row before the Luftwaffe switched to
a terrifying raid that flattened large parts of Coventry's historic city
center. The aerial Blitz continued until the end of December, when it
was suspended due to continued poor weather, but resumed in the
spring of 1941. As late as May 10 London was pounded again and
again by very heavy nighttime attacks. After that, and although German raids would continue off and on throughout the war (until replaced by the V-bombs), the Luftwaffe's bombing campaign would
never again be as strong. By the middle of May 1941 most of Goering's
bombers and fighters were either heading to the Eastern Front or being
diverted to the Balkan and Mediterranean theaters. Later, this had a
very significant but unforeseen result: because German planes were no
longer flying over England in daylight, there was no photo reconnaissance of the massive buildup of Allied forces in preparation for D-Day.
Aerial weakness meant that Germany was blind.

While the Luftwaffe's nighttime bombing offensive was much less dramatic than the great aerial duel of August and September 1940, there was a lot to be learned from it, had an acute observer bothered to do so. It showed how difficult it was to carry out a sustained night offensive, whichever air force was making the attack, for they all faced the task of finding their way in the dark. Once the German squadrons had passed over the coastal radar stations, finding them and then shooting them down became a game of hide-and-seek, so any losses inflicted were piecemeal rather than decisive. Over the next few years it galvanized the British into establishing a very sophisticated defensive system against aerial attacks at night—a combination of specially trained Spitfire and Mosquito squadrons, a much broader band of HF-DF detection stations, and interceptions of Luftwaffe Enigma messages (although, as with the Atlantic battles, detection and decryption breakthroughs did not automatically bring destruction of the foe).

But the Luftwaffe's task was much harder. To begin with, there were the usual hurdles of range and endurance, especially for twin-engine medium bombers; if RAF Blenheims and Wellingtons could not reach very far into France and Germany, then equally Dorniers and Heinkels could not reach far into Britain—or stay very long once they got there.* Aside from the gigantic target of London and some special maritime targets such as Harwich and Portsmouth, the greater part of British industry was located farther north and west. The main shipyards were all there, and so were many of the vital coalfields. Apart from lighter vessels in the south, the Royal Navy positioned its heavier warships and its Atlantic escort squadrons in northern Ireland, Liverpool, the Clyde, and the Orkneys. And while the Midlands contained many factories, the late Victorian structure of British manufacturing meant that they were scattered across a very large area. Even on a rare moonlit night (which also made the German planes most vulnerable to RAF fighters), the rigidly kept blackout meant it was difficult for German pilots to find out where their appointed target was; on cloudy nights, the natural temptation was to dump their bombs somewhere in the area and run

---

*Roughly speaking, the German medium-range bombers flew within an arc from Coventry to Exeter, the RAF's bombers from Bremen to, say, Lille or St. Nazaire; Cologne and the Ruhr were scarcely in range in 1940.

for home. Most of those bombs dropped harmlessly into fields, though not a few of them struck a school, a hospital, or a row of workers' houses.

Finally, although German air intelligence had assembled a portfolio of British industrial and infrastructural targets before the war, there hardly seems to have been much pattern in the Luftwaffe's nighttime campaign. The bombing of London eased the pressure not only upon RAF bases but also upon the manufacturing regions, yet the slightly later attacks upon Birmingham, Coventry, Bristol, Exeter, the Tyne, Plymouth, and South Wales, although inflicting damage, were too scattered and sporadic to have a strategic impact. During the spring of 1942 there occurred another case of mutual stupidity by each side's air commands. In late March the RAF planners inexplicably ordered a raid upon the ancient, wooden-framed Hanseatic city of Lubeck, so enraging Hitler that he ordered retaliatory raids upon British cathedral and university cities (the so-called Baedeker raids) such as York and Norwich, and on May 3, 1942, the medieval heart of Exeter's town center was bombed to bits. Such raids had no strategic purpose whatsoever. They simply wasted bombs (and crews) and inflamed hatreds. In retrospect, as will be seen in the rest of this chapter, the most important industrial targets by far were the Rolls-Royce engine factories in Derby and the Spitfire and Lancaster assembly lines (often with "shadow factories" built even farther away, in north Wales or northwest England). A flailing, retributionist bombing offensive against cathedral cities simply lacked strategic purpose and took the focus away from the need to cripple Britain's rising aircraft production. Total aircraft production figures tell it all: 15,000 planes were built in 1940, 20,000 in 1941, and 23,000 in 1942.[24]

The Blitz upon Britain's cities did two other things: it steadily inured people to the coming of indiscriminate aerial bombing, and it aroused a yearning to hit back at Germany's population. Churchill's most popular speeches in this period were those in which he warned the German people that if they continued to follow the heinous Nazi leadership (what choice had they?), they would in turn suffer from what had been done to Warsaw, Rotterdam, London, and Coventry. In recent years it has become quite fashionable to denounce the Anglo-American strategic air offensive against Germany as a special form of holocaust—and there is indeed severe criticism to be leveled at it, as will be discussed below. It is only proper to remember that "terror bombing" first took place when Japanese, Italian, and especially German bomb bays opened upon civilian populations below.

RAF Bomber Command's aerial offensive against German targets encountered many of the same problems that the Luftwaffe had faced beforehand. The early months of the war had not been auspicious for a service that for over two decades had advertised the benefits of independent strategic bombing. It carried out some early limited raids over western parts of Germany, was forced by bomber casualties to abandon daytime raids, and then found nighttime bombing over Germany full of difficulties; the usually thick clouds hid Wellington bombers from enemy fighters but also hid enemy targets, and detection aids, if any, were primitive. After May 1940, Italy's entry into the struggle meant that an increasing number of RAF squadrons, including modern medium bomber squadrons, had to be diverted to the Mediterranean. And everyone—Coastal Command, the Fleet Air Arm, Fighter Command, the tactical commands across the Middle East and India—was screaming for more trained pilots and aircrews.

Still, by the end of 1940, and for another two years, Bomber Command occupied a very special place in British grand strategy. With the fall of France and much of the rest of western Europe, Britain and its Empire stood alone against the Axis, with the United States and the Soviet Union watching on the sidelines. In almost all respects its posture had to be a defensive one: to protect the merchant shipping convoys from the U-boats, to head off the German surface raiders, to blunt the Luftwaffe's nighttime attacks upon British cities and factories, to hold out against vastly larger Italian forces in Africa and the Mediterranean, and (if possible) to try to send some reinforcements to India and the Far East in response to Japan's piecemeal aggressions. Repeated and ever larger aerial attacks against Germany, by contrast, were the proof that Britain could hurt the foe and was serious about winning. This was hugely important in Churchill's relationships with Roosevelt and Stalin, and immensely helpful in sustaining British domestic morale. Not only that, a successful bombing campaign *would* weaken Germany's fighting strength, perhaps not to the extent that dedicated airpower advocates believed (i.e., bringing enemy collapse through bombing alone), but certainly sufficient to make the reconquest of Europe—when it came—somewhat easier. Even those British admirals and generals skeptical of Bomber Command's claims had to agree that wrecking German shipyards, hitting aircraft factories, and disrupting gun production were jolly good things.

The question was, could a successful bombing campaign be accomplished? The answer was no in 1941, and no again in 1942. Bomber Command sorties went out night after night, sometimes diverted to attacks upon German battle cruisers in Brest or the U-boat pens, but always returning to the assault upon the enemy heartland. But the extraordinary courage of the crews in being willing to undertake those appalling journeys, to lose many friends and colleagues in the night and then to go out again, was no guarantee of operational success if the right tools were not at hand to match the strategic doctrine. If the RAF's aircraft had limited range (and even when they had more range), usually couldn't see the targets, and possessed inadequate navigational and target-setting instruments, then there was little chance of damaging the foe's massive industrial strength even if he had no night fighter and flak defenses with which to hit back. With increasingly enormous antiaircraft barrages driving the bombers to fly at ever higher altitudes, the level of accuracy tumbled further. By the spring of 1941 the Air Staff was assuming a theoretical average error of drop of 1,000 yards, which was dismaying enough. However, in the rigorously compiled internal Butt Report of August 1941, based upon new day-after photographic reconnaissance, it was found that in a series of raids on the Ruhr only one-tenth of the RAF's bombers found their way even to within 5 miles of their assigned targets. This evidence really shook Churchill's faith in bombing. While he might continue to boast of the air offensive publicly, he was withering in his reply to the Air Staff's plea for a war-winning force of four thousand heavy bombers, pointing out that increased accuracy *alone* would quadruple their capacity to damage Germany. He also, for the first time, privately admitted that "all we have learnt since the war began shows that [bombing's] effects, both physical and moral, are greatly exaggerated."[25]

The RAF's response was to concede that "the only target on which the night force could inflict effective damage was a whole German town," which clearly implied a shift toward indiscriminate bombing.[26]

It is from this time onward that increasing stress was placed upon weakening the enemy's morale, whatever that meant. Moreover, the means of mass destruction were increasing fast. What damage could hundreds of the newer, powerful Lancasters inflict if let loose over the Third Reich? Also, the British scientific establishment was beginning, with RAF funding, to develop a series of top-secret directional aids to navigation and target identification—code-named Gee, Oboe, and H2S—that were increasingly capable of getting the bombers closer to the target. So, too, were the new RAF Pathfinder squadrons, specifically trained to fly ahead of the main bombing force and identify the target with flares and explosives. The British authorities also decided, a little later, to authorize the dropping of odd strips of aluminum (called variously "window," "snowflake" or "chaff") that blurred the enemy's radar screen. And in the midst of these many improvements, on February 22, 1942, Air Marshal Sir Arthur Harris became C in C, Bomber Command, and the service received its most implacable advocate of the sustained and general bombing of Germany.

None of this actually solved the basic operational dilemma: how to get control of the air over Germany so as to be able to eliminate the Nazi industrial machine. The idea that this might *not* be possible was absent from Bomber Command's mind. All that was needed now, it was argued, was the systematic application of further force, chiefly by the RAF's own efforts, although the coming to Britain of the USAAF in 1943, with its own conviction of the centrality of strategic bombing, was hugely welcome. If the American air generals Henry "Hap" Arnold, Carl Spaatz, and Ira Eaker needed the Air Staff's strong support for their European strategy, the RAF definitely required senior American airmen to keep up the pressure for the strategic bombing campaign. At Casablanca, the Anglo-American air chiefs stood firm together.

Harris was a remarkable character who was both much loved and much hated; he was as strong and egotistical as MacArthur or George Patton and just as aggressive. Like those two, he recognized the need for display, hence the coming in 1942 of Bomber Command's much-acclaimed "Thousand Bomber Raids." By scraping together training squadrons and second-line units, he managed to dispatch 1,046 bombers against Cologne on the night of May 30—a chiefly symbolic raid, although the city also possessed light industries and occupied a key spot on the lower Rhine. Six hundred acres of that ancient city were flat-tened, at a cost of forty bombers (3.8 percent). Other such raids fol-lowed against Essen and then Bremen, though repeatedly cloudy conditions and the higher losses among the training squadrons forced Harris to abandon such spectacles for a while. Still, he had made his point: he had an independent tool for killing Germans and hurting the Third Reich. This gave him the elbow room for the RAF's three great campaigns of 1943–44: the Battle of the Ruhr, the Battle of Hamburg, and the Battle of Berlin.

The results of these three battles, taken together, show how difficult it was (and still is) to offer a balanced assessment of Britain's strategic bombing campaign and why Bomber Command was allowed to do what it did for so long. The Battle of the Ruhr consisted of no fewer than forty-three major raids between March and July 1943 and did massive, indiscriminate damage to Essen, Aachen, Duisburg, Dort-mund, Bochum, Duesseldorf, and Barmen-Wuppertal (90 percent of the last named was flattened in one attack in May). Helped by the Oboe directional system and by the Pathfinders, Bomber Command's accuracy had improved; its planes could certainly reach their target cit-ies, stay in the air longer, and scatter their incendiaries and larger bombs (by this stage some of the newer Lancasters carried 8,000-pound high-explosive bombs). And the losses of RAF aircraft and crews, although steadily rising, were matched by continual reinforcements.

Perhaps Harris should have stopped there and not sent his bombers farther afield. There were two good arguments for constraining his campaign in this way, of which the first, of course, was geography. Whereas Britain's main steelworks (Sheffield, Doncaster) were a long way north, much of Germany's heavy industry was in the west of the country. Thus Bomber Command's squadrons only had to run a rela-tively short gauntlet of antiaircraft fire and Luftwaffe night fighters. Second, as the historian Adam Tooze has recently pointed out, the Ruhr bombings really were hurting that part of the German war econ-omy, and much more than the later critics of the overall strategic air offensive have understood.[27] While there may not have been pinpoint bombing of specific factories, the RAF was unloading an awful lot of bombs onto great concentrations of strategic industrial manufacture. Spaatz, when he heard American visitors criticize Britain's limited war efforts in 1943, was very quick to remind them that Bomber Command was the *only* force in the western camp that was directly hurting Germany.

But Harris wanted to move forward. To his delight, the Battle of Hamburg was a further advertisement for mass bombing. Between July and November 1943 a staggering 17,000 bomber sorties were flown against that great port city and others in the western parts of Germany, under the name of Operation Gomorrah. The initial raid, on July 24, was terrifying: 791 RAF bombers, including 374 Lancasters, screened by the aluminum "window" strips, directed by Oboe, led by Pathfinders, and aided by clear weather, pounded the center of that historic Hanseatic—and, traditionally, very Anglophile—city. There was no rest for Hamburg over the next few weeks, since the USAAF joined in, as did Mosquito fighter-bombers that had been specially modified to carry 4,000-pound bombs. Then the main force of Bomber Command struck, again and again, at German targets as the summer unfolded. On August 17, Harris sent 597 Lancasters to attack the Reich's experimental flying-bomb station at Peenemuende.

This, surely, was the apotheosis of the Trenchardian doctrine (the old man was still around, closely informed and grunting approval). Around 260 factories in the Hamburg area were obliterated; so, too, were 40,000 houses and 275,000 apartments, 2,600 shops, 277 schools, 24 hospitals, and 58 churches; some 46,000 civilians were killed. The devastation of Hamburg totally shocked the German leadership. Speer warned the Fuehrer that six more such attacks would finish off the Third Reich, a conclusion Hitler rejected. Goebbels, however, privately called the Hamburg bombings "a catastrophe."[28]

But Harris's terrible success in the flattening of Hamburg soon turned out to be Bomber Command's downfall, or close to it. These nighttime attacks were so obviously destructive and produced such huge panic in the German civilian population that the Reich's leadership could no longer dismiss them as a marginal activity. The damage inflicted upon Hamburg, following as it did upon Stalingrad, Kursk, El Alamein, North Africa, and Sicily, constituted a trio of warnings to the Third Reich that the relatively easy early years were over. From now on, it could not coast along, offering the German people guns and butter. Speer for his part, and the Luftwaffe leadership for its—by this stage,

Goering was just an ineffective blusterer—had the talent and drive to do just that.

As soon as the Hamburg campaign was over, Harris immediately raised the stakes by moving to the Battle of Berlin, which involved sixteen major attacks between November 1943 and March 1944 (along with side attacks on Frankfurt, Stuttgart, and Leipzig—20,000 sorties in all). During the same time, however, and with great alacrity, the German air defense system had been strongly revamped and reinforced. A gigantic army of heavy antiaircraft battalions now covered Germany. Special rocket-firing night fighter squadrons had been formed. German detection systems had made huge advances. The Luftwaffe had also figured out a way to circumvent the blurring effects of "window." Finally, of course, there remained two obstacles to any strategic bombing campaign: the weather, which during these winter months was unrelentingly bad, and distance, because attacking Berlin took far longer than raids on Hamburg or Cologne. German night fighters based, say, near Hanover could attack the bombers once, refuel, then attack again on the enemy's return flight.

For all these reasons, the damage inflicted upon Berlin—a massive, sprawling city, like London—was far less than that achieved against smaller German cities such as Essen. By contrast, Bomber Command's losses of aircraft and crews spiraled sharply upward as 1943 led into 1944. The loss rate for each raid was averaging 5.2 percent, and the morale among these extraordinarily brave crews was plummeting. In all, Bomber Command lost 1,047 heavy aircraft and had a further 1,682 damaged in the Berlin campaign. Shortly afterward, the catastrophic losses the Pathfinder pilot Pat Daniels had observed during the Nuremberg raid of March 30, 1944, brought the RAF's strategic bombing campaign against Germany to a grinding halt. To lose 95 bombers out of 795 dispatched was a totally unsustainable casualty rate of 11 percent. At this stage, even the Air Staff in Whitehall (Air Chief Marshal Portal and planners) had abandoned Harris, and they probably were glad that Eisenhower had now directed both the USAAF and RAF to concentrate solely upon targets connected with the impending D-Day landings—that is, tactical targets such as French railway bridges rather than strategic operations. In April 1944 Harris himself admitted to the Air Ministry that "casualty rates . . . could not in the long run be sustained." With recent innovations such as "window" and the Pathfinders no longer so effective, remedial action was needed, including perhaps RAF night fighter escorts for his bombers. As the official British historians who examined these reports two decades later concluded, the Battle of Berlin was more than a failure—"it was a defeat."[29]

The third new element between late 1940 and 1943—that is, the coming of the United States Army Air Forces to join the RAF in the aerial campaign against Germany's capacity to fight—was the most important of all. This story also can be reduced to its few fundamental parts. To begin with, it was strategically significant because the arrival of a new strategic air force in Europe meant more pressure upon the Third Reich and more pressure from the west, which logically implied less capacity (fewer aircraft, fewer industrial resources, fewer trained personnel) for Luftwaffe allocations to the grinding battle in the east or the expanded fighting in the Mediterranean and North Africa, which nonetheless required a greater number of German squadrons to meet the steadily growing British Empire and American air forces. It is about this time (1942–43) that one can detect Luftwaffe air groups being shuffled to and from the Eastern Front, North Africa and Italy, France, and the defense of the Reich. Hap Arnold, though always complaining about the lack of aircraft and crews for the USAAF, actually was able to dispatch fresh streams of squadrons to Britain, North Africa, and the Pacific. Goering no longer possessed that option. By late 1943, new airfields were opening up every six days across the topographically convenient flatlands of East Anglia to receive the flow of brand-new American bombers and fighters.

The coming of the American squadrons was important not just because it was a major addition in the fight against the Axis in Europe and not just because it provided an inordinate boost to British morale, but also because it brought two critically important operational elements into the fray. The first was that the USAAF insisted on carrying out only daylight bombing; this had been their doctrine for years because it played to the strengths of the high-altitude, heavily armed B-17s and B-24s, equipped with the purportedly ultra-accurate bombsights developed by the Norden Company in the 1930s. For some time Harris and other senior RAF staff, even Churchill himself, pressed the Americans to join in the nighttime campaign, a foolish operational idea that was rightly resisted. The beauty of the American plan was that it brought the possibility of twenty-four-hour-a-day bombing of Germany. Londoners at least had the daylight hours to clear away the Luftwaffe's nighttime damage and to get some rest. But round-the-clock bombing by the USAAF in daylight hours and the RAF at night promised, in theory at least, unrelenting pressure upon German aerial defenses, air controllers, and damage and relief services, as well as the German workers themselves.

The second development was that the Americans also insisted on the other critical aspect of their prewar aerial doctrine, the "pinpoint" bombing of identifiable military-industrial targets. As we shall see, this was much easier said than done, and more often honored in the breach by confused and frightened crews. But the USAAF's insistence upon attacking targets such as a Messerschmitt factory or a railway marshaling yard was significant because it complemented the more general area bombing of the RAF's nighttime raids, and made a response more complicated for the German aerial defense planners—should they cluster their antiaircraft battalions around Cologne or around an important tank factory 30 miles away? After the war Allied assessors compared the performance of these two very different bombing doctrines. In the heated postwar controversies about the mass bombing of enemy civilians, the U.S. Army Air Forces could claim a somewhat higher moral ground—at least in their European campaign. But in practical terms, that is, in destroying German infrastructure, factories, and roadways, there wasn't that much difference.

All these apparent advantages were welcomed on the Allied side. The only problem was that these operational assumptions about being able to carry out pinpoint bombing on any selected enemy target *really didn't work* when put into practice between 1942 and early 1944, so in fact the American record was little different from Bomber Command's. And finding a solution to this particular challenge was, clearly, taking much longer than winning command of the Atlantic sea-lanes.

With all the wisdom of hindsight possessed by armchair strategists and historians, it is not difficult to see where the obstacles lay. The first reason must be that the Eighth Air Force's bombing assaults against Germany, especially in 1942 and early 1943, were a case of what one might call "too little, too *early*." Politically and institutionally, the

USAAF's leaders (Arnold, Spaatz, and Eaker) and their staffs worried that the existence of their service as an independent third arm depended upon proving itself as soon as possible. For obvious geographic reasons, and possibly because of memories from 1917–18, both the U.S. Army and the USAAF had always regarded the European theater as being much more important than the Pacific and Far East. To the army, distances across the Pacific were simply far too wide to allow it to play the lead role—the island groups were much too small to permit the deployment of twenty-five or fifty divisions in the tradition of the "American way of war," that is, the massive allocation of men and munitions to overwhelm a foe, as happened in the Civil War and again in 1917–18 in France.[30] There was no room in the Pacific for tanks, and little room for artillery; there was nothing that the army could do in those theaters that an expanded U.S. Marine Corps could not do as well, or perhaps better.

This was also the air force's problem. There was nothing to *bomb* in the Pacific, strategically, until one got close to Japan, probably after years of fighting. Not only was Germany the greater foe industrially and technologically, but its factories and railways lay just across the North Sea from Suffolk—just as the beaches of Normandy lay right across the Channel from Sussex. Without a German focus, Arnold feared, the air force's precious new bomber squadrons would be divided and scattered, from Newfoundland (anti-U-boat work) to North Africa (supporting Eisenhower and Patton's ground forces) to New Guinea (assisting MacArthur's laborious campaigns).

As he flew in for the Casablanca Conference, Arnold had every right to be fearful. The Japanese attacks upon Pearl Harbor and the Philippines had given Admirals Ernest King and William Leahy terrific ammunition to argue for a "Pacific first" policy, one supported by much popular sentiment, but one that would surely give the lead role to the U.S. Navy. At the same time, the increasing British reluctance to commit to a definite date for an Allied invasion of France weakened the precarious 1941 agreement that the defeat of Nazi Germany had first priority. Then there was the enormous, angry pressure that Stalin was putting upon London and Washington to do more than they were, and the fact that even America's enormous productive capacities could not meet the gigantic claims from all services for weapons and men. There were fierce interservice clashes over priorities all throughout 1942, causing the USAAF to fear for its bomber programs as cuts were made to every service's early demands—and so the need to show that the service was hitting and hurting Germany became overwhelming.[31] Like it or not, the American bombers had to take off, however few and poorly prepared, as had their British cousins in 1939.

Arnold and his colleagues came away from the Casablanca conference much relieved by the pronouncement that "the progressive destruction and dislocation of the German military, industrial and economic system" was second only to the winning of the Battle of the Atlantic in order of priority. They had quarreled with the navy right up to the eve of the conference and then encountered enormous disputes with Churchill and the British chiefs (normally their natural allies for the "Germany first" argument) regarding allocations in the Mediterranean. They had spent 1942 employing every conceivable argument to advance their case—for example, touting the October 9 raid upon Lille (with its mixed if not downright dubious results—see below) as an example of a powerfully armed bomber force executing its mission unescorted. And any and all RAF reports that year on the effectiveness of *their* strategic bombing were eagerly consumed and swiftly circulated to higher authorities. By January 1943, then, the USAAF hoped that at last it had demonstrated its significance and preserved its strategic mission. Years after the war, Albert Speer asked Eaker why he had committed his squadrons so early, in such relatively small numbers, and without long-range fighter escorts. The answer was that it was politically necessary. No wonder Elmer Bendiger got so furious when he learned of that reply thirty years after the event.[32]

There were other reasons the first two years of the American bombing campaign against Germany were disappointing. The USAAF was creating the single largest strategic air force in history (by 1945 the total American bomber force in Europe, including the new and separate Mediterranean air groups, would be twice the size of Bomber Command) and was expecting to do so in an exceptionally short time. Where were the thousands of pilots to come from, and who would train them? Whence the tens of thousands of crew, whence the vital ground staff and repairmen? The adjutants, the intelligence personnel, the staff for group headquarters? The airfields, control towers, hangars, and messes? The aircraft and the bombs? Arnold also knew that if he did not dispatch a lot of long-range bombers to the Southwest Pacific, the United States might lose the hard-fought Battle of Guadalcanal, which raged from August 1942 until January 1943, and he could not neglect the non-European theaters of war; thus the supplies of men and planes to England were bound to be limited. In the event, it was scarcely a surprise that the initially small Eighth and Ninth Army Air Forces got off to a rough start when they carried out their first flights across the Channel, because they had new commanders, new crews, new planes, new bases—and a new, sophisticated enemy.

There was also, inevitably, the usual array of practical problems: poor weather, aborted missions, disappointment with the bombsights, range limits, Luftwaffe counterattacks, the lack of escorts. All this would have tested a completely trained and prepared bomber force. There was simply nothing to do about the almost continually cloudy weather over northwest Europe for most of the months of the year, as the British had pointed out. Placing "a bomb in a pickle barrel" with a Norden bombsight may have been possible over the clear testing grounds of Texas, but it simply didn't work when there was ten-tenths cloud cover. At this early stage, there were no sophisticated directional tools to put the B-17s directly over their targets, and squadron leaders had often to order a return to base, mission unfulfilled. Whether one came back with the bomb bays empty or not was an unpursued question, but few pilots liked landing with a belly full of bombs. Even when the skies were clear, there were still strong winds to deflect the bombardiers' aim; the higher one flew to avoid the flak, the greater the margin of error.

Then there was the USAAF's vulnerability to enemy fighters. As noted above, by 1942–43 the German high command had at least partially woken up to the threat posed to the Reich's industrial production by Bomber Command's thousand-bomber raids and was hastening, impressively, to respond. This meant that German defenses had become much tougher just as the Americans were making their first tentative daylight raids. Finally, the Luftwaffe now enjoyed the home field advantage for daytime combat that had been so critical to Fighter Command during the Battle of Britain; the farther away the targets set by Spaatz and Eaker, the more opportunities the German fighters had to refuel and attack again while the bombers lacked the protection of the short-range Spitfires or medium-range Thunderbolts. (It should come as no surprise that Speer was now locating his new aircraft and tank factories as far from the English Channel as possible, well into Poland and Czechoslovakia.) The American bombers, like the Atlantic convoys, were now in an escort gap.

Given all these constraints, it was natural that the early daylight raids would be tentative and would keep close to home bases; for now, operations would take place over northwestern France and the Low Countries, and rely upon fighter cover. The weather throughout August 1942 was fine, the Luftwaffe noticeable by its absence, and the bombing remarkably accurate. This continued into September, and even when Fw 190s were thrown in, they chiefly tangled with the Spitfires. All this changed with the attack by 108 USAAF bombers upon the heavy industries of Lille on October 9, 1942. Rushing past no fewer than 156 Allied escorts, the German fighters concentrated upon the heavily laden B-17 Flying Fortresses and B-24 Liberators. As it happened, the Luftwaffe hadn't yet figured out its more destructive tactics against tightly packed bombers flying at around 25,000 feet, so the USAAF losses were four destroyed, four seriously damaged, and another forty-two needing repairs. But the bloodletting had begun. More significant was that the constant fighter attack led to a catastrophic falloff of accuracy—only 9 of the 588 high-explosive bombs fell within 1,500 feet of the targets, and many bombers aborted their load before even getting close. In such massive melees, it is not surprising that the gunners wildly overestimated the number of German fighters downed (they claimed fifty-six certain kills and twenty-four probables). Such estimates greatly helped Arnold and Spaatz in the air force's political campaigns in Washington but also showed how green and nervous these brand-new crews were and how poor the after-action analysis was. (Postwar analysis revealed that the Luftwaffe recorded only one definite loss that day, possibly a second.)

Even among those who believed that the USAAF's campaign was proving itself, the last months of 1942 and early months of 1943 brought a widespread slowdown in pace. More and more squadrons of American bombers and fighters were being diverted to support Operation Torch and the follow-up land campaigns in North Africa. The winter weather made daylight precision bombing virtually impossible against any target in northwest Europe. (The same storms that were preventing the U-boats from attacking Allied merchantmen were also stopping the U.S. planes from finding their targets in France.) There was gloom at the publicity about French civilian losses from these aerial attacks, especially when so many of them had occurred quite some way from the intended targets. For a while, therefore, American bombing focused on coastal targets such as German submarine pens. All of this disappointed RAF Bomber Command, which had been hoping for swift reinforcement of its own (mutually) punishing fight against the Luftwaffe over Germany. Most important of all, it gave the German high command time to figure out its aerial and antiaircraft defenses and to switch production more toward fighters.

The Eighth Air Force made its first raids upon targets in Germany at the end of January 1943, but they were virtually harmless, being much disrupted by bad weather. This dismal and unproductive period lasted well into the year, to Arnold's great frustration. Only with the coming of early summer, improvements in the weather, and the arrival of more squadrons (in February the Eighth had had a daily average of only seventy-four operating aircraft with crews) did the American offensive get under way again. Yet these improved operational conditions also meant that the new extent of the Luftwaffe's defensive capacities against daylight raids would be learned in full. In a raid on Kiel on May 14, 8 bombers were lost and 36 damaged, out of 126 dispatched; in a double raid upon Bremen and Kiel on June 13, 26 bombers were lost and 54 damaged, out of a total of 182; from the Ochersleben raid of July 28, only one-third of the 120 bombers returned to base unscathed, for 16 others had been lost and 64 damaged.[33] And those were operations where the bomber squadrons could count upon fighter escorts for a greater part of the way. Despite such ominous signs, the momentum was increasing for more longer-range raids.

Thus came the ghastly Schweinfurt raids and losses with which this chapter began. In addition to the 60 bombers lost during that October 14, 1943, raid, another 138 were damaged, which meant that a mere 14 percent of the 229 aircraft that actually attacked—many turned away early—arrived back at base unscathed. This seems a fantastical figure until one reads the detailed analysis in the official USAAF history's account of a Luftwaffe performance that, its authors observe, was unprecedented in its magnitude, cleverness, and severity of execution:

> Wave after wave of fighters attacked. Usually a screen of single-engine fighters would fly in from in front, firing normal 20-mm cannon and machine-guns until very close to the formation. Closely following the single-engine fighters, large formations of twin-engine fighters appeared in waves, each firing large numbers of rockets from projectors carried under the wings. . . . Meanwhile, the single-engine fighters refueled and attacked from all directions. Soon they were followed by re-formed groups of twin-engine rocket carriers. After expending their rockets, these twin-engine fighters frequently came in firing cannon and machine guns. The enemy aircraft concentrated on one formation at a time, breaking it up with rocket attacks . . . and then finishing off cripples with gunfire. One combat wing of the 1st Bombardment Division, which bore the brunt of the counterattack, was almost completely wiped out by these tactics.[34]

Despite these losses, the USAAF's attacks upon the German aircraft industry in this period *did* have a considerable impact: the selective targeting strategy, rather like the RAF's night attacks on the Ruhr, frightened Speer enough to spur him to summon even more of the Reich's massive resources for the campaign against the Anglo-American bombing offensive. The struggle was only going to get harder. Nevertheless, to the Allied air chiefs, this was a completely unsustainable rate of attrition, higher probably than that of any other aerial campaign in the war, and the offensive simply had to be closed down, with bombing temporarily restricted to short-range targets. The winter weather made all this justifiable, and provided time for efforts to rebuild shattered morale, which could be measured negatively by the growing number of missions aborted or of partly damaged aircraft gliding into neutral havens such as Sweden and Switzerland.*

*The classic postwar movie about this problem is *Twelve O'Clock High* (1949), in which an extremely tough air force commander, played by Gregory Peck, restores the morale and effectiveness of a hard-luck squadron. The real story was altogether more complicated.

It is not possible to better the conclusion drawn in the air force's own official history on this dismal chapter of the struggle:

> By mid-October 1943 the daylight bombing campaign had reached a crisis. Its cost had risen alarmingly while its successes remained problematical. The assumptions underlying it therefore came up for reconsideration. . . . The fact was that the Eighth Air Force had for the time being lost air superiority over Germany. And it was obvious that superiority could not be regained until sufficient long-range escort became available. . . . [C]learly, also, fighter range would have to be extended.[35]

Fortunately for the battered Allied bomber crews, there were solutions just around the corner.

The Merlin and the Mustang

The Americans' decision to halt their daylight bombing of the enemy's industries occurred approximately twenty-six years after the Smuts Report, a fact disregarded in the more teleological accounts of the rise of modern airpower. This offers to the student of war a superb example—perhaps the best we have—of how a novel strategic doctrine was promulgated in theory well before it could be realized in practice. In no other chapter in this book is there such a need to explain in detail how and why it took so long for the Allies to close this critical gap between the concept and the accomplishment of their strategy. This is because the other forms of warfare discussed here (such as convoys, amphibious landings, and mass land battles) had already existed for many centuries. Yet a detailed retelling of this particular story is also justified, surely, by the enormous claims for the efficacy of airpower made by its advocates both before and well after the Second World War.

The "solution" to the problem of the bomber's vulnerability, when it emerged, was developed very swiftly indeed, although, as we shall see, it could have arrived in Europe's skies even faster than it did. What was needed was a technological breakthrough for the military problem succinctly stated by the U.S. official historians: a long-range Allied fighter capable of beating off aerial attacks upon the vulnerable bomber groups.

Mechanically, this could be defined by a set of operational and engineering specifics for a new flying machine. It had to be a single-engine fighter that was faster and more maneuverable than anything the Luftwaffe possessed or would be able to field in the next year or so, and it had to be able to maintain its performance at all altitudes from 5,000 feet to 40,000 feet, something experts thought was aerodynamically impossible given existing aircraft-design assumptions. Most important of all—and equally impossible—this aircraft had to have the fuel capacity to permit it to fly with the USAAF bombers from East Anglia to beyond Berlin and back, and to protect its "convoy" even 600 miles from their bases. These requirements appeared to defy the laws of physics: how could a fighter be capacious enough to carry enough fuel to, say, Prague, yet nimble and fast enough to shoot a locally based Fw 190 out of the air? Yet the aircraft was made. Sir Henry Royce (1863–1933) was one of the greatest inventors and engineers of the Western world since Macadam, Brunel, Stephenson, Bell, and Edison. Royce had only a year of formal education, and survived his youth by selling newspapers. Fascinated by machines, especially the new automobiles, he began building them himself even before the First World War; when he joined with his business partner, Charles Rolls, and the firm expanded, he still insisted on very high quality control. The man had a mania for exactitude. Late in his career, this creator of the finest and most reliable automobiles turned his attention to the novel though related field of airplane engines. It was a logical move: both cars and planes rely upon the inanimate power of gasoline to propel their heavy frames, and their human occupants, along winding country roads or into the wide blue skies above. Both vehicles consist of a solid frame that houses hundreds of moving parts, many of which have to turn and change and ignite with precision hour after hour, day after day. One reason for the "rise of the West" after 1700 was the development of the exact sciences, that is, manufacturing and technological breakthroughs, peaceful and military, that were not occurring in other parts of the world.[36] A Rolls-Royce engine was an apotheosis of an astonishing human achievement: the ever-greater conquest of the obstacles to time and speed and space that had always existed in a world of animate and wind energy.

The late 1920s and early 1930s were an exciting time for all those who were devoted to newer and better airplanes and airplane engines.

The American-designed Curtiss V-12 engine was way ahead of its competitors at this time, and Royce and his team had no qualms about buying one in the United States, shipping it over, stripping it down for analysis, and then rebuilding it as, variously, the Rolls-Royce Kestrel, Griffon, and Merlin. Malcolm Campbell used a Rolls-Royce engine in his record-breaking 300-mile-an-hour drive across the Bonneville Salt Flats in 1933. And Supermarine Aviation used these engines to become permanent owner of the famous international Schneider Trophy Race (for seaplanes) after three successive wins; in 1931, the final year of the competition, a Supermarine plane achieved the then astounding speed of over 400 mph.

In the midst of all this, Royce began to conceive of placing one of his engines in a new type of very fast fighter for the RAF. On a memorable occasion in 1931 near his country house at West Wittering, Sussex, he walked along the sandbars with his chief engineers and sketched out the design in the wet granules. Already the familiar biplane-shaped, canvas-covered planes of the First World War era were giving way to sleeker, aluminum-covered monoplanes carrying a single pilot; also on the drawing board were fast twin-engine (and even some four-engine) longer-range bombers. All would require much greater propulsion. And that meant creating an engine that was of the most exact engineering standards, converting fuel into thrust as efficiently as possible, which was what the firm of Rolls-Royce specialized in. Naturally, they were not alone in this passion. Companies in America, Germany, Italy, Japan, and France were driven by the same motive, each of them seeking to squeeze more and more power out of an elaborate design of pistons, cylinders, spark plugs, wiring, and steel chambers.*[37]

The Rolls-Royce Company had the habit of naming its various airplane engines after the swift raptors of the skies, the hawks, falcons, and eagles. In the case of the engine in question, the designated name was Merlin, which was a reference not to the legendary wizard but to

---

*Near the end of the war, there would appear truly transformative weapons such as the jet fighter, the V-2 rocket, and, of course, the atomic bomb. In all of the middle-of-the-war narratives we are examining in the present book, however, it is more a tale of *incremental* changes in technology, organization, and war fighting—and the difference such changes cumulatively made.

the smallest of the falcons, a fast, aggressive aerial wonder that could attack but not be attacked.

It should be stressed that the increasingly powerful engines Royce forced ahead could be embedded in a whole array of aircraft; the official Rolls-Royce history of the Merlin, for example, shows it in about forty different planes that ranged from one to four engines.[38] Variants also powered hydroplanes, speedboats, racing cars, and, of course, the Rolls and the Bentley. It was not the case, therefore, that the Merlin was designed for the Spitfire, and the Spitfire for its Merlin engine, although popular legend has it that way. Had the Spitfire never been created, the Merlin would have had its place in aviation history. Yet it is not surprising that, just as Rolls-Royce was developing this particular thrust machine, one of its recipients should be a new, single-engine monocoque plane, based upon the great aircraft designer J. R. Mitchell's elegant racer for Supermarine. And, despite all understandable interwar restrictions and difficulties, the impoverished Air Ministry pushed Vickers (which had bought Supermarine) toward the production of a slim, swift PV-12-powered fighter. The Merlin and the Spitfire had been brought together. But it was a close-run thing: Royce had died in 1933, still working on the newer engine design; he never even saw a prototype Spitfire, though he left an accomplished engineering team behind him. Mitchell, battling cancer, died in 1937 at the age of forty-two, also working until he dropped dead. He, at least, saw the first flying models.

The role of the Spitfire in the Second World War in Europe is legendary. From the closing stages of the Battle of Britain to the dogfights over Malta, from the great daylight sweeps over France to aerial patrols over the D-Day beaches (the Polish Spitfire squadrons were given this honor), the Spitfire became the RAF's most important single-engine fighter. Its elegant frame and tapering elliptical wings created thousands of devotees, even on the other side: when Goering asked Galland around 1942 what more he needed, the latter replied, "A squadron of Spitfires."[39] Its evolution from the Supermarine seaplane into an urgently needed addition to RAF Fighter Command's armory has been the subject of movies, books, and documentaries.[40]

The place of the Merlin-powered Spitfire in this history is, however, that of an intermediary weapons system. By late 1940 it had proven itself as Britain's best and fastest fighter, even better than the Bf 109. But the Luftwaffe was constantly upgrading the horsepower, armor, and armaments of its own fighters, as well as bringing newer craft into play, so both the Spitfire itself and the Merlin engine had to be upgraded also. In particular, the development engineers at Rolls-Royce had to create ever more powerful versions of the engine if British fighters were to match the fantastic rate of climb of the Fw 190s. By 1942 they had done just that, with the enormously successful Merlin 61 variant, which more than doubled the engine's original horsepower.[41]

Unsurprisingly, everyone wanted the enhanced engine—Bomber Command for its new heavy four-engine Lancasters, Fighter Command for its Spitfires, Reconnaissance and Pathfinder squadrons for the high-flying Mosquitos. Ten thousand engines would not be enough. Even the highly productive Packard factories in the United States (brought in to treble output, as they did so well) could not match the demand. Getting one's hands on a batch of Merlin 61s was like getting a shipload of East Indian spices in seventeenth-century Amsterdam.

The person most responsible for allocating these scarce resources was Air Marshal Sir Wilfrid Freeman, little known to the public then or now, but perhaps the single most influential figure in the Air Ministry—Churchill, for one, regarded Freeman with the greatest respect.[42] As air minister for supply in the late 1930s, Freeman had been responsible for pushing ahead with the development of virtually all the RAF's newer planes—the Hurricane, the Spitfire, the Wellington, the Lancaster and Halifax heavies, and the Mosquito. It was Freeman who realized that the miserably underperforming Avro Manchester bomber of the late 1930s could be transformed by giving it Rolls-Royce engines (Merlins) and turning it into the powerful Lancaster. In the same years, it was Freeman who pushed for the development of the twin-engine Mosquito when everyone else thought the design absurd: how could one invest scarce resources in a plywood-framed, unarmed aircraft and expect it to fly unscathed across Germany and back? Critics called it "Freeman's Folly." It turned out to be the most versatile plane of the Second World War—it could be modified (that is, armed) to be a night fighter, submarine killer, high-level reconnaissance plane (in a version that flew too high for any German fighter to reach it), and bomber (modified, it carried a greater weight in bombs than a B-17). By 1942, Freeman was deeply respected by the British airframe and airplane engine manufacturers, got on fabulously well with his American counterparts, and could go directly to Churchill. If he had a new, sensible idea, he would seek to make it happen. In the spring of that year, he got another one.

There now enters another key figure in this remarkable history, although this one is even less well known. In late April 1942 the RAF liaison test pilot for Rolls-Royce Engines in Derbyshire, Ronnie Harker, received a phone call from the RAF's Fighting Development Unit, asking if he would come down to the Duxford airfield to test a problematic U.S. plane, designated Pursuit Fighter 51 (P-51). This was not a novel task for Harker, who had flown in the RAF before being hauled back to his current critical job; he methodically tested all new aircraft, including captured variants of Messerschmitts and Focke-Wulfs, to report on their relative performance. The airplane he was asked to fly this time was a single-engine fighter that had been produced by the North American Company—one of the dozens of aircraft types that had been hastily commissioned by the U.S., French, and British air forces after 1938, when the Luftwaffe's much-touted numerical advantage had become alarmingly clear. The problem was that the Alison-engined P-51 was not a great performer. Its original specifications had called for the design of a low-altitude interceptor, and it actually flew well in that respect, but the USAAF was (understandably) heading toward the much more powerful P-38 Lightning and P-47 Thunderbolt fighters, both of which were a match in the air against a Zero or a Bf 109, and the British were heading toward improved Spitfires, Typhoons, and Beaufighters. Thus, when in early 1942 RAF Fighter Command received its first batch of P-51s, it didn't quite know what to do with them and was contemplating scrapping the order.

Harker flew the P-51 for the first time on April 30, 1942 (the Rolls-Royce archives have, happily, a photo of that very plane after he brought it back to base). It clearly puzzled him. It turned easily, never stalled, and was fine at low to medium altitudes. Aerodynamically it was superb, that is, it had very low drag, although neither he nor anyone else at the time could quite figure out why.* Harker's report finished with a sentence that, though laconic, caught the attention of everyone who read it: "The point which strikes me is that with a powerful and good engine like the Merlin 61, its performance should be outstanding, as it is 35 m.p.h. faster than a Spitfire 5 at roughly the same power."[43] A few days later, a team of Rolls-Royce mechanics pulled out the Alison engine and carefully lowered a Merlin 61 into the front of the machine Harker had flown. As the sharp-eyed test pilot noted, the distance between the front edge of the cockpit and the nose of the P-51 chassis was the same as that on the latest Spitfire, so it was a perfect fit.

At the same time, Rolls-Royce's outstanding Polish mathematician turned performance engineer, Witold Challier, equipped with the comparative aerodynamic details of both planes, produced a chart suggesting that a Merlin-powered P-51 should be able to outperform the Spitfire at all levels up to 40,000 feet and reach the astounding speed of 432 mph. (All Challier's calculations proved to be absolutely correct.) Both men, supported by the dynamic general manager of Rolls-Royce, E. W. Hives, began agitating for the new hybrid version as soon as possible.

When this information got to the resourceful Freeman, he responded immediately. Although Fighter Command and Bomber Command wanted all the Merlin engines for their own projects, Freeman ordered the engine conversion on another five P-51s, directing that two of them be sent to Spaatz in the United States as soon as possible for USAAF testing. When Hives shortly afterward proposed the conversion of 250 aircraft, Freeman doubled it to 500. Freeman was swiftly in touch with the managers at Packard (who put one of their Merlins into a P-51, which by now had received its more familiar name, the Mustang); with the influential U.S. ambassador to Britain, John Winant; and with the well-connected assistant air attaché Thomas Hitchcock, a former Lafayette Escadrille flier and a great Rolls-Royce enthusiast who had family links to the White House. Freeman also started nudging Churchill to write to Roosevelt, since everyone in the United Kingdom realized that if this Anglo-American hybrid was to be produced speedily

---

*Later aeronautical studies, based on wind tunnel experiments, point to the slightly inward-turning curves of the sides of the P-51's chassis as the reason. Harker received a £1 increase in his weekly salary as a tribute to his work.

and in sufficient numbers to alter the aerial balance, it had to be done in American factories. It was like the cavity magnetron story all over again: Britain was industrially overstretched, but the United States still possessed enormous capacity for additional aircraft and engine production.

Then, inexplicably to Freeman, the scheme foundered. Sheer obstructionism on the American side now slowed down the mass production of the Merlin-engine Mustang. In the United States there were genuine rival claims being made on resources, and it was always going to be hard to argue that the output of American aircraft already in production should be curtailed for an unknown and essentially foreign newcomer. There was continuous misunderstanding of Harker and Challier's point that the P-51 was good at *all* altitudes, and certain area commanders kept insisting that the plane was to be used according to the original specifications as a low-altitude tactical fighter, in which capacity it was just one of many. Then there were USAAF leaders who could not accept any claims of the Mustang's superiority because they were devoted to the P-38 Lightning and the P-47 Thunderbolts, both of which had well-tried combat records. There were also powerful objections in the air force's procurement offices and among rival manufacturers. Freeman, who followed American production figures as anxiously as he did Britain's own, was warned by Roosevelt's personal and very Anglophilic advisor Harry Hopkins in September 1942 that the USAAF had on order no fewer than 2,500 P-40 Kittyhawks, 8,800 P-39 Airacobras, and 11,000 P-63 Kingcobras, all of them hopeless against the formidable Focke-Wulf 190 fighter that was beginning to dominate the European skies, but each of these underperforming aircraft had its own significant backers.[44] Additionally, the Mustang surely needed further testing and improvements.

This was all understandable, if regrettable. What was less understandable was the sheer, relentless anti-Britishness of key members of the all-powerful Air Production Board under the stiff-necked Major General Oliver Echols. Hopkins was only half joking when he told Freeman that many Americans believed they naturally flew better than the British and always built better planes than the British. Since the P-51 had been first ordered by the RAF, it had not gone through the normal American channels of engineering scrutiny, and many officers devoted to that system fed Echols disparaging details when they inspected some of the early models. Essentially, the attitude was "not invented here."[45] The attaché in London, Hitchcock, was particularly scathing about all this: "Sired by the English out of an American mother, the Mustang has had no parent in the Army Air Corps or at Wright Field to appreciate and push its good points. . . . [I]t does not satisfy important people on both sides of the Atlantic who seem more interested in pointing with pride to the development of a 100% national product than they are concerned with the very difficult problem of rapidly developing a fighter plane that will be superior to anything the Germans have." The more restrained official historians merely point out that "the story of the P-51 came close to representing the costliest mistake made by the AAF in World War Two."[46]

At this stage, there was little more that Freeman or his business allies at Packard and North American could do. There was little more that even Churchill could do, though he again used every one of the "usual suspects" (Ambassador Winant, special envoy and later Churchill family member Averell Harriman, and Hopkins, plus his private messages to Roosevelt) to have Mustang production given priority. The push had to come from the USAAF itself. Eventually the air force did come around, for two reasons. The first was the Schweinfurt-Regensburg catastrophe of October 1943. Although the Air Force told the press that the lull in bombing Germany was simply due to the winter weather, as well as to a shift to bombing the Reich from newly acquired Italian airfields, both Arnold and Spaatz now privately recognized that their basic operational assumption was faulty: the bombers could not get through without fighter protection, and the existing fighters were too short-legged. By now, Arnold knew that even members of Congress were agitating. In his so-called Christmas address (given December 27, 1943) to air force commanders in Britain and Italy, Arnold gave out his bluntest message: "OVERLORD and ANVIL [a landing in the south of France] will not be possible unless the German Air Force is destroyed. Therefore my personal message to you—this is a MUST—is to 'Destroy the Enemy Air Force wherever you find them, in the air, on the ground, and in the factories.'"[47] But only an extremely nimble and long-range fighter, of a design recognized by Harker nineteen months earlier, could do that.

The second reason was the continued pressure exerted by a small and remarkable cluster of individuals in the middle levels of the Allied effort. There was the irrepressible Tommy Hitchcock, who possessed the characteristics of an Ivy League playboy—he was acknowledged to be the world's best polo player before the war—and, in addition, possessed a great flying record, a fluent pen, and a noted set of connections. Hitchcock was one of the few people who had no fear of browbeating Echols and the Air Production Board, or of privately approaching his neighbor Eleanor Roosevelt about the matter. Then there was the assistant secretary for air, Robert A. Lovett, who had flown in the legendary First Yale Unit in 1917–18 and in the Royal Navy Air Service, and would finally complete his public career as secretary of defense in the early 1950s. Lovett came from a patrician New England family, was not drawing a salary, and was much respected by Arnold. Like Hitchcock, he was not daunted by the Air Production Board and as early as 1940 had regarded the Alison-powered P-51 as a "washout." Lovett carried out a major inspection of the USAAF bombing campaign during a June 1943 tour in England and made a number of recommendations in his report to Arnold, the two most important of which were (1) the absolute necessity of increasing the production of auxiliary drop tanks as swiftly as possible, to enhance the range of all aircraft in Britain, and (2) the greater urgency that needed to be given to the development of long-range fighter escorts, with the Merlin-powered Mustang being particularly promising in this regard. These were the keys to unlocking the aerial deadlock. When Lovett returned to his office in Washington, by no coincidence right next door to Arnold's, obstacles began to fall. Four days after reading this report, Arnold composed a memorandum that stressed "the absolute necessity for building a fighter plane that can go in and come out with the bombers."[48]

There was also strong pressure from air force officers stationed in England, including the irrepressible Major Donald Blakeslee, an American who had fought through the Battle of Britain with a Royal Canadian Air Force squadron, then, to avoid being appointed a training officer, switched into the volunteer American Eagles squadron. Blakeslee flew Spitfires and adored them, but he knew their limitations of range. When he flew the first RAF Mustangs, he agitated incessantly to have the fighter groups attached to the Eighth Army Air Force (based in eastern England) equipped with the same planes. When the pugnacious Lieutenant General Jimmy Doolittle was pulled back from the Mediterranean and put in charge of the Eighth, he immediately pressed for Mustang squadrons, and Arnold and Spaatz, now persuaded of their virtues, found them. Portal in turn ordered four RAF Mustang squadrons to fly with the Americans. All such squadrons from the Ninth were transferred northward and placed under Blakeslee's boss, Major General William Kepner, the forceful head of the fighter groups. In fact, all USAAF pilots with *any* Mustang training were ordered to the Eighth. The logjam had broken, and not a moment too soon.

Air Supremacy, at Last

The air war in Europe was not, of course, transformed by a single "wonder weapon." Rather, the crucial developments came about as a result of tactical, technical, and operational complementarities that, coming together after many setbacks, allowed a small group of British and American individuals to solve a problem that had plagued strategic bombing for many years: how to realize the early visions of weakening an enemy's capacity to fight through the steady application of modern airpower.

What was critical were the interactions and feedback loops: the impact of one operational failure or victory upon another, for example, or the changes a new weapon could bring about at the tactical level that in turn affected the larger outcome of battles.[49] One such complementarity was mentioned earlier, namely, the fact that the RAF and USAAF bombed around the clock and thus gave the enemy little respite. Much more could have been done in this respect, and the Combined Chiefs of Staff repeatedly urged a greater integration of targeting, but the mere fact that British bombers flew over Germany each night and U.S. bombers did so each day placed enormous stress on the Reich's defenses. Neither air force could claim that they did it alone.

The other great complementarity was the use—perhaps a better word is *placement*—of the Mustang as a long-range daylight escort. It did *not* make redundant the Spitfires, Thunderbolts, and Lightnings, but augmented their efforts, especially at the greater flight distances. The tests by Rolls-Royce had shown that while the P-51 was considerably heavier than the Spitfire, it needed far fewer revs per minute to attain the same altitude and speed (this was the aerodynamic puzzle); it got to where it needed to go with much less effort. But there was another aspect to this aircraft that had intrigued Freeman, namely, its astounding fuel capacity, which when combined with its aerodynamic economy of consumption produced a miracle. "In terms of US gallons, its normal internal fuel tanks held 183 gallons (269 with a full rear tank) compared with 99 for the Spitfire, and it consumed an average of 64 gallons an hour compared with 144 for the P-38 and 140 for the P-47. With full internal tanks, including an 86 gallon rear fuselage tank, and two 108 gallon drop-tanks, its combat radius was 750 miles."[50] That must have been more than twice a Spitfire's range.

The drop tanks—the second enhancement pressed by Lovett after his visit, and being toyed with by most air forces by the middle of the war—were thus also a critical part of this tale, for while they were the factor that most enhanced the P-51's range, they automatically gave *every* British and American fighter a much wider range. Once the full potential of drop tanks to increase an aircraft's range was appreciated, everyone wanted them, not just the fighter squadrons. The British were so desperate for them that they manufactured a 108-gallon fuel tank out of stiffened paper; it was actually lighter and more capacious than the later U.S. metal versions, and was preferred by many American fliers, who scrounged batches of them. It also denied the enemy the reuse of the discarded aluminum casings.[51] With attached drop tanks, all the Allied fighters—Spitfires, Lightnings, Thunderbolts, Mustangs—could fly farther and stay longer in the air.

The newly equipped Allied strategic bombing campaign (called Operation Pointblank) resumed with a vengeance in early 1944, at last putting into practice its declared goal of the "progressive destruction and dislocation" of the enemy's capacity to resist. The breakthrough was a consequence of the resumption of the USAAF's daylight raids upon the great industries of the Third Reich, this time increasingly escorted by long-range fighters. It didn't happen all at once, of course. The first Mustang squadron (Blakeslee's) had joined the Eighth at the end of 1943, but there were never enough planes in the early raids, nor were the new Mustangs themselves free of imperfections. But the American war machine was now in full swing, and hundreds of Thun-

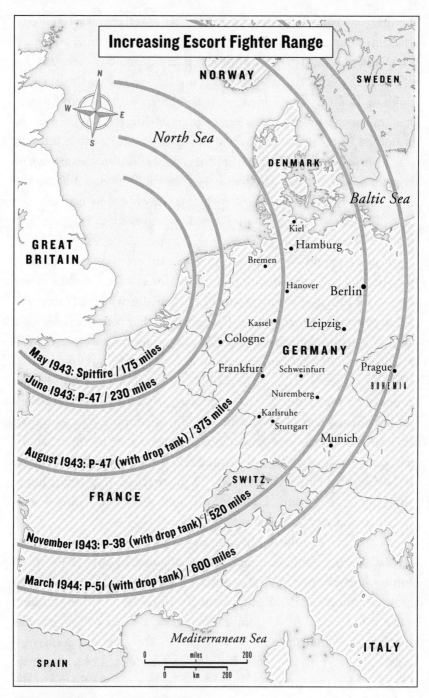

INCREASING ESCORT FIGHTER RANGE
Providing long-range fighter escorts in daytime for the American heavy bombers was the critical component in gaining air supremacy between 1943 and 1944.

derbolts and Mustangs were arriving in the United Kingdom each week—transported on the decks of the escort carriers no longer needed for that role because of the U-boats' setback—to join the hundreds of new Flying Fortress and Liberator bombers that were being flown overhead by the male and female shuttle crews of Allied Transport Command.

The breakthrough was the decimation of German fighter squadrons, and it worked in the following way. Spaatz stuck to specific strategic targets: Allied intelligence had identified Germany's shrinking oil production as a key weakness and, slightly later, Eisenhower ordered concentrated bombing of the railway lines and bridges of most of western Europe as preparation for the D-Day landings. North German shipbuilding yards for U-boat construction were also on the short list. As the Americans recognized, such targets *had* to be defended by the Luftwaffe. But the new Allied long-range fighters could now neutralize Galland's hitherto successful policy of waiting until the Allied bombers were unescorted. If Luftwaffe squadrons sought to disrupt the bombers before they reached the Rhine (which made a lot of sense), they would have to tangle with the advanced Spitfires and the remarkably tough Thunderbolts; if they waited until the aerial armadas were attacking key inland German industries, communications, and refineries, they would find Mustangs high above them, coming down out of the sun. More than that, Doolittle made the bold decision to release the Mustang squadrons from close escort duty to go hunt German fighters all over the skies, if necessary driving them down to ground level, where the Mustangs' astonishing aerodynamics would prevail.

The Americans lost heavily in the ferocious battles of that critical spring of 1944, and many of their best bomber and fighter pilots were killed, maimed, or captured by a desperate German resistance. But they always had reinforcing squadrons, while the Luftwaffe suffered a catastrophe from which it never recovered. To save the Reich, it pulled vast numbers of aircraft and crews from the Eastern Front, which gave the Soviet air force a major advantage in the advance on Germany. It also pulled most of its remaining squadrons from the Mediterranean. But it availed Goering nothing: seeing Mustangs flying in broad daylight over Berlin in mid-1944, he is reported to have said, "We have lost the war," perhaps one of his few honest statements. By March of that same year

Mustangs were downing proportionately three to five times more German fighters than the more numerous Thunderbolts, although the latter were certainly inflicting their own heavy damage.[52] And the giant British bombers still came during the night, intensifying the pressure upon Germany's aerial defense systems.

Around this time, in late spring 1944, the Luftwaffe cracked; its exhausted fighter pilots could take no more. "Monthly losses, which included most of the experienced German fighter pilots, averaged 450 in the first five months of 1944; by the end of May, 2,262 had been killed. By 24 May 1944, only 240 of Germany's single-engine day fighters remained operational."[53] Actually, many more aircraft were in the pipeline, but their completion was hampered by disrupted communications, slowdowns in ball-bearing production, and completely inadequate fuel production. Above all, there was the loss of fighter pilots and crews. The newly trained men had far fewer hours of flying experience than their American and British equivalents, and suffered accordingly. Mustang pilots who chased their prey to the ground reported that many German planes were unable to avoid crashing into a pylon, a tree, or a high building. Even the finest German aces, perhaps softened by an easier tussle on the Eastern Front, met their match, great and revered names simply exploding into the sky. In March 1944 the Luftwaffe lost more than a dozen veterans of the air war, including two group commanders, one with 102 kills, the other with 161. They were irreplaceable. As the U.S. official history diplomatically acknowledges, the Luftwaffe was hurt more through these aerial battles than through even the most relentless and expensive daylight and nighttime bombing of aircraft factories.[54]

In this campaign, as in several others described in this book, it was the numbers of *trained* crews that were relevant. Were there enough good pilots left to fly the Hurricanes and Spitfires of Fighter Command in 1940? Yes, just. Were there enough well-trained U-boat captains and chief engineers to execute Doenitz's resumed offensive after the autumn of 1943? No. Were there enough experienced midlevel British Empire officers in the Eighth Army to handle Rommel's explosive form of ground warfare in 1941–42? No. Were there enough competent Japanese fighter pilots left after the "Great Marianas Turkey Shoot" of June 1944? No. Were there enough rock-hard major generals left in Stalin's army after both the purges of the late 1930s and the first year of Operation Barbarossa? Scarcely, but yes.

There were certainly not enough competent German airmen left to handle the increasing waves of Spitfires, Thunderbolts, and Mustangs over western Europe from late 1943 and early 1944 onward, despite the more than 80 percent of the German fighter force that was being deployed against Allied bombers by this time. One of those aces, Heinz Knocke, reported in his early 1944 diary the "awe-inspiring spectacle" of going against a thousand heavily armed bombers surrounded on all sides by American escorting fighters. After flying two sorties in one day, he watched his own Bf 109 machine-gunned and destroyed while refueling at a makeshift base; a few months later he himself, leading a group of only five planes patrolling the Reich's western borders, was pounced on by forty Mustangs and Thunderbolts and shot to the ground.[55] Hitler and the Wehrmacht leadership now railed at the "cowardice" of the Luftwaffe pilots for letting the enemy's bombers get through. Nothing could have been further from the truth, but truth was in short supply in Berlin. The German fliers were as brave as the Tom Blakeslees and Elmer Bendigers and Pat Danielses and Guy Gibsons who faced them month after month, but the plain fact was that, as in the Atlantic convoys campaign, the Allies had come up with a better way of getting things done.

ATTRITION OF GERMAN FIGHTER ACES IN WEST AND REICH		
ACE	VICTORIES	DATE LOST (1944)
Egon Mayer	102	March 2
Anton Hackyl	192	March
Hugo Frey	32	March 6
Gerhard Loos	92	March 6
Rudolf Ehrenberger	49	March 8
Egmont Prinz zur Lippe-Weissenfeld	51	March 12
Emil Bitsch	108	March 15
Heinrich Wohlers	29	March 15
Johann-Hermann Meier	77	March 15
Stefan Litjens	28	March 23

Wold-Dietrich Wilcke	162	March 23
Detler Rohwer	38	March 29
Hans Remmer	26	April 2
Karl Willius	50	April 8
Josef Zwernemann	126	April 8
Otto Wessling	83	April 19
Franz Schwaiger	67	April 24
Emil Omert	70	April 24
Kurt Ubben	110	April 27
Leopold Moenster	95	May 8
Walter Oesau	123	May 11
Gerhard Sommer	20	May 12
Ernst Boerngen	45	May 19
Hans-Heinrich Koenig	24	May 24
Reinhold Hoffman	66	May 24
Horst Carganico	60	May 27
Friedrich-Karl Mueller	140	May 29
Karl-Wolfgang Redlich	43	May 29

Source: From To Command the Sky *by McFarland and Newton, from Obermaier.*

LOSSES OF LUFTWAFFE FIGHTER PILOT ACES, MARCH–MAY 1944
Luftwaffe pilots who had gained an enormous number of kills on the Eastern Front were no match for the aerial offensive in Western Europe in early 1944. Without these aces, German aircraft production was valueless.

This relentless elimination of Luftwaffe pilots, plus the disruption of Germany's aircraft industry, supply lines, and, increasingly, aircraft fuel refining, created a favorable feedback loop that affected Bomber Command's own campaign. By this stage it too had started sending night fighter escorts to accompany the bombers, but they were generally rather slow (Beaufighters), and even when more sophisticated night fighters were introduced they did not achieve a high rate of kills. Still, Germany's own night fighter strength was already ebbing when, in late spring 1944, Eisenhower, supported by the Combined Chiefs of Staff, demanded that Bomber Command redirect its enormous bomb-carrying capacity to join the USAAF in paralyzing all German rail and road communications that reached westward to the Channel. Attacking French rail yards was much less dangerous than round-trip assaults on Berlin, but the pressure was coming off Bomber Command in any case. As one of the British official historians (himself a former distinguished flier) put it in his later reflections on the whole bombing campaign, the greater the success of the American long-range fighters by day, the greater the chances of Bomber Command by night. Devastated by their enemies' low- and high-altitude search-and-destroy policies, and by the destruction of railways and bridges needed to ensure delivery of supplies, the Luftwaffe had fewer and fewer planes to put up at night.[56] Too, in the weeks before D-Day, Bomber Command began to make daytime raids for the first time since 1939–40, discovered the loss rate considerably reduced, and found the accuracy of the Lancasters (despite Harris's gloomy forecasts) devastating. By September 1944, the RAF/USAAF round-the-clock pattern had resumed against a weakened Third Reich, which now lacked the advantage of its advance radar stations in France and the Low Countries. By this time also, Blakeslee's P-51s with drop tanks could accompany the B-17 Flying Fortresses all the way to western Russia, a distance of a thousand or so miles. The same Mustangs would then escort the bombers back from Russia to Italian bases and hop over the Alps home to East Anglia. In that triangular flight pattern alone, one sees the Third Reich cornered.

Throughout this entire period of February to September 1944, the Allied chiefs quarreled among themselves and across national lines. Harris opposed any selective strategy that targeted enemy oil, transportation, and electricity grids (as opposed to his own weird panacea of blasting cities), though he was overruled. Spaatz struggled against being under army direction and argued for continued attacks upon the German oil industry. Air C in C Sir Arthur Tedder and his guru Solly Zuckerman joined the recently appointed RAF air marshal Trafford Leigh-Mallory in pushing for the transportation plan, that is, the interdiction of all rail lines and roads going to western France, though Churchill feared that this would kill many French civilians (he was right). In the end, Eisenhower insisted that both the RAF and USAAF focus upon preventing German army divisions from reinforcing their forward divisions in Normandy by taking out all the marshaling yards, Seine bridges, railway lines, and whatever else moved trains (though he was willing to release the bombers elsewhere once it became clear that the enemy's rail and road links were cut).

Eisenhower's decision to cut off the Wehrmacht's reinforcement routes was one of the single most important policy calls in the entire war. To avoid giving any indication of a Normandy landing, the two air forces bombed the length and breadth of the French transportation system. By June 6, 1944, French rail traffic was a mere 30 percent of what it had been in January; by early July, it was only 10 percent. The Wehrmacht could neither get reinforcements into western France nor get its forward divisions out.[57]

It is striking that in these furious debates—with Eisenhower on occasion even threatening to the Combined Chiefs to *resign* if the air marshals would not obey his directives—few of the protagonists, apart from the levelheaded Tedder, seemed to pull back and recognize that they were now discussing targeting choices. They were not worrying about whether they were winning the air war: they were on top. The point at issue now was the swiftest way, in the aerial dimension, to bring down the Third Reich. It was no longer a matter of how one got the bombers through, but of what targets they would bomb. Perhaps it is not surprising that group and air force commanders who had lost dozens of crews each month needed some time to recognize this fundamental strategic change, but the Combined Chiefs should have done so. The battle for air superiority over Germany and Europe was basically won, even if the Reich fought back with astounding tenacity in the final year of the war.

A single witness makes this point better than any statistic. Early on the morning of June 6, 1944, Flight Lieutenant Owen of 97 Squadron recorded in his diary a late-night raid on a small enemy gun battery on the Normandy coast. His squadron was, unusually, ordered not to fly below 6,500 feet, not to use the identifying friend/foe transmitters, and not to off-load any bombs in the Channel. What did all that mean? His Lancaster duly dropped its bombs on the target at 0500 hours and turned for home; then, amazed, he and his crew looked down and witnessed the entire D-Day landings beneath them, with "a grandstand view of the Americans running in on the beach."[58] As the British aircrews flew back to their Lincolnshire base, they were staggered by the sheer number of American B-17s and B-24s flying overhead toward the continent. The Allies were finally invading France, and their troops were protected above by a gigantic aerial armada, safe from Luftwaffe attack. Eisenhower had told his troops, "If you see planes overhead, they will be ours"—what a difference from Dunkirk, even from Dieppe. And they were protected at sea from any U-boat assault. The whole thing—sea, air, and land—had come together.

On that day, the American, British, Canadian, and other Allied air forces put an astonishing 11,590 planes into the air. There had been nothing like it in world history, nor has there been since. A total of 3,700 fighters, including those Polish-flown Spitfires, covered the invasion beaches and patrolled all the way northward to the middle North Sea and westward to the western approaches. There was no chance for the completely diminished Luftwaffe to do anything except lose more and more of its planes and pilots whenever they rose into the air. The German air force would soon receive many additional aircraft, but they would have little overall impact, for they were steadily losing their bases, their fuel, and their crews. (The new pilots were so unprepared that their experience was grimly referred to as *Kindermord,* "infanticide"; 13,000 German pilots and crewmen were lost between June and October 1944 alone.) Their flight commanders had been shot out of the sky, chiefly by the Mustangs, at the critical point in the aerial war, and from that blow they could not recover. For the Allies to gain command of the skies over western Europe only two or three months before D-Day had been, as the Duke of Wellington said of the Battle of Waterloo, "a damned close-run thing." But great battles often are. The defeat of the Luftwaffe between February and May 1944 was a close run, but also one of the most decisive campaigns in history.

After Normandy

A detailed analysis of Allied strategy is beyond the scope of this book. A few aspects are nevertheless worth noting. In the first place, Harris's resumption of "strategic" attacks upon German cities after June 1944 did not do much to reduce the Reich's aircraft production—output actually rose to its height (39,800 planes) in that year. Germany's massive industrial capacity, spread into its many captured lands, was far too big to be brought down by a general, nonspecific bombing campaign even if there was much less opposition from Luftwaffe fighters; as late as January 1945 Arnold was expressing his amazement at how resilient the German war economy had turned out to be. By the spring of 1945, it is true, Hitler's empire was being smashed each day and night by thousands of Lancasters, Liberators, and Flying Fortresses. Four-fifths of the total Anglo-American bomb tonnage (out of a colossal 1.4 million tons) was dropped in the final year of the war. But the RAF and AAF were able to inflict that damage only because other, more important things had happened first.

Second, the strategic aerial offensive did not break the German people's morale, at least not to the extent of stopping the fight or turning its battered inhabitants against their regime. There is ample evidence of the citizens of Darmstadt, Hamburg, Dresden, and other cities being unable to believe the intensity of the destruction around them and yearning for relief and an end to it all. But there is absolutely no evidence that such desperation ended the war in the way that the meeting of American and Soviet soldiers on the Elbe did. Rather, the continuation of the West's area bombing (or "carpet" or "blanket" or "indiscriminate" or "general" bombing) stained its reputation and produced a moral equivalent to what the Luftwaffe had done to Warsaw, Rotterdam, and Coventry.

Third, the Allied air forces did carry out some clever bombing campaigns after Normandy, especially in their attacks upon oil-refining installations, transportation bottlenecks, and electrical grids. Here again, one is talking about *choices*. It doesn't make sense to get into a messy technical debate as to whether those massive bomber fleets should by then have concentrated more or less of their explosives upon marshaling yards or oil refineries, or more generally upon Germany's cities. The fact was that collectively they were bringing their enemy to his knees and were doing so with directional aids, Pathfinder forces, and a bombing accuracy that they had not possessed in previous years. If the German mines were still producing masses of coal but the rail lines were destroyed, that increase in coal output meant nothing. If fantastic new U-boats were being assembled at Kiel and Bremen but the diesel engines could not be transported from the Ruhr, they were of no use. The Battle of the Bulge in the winter of 1944 saw German panzers run out of fuel. If Speer and his able managers were cranking up fighter production in factories hidden in mountainsides but those aircraft had no fuel, that also meant nothing—or, rather, it meant that the Allied bombers were opposed by fewer and fewer German planes in the air. If extraordinary German ingenuity produced the world's first combat jet fighter in the Messerschmitt Me 262—by late 1944 Mustang pilots were reporting some extraordinary non-propeller-driven plane sweeping past them at unmatchable speed—then it was a further example of a Nazi "wonder weapon" (like the schnorkel and type XXI "elektroboot" U-boats and the V-2 rocket) that could not affect the outcome of the war because the Germans had run out of time.[59] The greatest of all the wonder weapons, the A-bomb, was of course unknown to all the European commanders, who had to fight with the tools at hand. But such tools were enough to deliver an Allied victory.

This epic struggle had to be, and was, won between 1943 and 1944; it was virtually impossible for Germany to turn back its course in the twelve months following June 1944. The shrinking German aviation fuel production figures after they lost air supremacy say it all: in March 1944 the totals were at their highest output (185,000 tons), but after the late May assaults the totals tumbled fast, to 56,000 tons in June, 17,000 tons in September, and a mere 1,000 tons by February 1945.[60] When British paratroopers landed at Arnhem (its own sad story), some Luftwaffe squadrons could not rise to join the fight at all. Germany had run out of gas, as Ultra decrypts of the desperate messages of its high command readily confirmed. Spaatz and his planners had, at last, got it right. In any case, by the beginning of 1945 the Allied generals no longer cared all that much for strategic bombing. Their ground armies were to close in on the Third Reich, its factories, rail yards, missile launching pads, and U-boat pens. The sort of airpower the generals wanted in order to help them reach the Fuehrer's bunker in Berlin was now tactical, not strategic. Turning German cities into heaps of rubble and destroying all the bridges actually slowed down the advance of Allied armor.

The two great breakthroughs in the strategic air offensive against Germany, or so Hastings argues, were American: the introduction of the long-range Mustang fighter to weaken the Luftwaffe and then, near the end, the campaign against enemy fuel production and distribution.[61] This is true, but it is only a partial explanation, rather like arguing for the decisiveness of MIT's Rad Lab in the radar war without mentioning Birmingham's invention of the cavity magnetron in the first place. As the narrative above has tried to show, the greatest and most successful of those long-range fighters would not have been produced without Rolls-Royce, the Merlin engine, the P-51 airframe, the drop tank, and a small, dedicated group of first British and then American individuals who helped further advance this project. Counterfactuals such as "Without the long-range Mustang, how long would the war have gone on?" are always of limited usefulness, but it is surely incontestable that only the destruction of Germany's fighter defenses in early 1944 made possible the fulfillment of that Casablanca directive to place an invasion army into western Europe shortly afterward.

The meaning of the Allied strategic air campaign was fought out long after the Second World War ended. Predictably, Harris and his Bomber Command supporters rushed in early to claim that the aerial offensive was key to the defeat of the Third Reich.[62] In retrospect, we can see it was not. At the other extreme, there has been a tendency to describe it as an unwise and excessive use of scarce resources (especially Britain's), a point reinforced by the recent slew of books and films about the horrors of indiscriminate bombing. Strategic bombing, from this standpoint, was thus both evil *and* an unnecessary waste.

This criticism conflates two separate points. To the extent that any bombings of nonmilitary targets were intended to terrorize the populace below—to deliberately "undermine enemy morale" rather than to keep people from returning to their shipyards or their ball-bearing factories—that form of warfare was indeed morally wrong. Whether saturation bombing took place in China, over Japan, or against Warsaw, London, or Dresden, it stood in contradiction to the generally accepted codes of proportionality and discrimination; it offended the doctrine of "just war." By the same doctrine, though, if a bomb intended for a Tyneside or Bremen shipbuilding works missed its target and hit the row houses next door, it did not have that moral stain. The real problem, as we have seen above, is that *most of the time* strategic bombing by both British and American squadrons missed the targets due to awful weather, altitude, wind speed, and poor navigational tools. Such obstacles, together with the rise of enemy counterattacks (flak, fighters), caused Allied planners to move to more general bombing, and caused Allied aircraft to jettison their bombs when under pressure. Yet the latter responses do not deserve to be described as indiscriminate bombing. Moving to bomb "a whole German town," as the RAF did following the Butt Report (and its analogue by the USAAF against Japan in 1945), crossed a moral watershed.

Drawing a military and strategic balance sheet of the effects of the Allied bombing campaigns brings its own important conclusions, well treated in the better studies.[63] The Allied bombing campaign was extraordinarily effective both in its infliction of direct damage upon the Third Reich (and Italy and Japan) and in its many indirect effects—forcing the Germans to cede aerial control over the Eastern Front, diverting vast numbers of guns and crews into antiaircraft duties, consuming millions of workers in the task of rebuilding wrecked communications and relocating damaged factories, and slowing down U-boat development and training. To his credit, Harris did keep up a campaign of aerial mining in Baltic waters. Bomber Command, seduced by the doctrine that theirs was an "independent" service, exaggerated before the fighting what they could achieve by themselves and exaggerated afterward what they had achieved. Yet where their squadrons really worked best was in *conjunction* with the land and naval campaigns—in helping to get aerial command over the D-Day beaches, in pulling the Luftwaffe back from the Eastern Front, in paralyzing any German counterattacks into Normandy, in slowing down assembly of the newest U-boats, and in depleting the Reich of its fuel supplies. That is a considerable tally.

It is difficult to see where intelligence played much of a role here, except to observe that German knowledge of Allied aerial defenses—or offenses—was miserable. Secret agents were of no value to Harris as he pursued his carpet bombing of German cities. Reading codes helped if they could tell the Air Ministry that this or that Luftflotte was being moved from Calabria to the Russian front. But the most valuable data source of all was aerial reconnaissance, because it was consistent, technical, and objective. Bletchley Park sources, by their episodic nature, had no way of capturing how shocked the German leadership was by the bombing of Hamburg—or, for that matter, of spotting the secret development of the Luftwaffe's jet fighters.

The Allies were helped by mistakes on the German side, especially those made by Goering and the Fuehrer himself. Hitler paid no attention to the lessons of the Battle of Britain about the importance of careful, calibrated defense. Germany gave inadequate priority to aircraft production as a whole between 1940 and 1943 and very little heed to fighter-plane production. Hitler hated the argument that a greater number of defensive weapons platforms was needed, and he was still screaming for the renewed bombing of England as late as 1943–44. During the final few months he was insisting that the Me 262 be flown as a jet *bomber*, with a 250- or 500-pound load that would totally unbalance it and leave it vulnerable to the Spitfires and Mustangs. The increases in fighters that Erhard Milch and Speer achieved were done despite him. The V-2 rocket was truly impressive, but such a wonder weapon consumed staggering amounts of scarce resources—one estimate believes it was the equivalent to the United States' Manhattan Project (America could afford its superweapon, but Germany couldn't), another that it diverted the manpower and materials that could have built around twenty-four thousand regular fighter aircraft.[64] In any case, it was too late, like the jet fighter; bringing it into battle in such limited numbers when the Allied armies were advancing with increasing pace toward the Rhine and the Vistula was of little use. Such weapons had been needed during the critical year of 1943, or at the very latest during the early months of 1944, for that was when Germany lost the fight.

Looking East

The first half of June 1944 was also an important turning point in the strategic bombing campaigns in the Pacific, as will be detailed more fully in chapter 5. On June 15, the first wave of U.S. Marines landed on Saipan, in the Mariana Islands, and America burst right through Japan's vital perimeter. The seizure of the Marianas was critical because it was from swiftly built air bases on those islands that the USAAF could directly attack the Japanese homeland. By late November 1944, a full two years after the invasion of North Africa and the subsequent Casablanca resolutions, American airpower in the form of the massive new high-altitude B-29 Superfortresses was at last able to hit Japan in the way in which its heavy squadrons in Europe, along with RAF

Bomber Command, had been pounding the Third Reich over the preceding months.

The story of the strategic bombing offensive against Japan is beyond our narrative.[65] It had, to be sure, a common Trenchardian aspect, that of devastating aerial attacks upon the enemy's industries, cities, communications, and people. But, simply because of the imbalance of power, it lacked the dramatic ebb and flow of the prolonged, desperate fight against the Luftwaffe. Germany really was a tougher opponent in all respects, but especially in the duel for air superiority. By the time that the B-29s began their attacks, both the Japanese naval and army air forces—especially, once again, their pilots—had been greatly reduced in numbers. The long-range Mustang escorts, when brought into service in the Far East, had very little dogfighting to do. In any case, the B-29s could fly so high that it was difficult for Japanese planes to reach them. By the spring of 1945, their pugnacious commander, Curtis LeMay, had many of the bombers stripped of their gun turrets and ordered them to bomb their targets at night (to further confuse Japanese defenses) and from a mere 8,000 feet, unloading vast numbers of smaller incendiaries upon the timber-framed buildings below. The results of the Tokyo and other fire raids were terrifying, with losses unimaginable to the Anglo-American peoples' experience of war, or at least war upon civilians as distinct from war in the trenches. The entire Blitz on Britain took approximately 43,000 lives, with another million wounded.[66] The obliteration of 150,000 inhabitants in the firestorms of Tokyo was not the end of the story. Once the B-29s had replenished their incendiaries—they ran out of supplies in late March—it was the turn of many other Japanese cities, towns, and harbors to be totally crippled. Then came the two atomic bombs, which killed at least 200,000 people. The theory and practice of strategic air warfare had reached its apotheosis. And a new, threatening age of bombing civilians had begun.

The years after the war brought mixed blessings to the characters described above. The glamorous, casual, and very influential Tommy Hitchcock, rejoicing that the Mustangs were coming to England, died in 1944, just before D-Day, when a plane he was trying out plunged mysteriously to the ground. Jimmy Doolittle "retired" to a busy life that mixed a business career with government service, for which he deservedly received much recognition; in 1985 President Ronald Reagan pinned a fourth star on his epaulets, making him a full general at last. Lovett had an even more distinguished postwar career, being undersecretary of state to George Marshall from 1946 to 1949 and then secretary of defense himself during much of the Korean War (from 1951 to 1953). Don Blakeslee stayed in the air force and served in both Korea and Vietnam until a gentle retirement to Florida. Elmer Bendiger, like so many of the "average" fliers, returned to civilian life; thirty-five years later he began writing his account of the Schweinfurt raids.

Sir Wilfrid Freeman retired as air chief marshal and took a senior management position at the British company Courtaulds; he died in 1953, by one account of "overwork," which sounds plausible. He could be deeply satisfied that all the aircraft he had authorized, protected, and then put into service from the late 1930s had done so well. Arthur "Bomber" Harris was a less satisfied man. Already by the closing months of the war the British government was uneasy at the rising criticism of Bomber Command's "shock and awe" area-bombing policies. Harris was the sole British commander in chief not made a peer at the end of the war. In 1948 he left, disappointed, for a business job in South Africa, although he eventually returned to a quiet life in the Thames Valley until his death in 1984. Eight years later his statue was unveiled by the Queen Mother beside the RAF's church, St. Clement Dane's, but even that provoked protests, and the statue had to be guarded for some time. Regrettably, this moral and political unease hurt the reputation of Bomber Command itself, the only service not to be given a campaign medal—a lasting shame, because it confused the arbiters of a bad targeting policy with the gallant crews who carried it out, tens of thousands losing their lives.

Long after the war, Ronnie Harker retired to New Zealand, where he indulged in his two favorite hobbies, fly-fishing and taking a Mustang up, higher and higher, into the skies he loved. A large number of P-51s had been acquired by the Australian and New Zealand air forces and performed wonderfully during the Korean War before enjoying a long retirement in the antipodean climate. There were still many around as the second half of the twentieth century wound to a close, and one of them was always available to Harker. In 1997 he took a Mustang into the air for the last time, his final solo flight. One doubts if he pushed it around as much as he had done to the Alison-powered P-51 on that raw morning of April 30, 1942. But no matter: he was back in a machine that he had helped to transform from an ugly duck-ling into the most formidable long-range fighter of the Second World War. He, and the others mentioned here, had created not so much a war-winning weapon as a war-winning system. One wonders whether Harker, a mere eighty-eight years old, reflected on all this as he handed over his oxygen mask for the last time. Two years later, he himself went to the central blue.[67]

The historian of great-power politics and of the force of economic change in humankind's rise cannot but be pulled up by the role of chance and serendipity in this story of the Mustang's rescue from the scrap heap. What if Ronnie Harker had not been invited to fly the P-51 by the Duxford Testing Station in April 1942? What if another test pilot had not been so perspicacious or had not known about the new Merlin 61 engine? What if the Rolls-Royce manager had not had such a close relationship with Sir Wilfrid Freeman? What if Lovett had not made his important 1943 visit to the air bases in Britain?

So many conjunctures and hypotheticals. Perhaps other solutions might have arisen to unblock the late 1943 impasse in Allied strategic bombing. Still, without the breakthroughs described above, it remains difficult to see how Overlord could have taken place in June 1944 and the Third Reich could have collapsed less than a year later. Serendipity counts.

CHAPTER THREE

HOW TO STOP A BLITZKRIEG

German success between 1939 and 1942 owed as much to the German armed forces' better understanding of the balance between offensive and defensive firepower as it then existed as to any material consideration. Opposed by a number of enemies of limited military resources and inferior doctrine, the *Wehrmacht* had been able to defeat opponents lacking adequate anti-tank and anti-aircraft defences and—crucially—the space and time in which to absorb the shock of a *Blitzkrieg* attack . . . yet by 1943 the Soviet Army had survived two *Blitzkrieg* attacks and in the process had learned to counter this form of warfare.

—H. P. WILLMOTT, *The Great Crusade*

By that time [El Alamein, October 1942], the British superiority in strength . . . was greater than ever before . . . the Eighth Army's fighting strength being 230,000, while Rommel had less than 80,000, of which only 27,000 were German. Moreover, the Eighth Army had seven armoured brigades, and a total of twenty-three armoured regiments, compared with Rommel's total of four German and seven Italian tank battalions. . . . In the air, the British also enjoyed a greater superiority than ever before.

—B. H. LIDDELL HART, *History of the Second World War*

On February 20, 1943, little more than a month after the Casablanca conference, units of the U.S. Army a few hundred miles away had their first serious encounter with the Wehrmacht. The grinding battles that followed took place in and around a strategic, stony mountainous route in southern Tunisia known as the Kasserine Pass. If anything could be termed the American army's baptism by fire in the Mediterranean/European theater of war, it was this. The U.S. Army II Corps had, unfortunately, bumped into the Wehrmacht's most aggressive panzer general, Field Marshal Erwin Rommel, a person who held to the tactical principle that when one was being pushed backward by superior enemy forces, the only sensible response was to retreat a little, regroup, then ruthlessly counterattack, relying on his men's sheer professionalism and superiority of fighting experience to intimidate and overcome more numerous but (he could assume at this point) half-green troops. Most other German generals would have agreed, and it was their long-standing operational doctrine: fight as hard as possible, fall back if the position becomes untenable or is in danger of being outflanked, reorganize and combine depleted units, and then strike again, just as the foe is taking a rest. Stunningly bold in their aggressive blitzkrieg attacks between 1939 and 1942, the Germans also showed themselves to be the world's best defensive fighters for the last three years of the war.*

The newly recruited American troops had never before encountered this form of warfare, nor had many of their commanders—and they were facing a mere six-battalion attack. The historian Rick Atkinson's laconic but brilliant account captures their collapse so well. We do not know whether the cynical lieutenant who muttered "Into the valley of death rode the six hundred" himself survived the subsequent slaughter, but the subaltern had it right and his overconfident superiors had it wrong. The GIs, along with an unfortunate battalion of the Leicester Regiment nearby, paid the price.[1] From the enormous number of immediate postbattle analyses and the many later histories, it is clear that

---

*The term *blitzkrieg* gets used, often sloppily, to cover different though related things. Literally, it translates as "lightning war." As such, it could be applied to many campaigns (that of Frederick the Great; Israel's 1967 fighting), although the word popularly arose to describe the so-called German way in warfare in 1939–41—fast battleground movements by armored and motorized infantry units to take the enemy's army off guard, sometimes followed by a pulling back to regroup and then strike again. German tactical airpower (dive-bombers, medium-range bombers) gave it a new touch. It has nothing in common with the London Blitz of 1940–41 (although obviously the Luftwaffe was once again to the fore), which was a lengthy aerial bombing campaign against the capital city.

much was wrong in the Anglo-American armies in Tunisia. At the top, chains of command were far too entangled, midlevel generals were absurdly optimistic one day but then lost their nerve and blamed their subordinates the next, and there were extensive failures in communication. At the bottom, there were far too many green troops who sited themselves in the wrong positions, were shocked by the astounding noise of enemy howitzer fire, and, when they observed a sister unit running away, joined in the retreat, abandoning their heavier weapons and staggering through the desert thornbushes and stones. They were badly beaten.

At the end of the week, the Allied troops, though pushed well back, held off Rommel's assaults. Overall, 30,000 GIs had been committed to the Kasserine fight, and 6,000 of them were lost, a majority of them captured, to spend the rest of the war in camps in Germany. General Lloyd Fredendall's U.S. Army II Corps lost 183 tanks, 104 half-tracks, more than 200 guns, and 500 jeeps and trucks. The Germans suffered a mere 201 dead.

But this was not the fall of France; nothing like it. Eisenhower moved more units and better commanders in from Algeria, Montgomery's Eighth Army was coming in from the east, and Rommel's forces simply ran out of fuel and ammunition. But it had been a shock, and a preview for the Americans of how hard this foe would fight. Fredendall was very soon replaced by the more aggressive and highly ambitious George Patton.

Apart from MacArthur's calamitous defeat in the Philippines in early 1942, it was probably the most humiliating smack in the face that the American army received in the Second World War. And yet, in some respects, it was unsurprising. Man for man, the German troops were simply much more experienced, and this was true when they fought more battle-hardened British troops, too. Shortly before the fight for the Kasserine Pass, a considerable force of Argyll Highlanders and the West Kent Regiment had moved into a valley at Jefna, also in Tunisia, only to encounter the heavily camouflaged 21st Parachute Engineer Battalion under Major Rudolf Witzig in the hills above. The British contingent, not spotting the formidable enemy positions, was routed by a German contingent one-tenth its size. Witzig had already fought with distinction in France, on Crete, and in Russia, and had led the famous paratroop drop that captured the supposedly impregnable Belgian fortress of Eban Emael in May 1940, allowing the panzer breakthrough.*[2] Compared to Stalingrad, fighting in North Africa was easy. No one in the Western armies had forces with battle experiences comparable to Witzig's.

General Fredendall's units were simply the fifth—or was it the seventh or tenth?—national army to experience the most uncomfortable military situation of the Second World War: being attacked by Wehrmacht forces who struck hard at first, and if necessary struck hard again in a later counterattack, to shock the enemy, dislocate his communications, weaken his morale, and paralyze his high command. On most occasions, the first, blindingly fast assault was enough. If resistance on the ground began to grow, the German units pulled back—but only to prepare to strike again. This, at least, had been the story from 1939 until the end of 1942.

Even knowing of the Third Reich's final and total defeat, a historian writing about these events seventy years later can only wonder at what Williamson Murray termed in his book "German military effectiveness."[3] The Wehrmacht blew apart a large, gallant, but highly disorganized Polish army within two weeks, in September 1939. There are reasons to discount the significance of this extremely lopsided fight, *except* that all of the components of this new way of warfare were deployed: the Luftwaffe immediately took control of the air, destroyed Polish air bases and scattered army columns, then proceeded to devastate Warsaw; the fourteen mechanized divisions swept past the badly emplaced Polish infantry, brushed aside the cavalry, and raced toward all the major targets—Lodz, Cracow, Lwow. While masses of Polish soldiers were being steadily herded into encircled positions, one of the armored divisions of Walther von Reichenau's Tenth Army raced to the outskirts of Warsaw in a week. Soon it was all over.

After spending some time during the so-called Phony War assessing how they could do even better during their next battles, the German

---

*Witzig and Otto Skorzeny (whose daring raid captured Mussolini and brought him out of captivity) compete for being the model for the Michael Caine character (a paratroop colonel who fought in the Low Countries, Crete, and Russia) in the celebrated novel and movie *The Eagle Has Landed.*

panzers, infantry, and air forces struck westward and northwestward in the spring of 1940. The smaller armies of Denmark, Belgium, and the Netherlands were engulfed, perhaps understandably, but the swift Nazi takeover of Norway—in the Royal Navy's front yard—was astounding. Even more astounding, in one of the epic battles in the long military history of western Europe the massive French army was crushed and a half-trained British Expeditionary Force bundled out of the continent only a month or so later. Evidently the Wehrmacht's swift defeat of Poland was no fluke. After all, the French had larger ground forces than Germany (sixty-five active divisions for the French, compared to fifty-two active divisions for the Germans) and more tanks, including some heavier ones.* The French had been preparing for two decades to counter a German attack in the West, and they would be joined by the British Expeditionary Force and a Belgian army reluctantly sucked into this war. But the French air force was weak and outdated, and so the Luftwaffe dominated the air. The unorthodox German panzer thrust through the Ardennes dislocated the French high command, which simply could not keep up with the pace and boldness of Heinz Guderian's advances toward the Channel—even his Wehrmacht superiors and Hitler himself were unnerved as they watched this unexpectedly fast victory, fearing the panzers might go too far and get trapped.

By June 1940 France was done for, and Britain stood alone. The whole geopolitical and military shape of the war was transformed. Stalin was amazed and anxious, knowing that his purge-weakened armies were nowhere near ready to fight. The American government was transfixed. Mussolini rushed to join Hitler. The Japanese recalculated their

---

* As this chapter unfolds, it will be clear that the standard units of military size—army groups, divisions, brigades, regiments, battalions, et cetera—only help us a small way in understanding the effective power of one protagonist vis-à-vis another in a campaign. American divisions were huge; Soviet divisions were half their size. German divisions on the Eastern Front had shrunk to one-quarter of their size by 1944, but Hitler still insisted on calling them divisions. Tank numbers were similarly confusing; if it took four or five T-34s to knock out a Panther during the 1943 Battle of Kursk, what did raw numbers matter? The same was true in the naval balances, though in a reverse form—British cruisers and destroyers were far better balanced in the North Atlantic than their massive, top-heavy German equivalents, which yawed so much that their crews became seasick.

options. The European war was not even ten months old, and the world had been turned upside down. Little wonder the term "lightning war" seemed so appropriate.

For Britain and its partners, 1941 was no better. Yugoslavia's political tilt against the Axis that spring had been punished by an enormous, swift invasion that carried on through the southern Balkans to overwhelm Greece and then capture Crete in a bold parachute attack. Was there a particular British regiment that was pushed out of Norway, then pushed out of France, only to be pushed out of Crete in May 1940, a fate magnificently imagined by Evelyn Waugh in his great wartime trilogy?[4] If so, that regiment might well have been pulled back to Egypt, where soon it would have encountered the fast-moving and highly dislocating assaults of Rommel's Afrika Korps. By that time, of course, the much larger Operation Barbarossa had begun, and German panzers were slicing through the Ukraine, leaving millions of Russian soldiers to be rounded up and herded to their fate. Really, the American soldiers crushed in the Battle of the Kasserine Pass had little to apologize for; it was simply their turn to get beaten up.

What follows in the rest of this chapter are two related questions. The first is why the Germans were so good operationally and tactically, and if they were that good, how on earth did one defeat them? The second is more general: did offensive warfare stand a better chance of claiming victory than any defensive strategy whatever? Assuming bold leadership and well-trained troops in the attacking army, would this phenomenon of lightning war almost always succeed—or was it critically affected by other factors such as time, space, and sheer numbers? It makes sense to treat the second, more general question first before looking at the reasons for the Wehrmacht's performance in the Second World War; by doing so we can arrive at a better understanding of why the land battles in western Europe, Russia, and the Mediterranean unfolded the way they did. The history of warfare is littered with examples of swift and spectacularly successful campaigns—probably nothing in modern times equals the achievements of Alexander the Great and Genghis Khan in the rapid overthrow of enemies and the extensive conquest of lands. No doubt the wide-open spaces across which they chiefly fought explain a great deal; with a decisive leader and mobile shock forces, an army could travel a long way in a week. This may also explain, conversely, why Europe's topography of mountain chains, dense forests, extensive swamps, and numerous rivers made complete control by any one power so very difficult.[5] Aware of these constraints, the Romans kept to their limits, while the later Holy Roman Empires of Charlemagne and the Hapsburgs were those of a large regional power. The wars of the Middle Ages were chiefly slug-it-out affairs, and the arrival of newer defensive fortification designs after 1500 put the emphasis upon laborious siege warfare.

Even in Europe, there were historic exceptions: dramatic and swift campaigns that threw the enemy off balance, because the attacking army was so well trained and motivated that geographic obstacles seemed to shrivel. The Duke of Marlborough's dramatic march up the Rhine from the Netherlands to upper Bavaria (the Battle of Blenheim, 1704) is a good example. A half century later, Frederick the Great often stunned his enemies by the speed at which he switched his armies from one front to another, and sometimes divided his forces so that while one half contested the field of battle, the other was making a flank attack obscured by hilly terrain. Napoleon's capacity for moving armies—very large armies, and at high speed—is legendary, and in 1866 and 1870 Helmuth von Moltke the Elder hit the Austrians and French so fast and decisively that those wars ended very swiftly.[6]

Yet these successes by lightning strikes were exceptional. Most of Marlborough's other battles (Ramillies, Malplaquet) were great, static bloodbaths in the Low Countries. Surrounded by enemies on all sides in the critical period of the Seven Years' War, Frederick had on many occasions no choice but to stand and fight. Try as they might, Napoleon's great marshals and battle-hardened armies could never achieve a decisive victory in Spain: the terrain was too broken and harsh, giving the advantage to the many Spanish guerrilla groups that sprang up to conduct irregular warfare, and giving the advantage also to the British-led coalition under Wellington, that cautious master of situational battles. The most spectacular exposure of the weaknesses of the Napoleonic way of warfare came, of course, in France's catastrophic defeat in the War of 1812, where the weather, sheer distance, and the Russian willingness to pull back hundreds and hundreds of miles made nonsense of the very idea of an early, decisive victory. Moltke concluded a decade or so after his 1871 victory over France that fast campaigning across industrialized western Europe by his Prussian armies would be impossible in the future: there were too many new urban zones, too many canals, too many railway embankments.

But the old field marshal was almost alone in this conclusion. His successors, Alfred von Schlieffen and Moltke the Younger, were mesmerized by the coming of railways and the telegraph, as were the generals in neighboring countries as 1914 approached. The advent of newer military technologies, such as modern aircraft and the tank in the years before the Second World War, saw another resurgence in the belief in swift military victories.

If we are looking to find the conditions, logically and logistically, in which lightning warfare doesn't work, the first answer has to do with topography. A decisive victory isn't achievable if the fighting has to take place across large mountain ranges, as the Wehrmacht discovered in trying to crush the partisans of Yugoslavia and Greece after 1941, and as, over centuries, successive waves of invaders have discovered in the high peaks of Afghanistan. Dense jungles, such as those in the southwestern provinces of China and across all of Southeast Asia, confirm this obvious military point: difficult physical circumstances tend to equalize the contest, even if one side possesses much more fighting power than its opponent (as was evidenced in Vietnam). And wide deserts, with hundreds of miles of shifting sands, definitely restrict the bolder Rommels of the world and give advantage to the more cautious Montgomerys. Great rivers, miles across from one bank to the other, slow down offensives; even if the attacking army has pontoons and other bridging equipment, such heavy stuff may have to be brought considerable distances to reach the water's edge. Topography does not remove human agency, but the strategic planner who sees how best to exploit it will aid his commanding generals greatly.

Two further reasons, also very obvious, complete the picture. Fast mobile warfare leading to a swift and decisive victory in the field will, logically, not occur when the defensive armies themselves are too strong, too entrenched, too numerous to permit it. This would be true whatever the geographic extent of the battle zone. Yet it is also true that fast, aggressive forms of operation will most likely falter, then fail, if the attackers have to fan out over an increasingly broad area of land, with the leading troop units ever farther from the home base, each other, and their supply lines.

So much for the historical and abstract theory of blitzkrieg, of swift, aggressive, and shocking warfare conducted by forces that usually were also very good at smart counterattack. But what exactly did it mean in the case of the Wehrmacht's operational performance during the Second World War? In the context of modern, industrialized warfare, the internal combustion engine, armored and armed vehicles, railways, and aircraft were melded into a form of fighting that appeared to contemporaries to be totally new and could transcend the larger geographic constraints of earlier eras. If proof were needed, one need look no further than the German tanks bursting through the forests of the Ardennes in May 1940.

And why was everyone so astonished at these very short campaigns? It was because almost everyone's image of great-power land warfare had been shaped by the experiences of the grinding, static, slogging-it-out battles of the First World War, particularly in the campaigns along the Western Front and in northern Italy. At the Somme, at Verdun, at Passchendaele, and along the Isonzo, hundreds of thousands of soldiers were killed for a few miles gained. Sometimes the attackers, after months of fruitless assaults, ended up right back where they had started. *This* was the reality of modern industrialized war, as confirmed and chronicled in the histories, memoirs, and literary writings of the 1920s and 1930s.

What almost all of this literature missed was that in the closing years of the war certain military staffs had privately figured out how to break the deadlock of static trench warfare. The first changes came, unsurprisingly, in eastern Europe, where the lengths of the fronts were far greater and thus less densely held than along the Western Front. The Brusilov Offensive of summer 1916, when the Russian army overran the flaccid Austro-Hungarian forces, had succeeded because the attackers did not use massive Haig-like bombardments (they had too little ammunition) against the sluggish defenders; instead, they employed surprise, and in various places attacked shortly after they commenced firing. A German offensive around Riga in the next year used the same principles of shock, swiftness, and going around obstacles. In 1917 also, Italy suffered its biggest defeat at Caporetto, when fast-moving German units (including a young Erwin Rommel), sent to reinforce the Austro-Hungarian armies, made a spectacular breakthrough and compelled the Italian high command to call for Allied assistance. Perhaps the most important point about all this, as Timothy Lupfer argued, was that the Prussian General Staff allowed midlevel officers to circulate impressions, ideas, and experiences from their respective fronts, to stimulate initiative.[7] If surprise assaults by well-trained infantry had worked in the Baltic, why not see if they might also work in Italy, or even in the West?

Slowly, roughly, the German army stumbled toward newer tactics and, of equal importance, newer types of troops: shock troops, better-trained, equipped with a new package of weapons (machine guns, grenades, wire cutters), and encouraged to press forward swiftly and move around enemy strongpoints. Meanwhile, on the Western Front, the British Army was at last coming up with its own way of cracking the trench warfare deadlock, a mechanical way, in the form of the first tanks; despite many early setbacks, they intervened to great effect in the August 1918 offensives. In the age-old tale of the ebb and flow contest of offensive and defensive warfare, the offensive had once again taken over.[8] Here, in mobile strikes and fast-moving incursions, was the future of war, and the intellectual high priests of this latest "military revolution," such as Liddell Hart and General J. F. C. Fuller, were to expend all their energies after 1919 in preaching the new gospel. Technology had conquered topography, newer types of armies could go anywhere, and if the forward columns didn't stop, the enemy's nervous system would be paralyzed. Yet none of the armored-warfare enthusiasts considered that the greatest attempt at a decisive, swift victory— Ludendorff's massive offensive in the West in spring 1918—had achieved significant advances initially but in the end could not get through the opposing defenses because the Allies' firepower was too deep and the room for maneuver was far too small.

The lessons drawn from the First World War by the military experts about the potential of offensive mechanized warfare were therefore very mixed. The French high command, conservatively but rather logically, drew the 1918 lesson that only strength in depth had saved the Republic and that if it was attacked by Germany in the future, the best strategy would be a defensive one. This led it to create an even greater physical barrier against invasion in the form of the Maginot Line. In Great Britain, the armored-warfare enthusiasts, Fuller perhaps in particular, argued for all-tank forces, expanding outward from the breakthrough points and paralyzing the enemy's nervous systems; Liddell Hart pushed for the indirect approach to battlefield victory, which also involved swift-moving attacks but by mixed military units making flank approaches or landings from the sea. Large-scale tank maneuvers on Salisbury Plain in the late 1920s showed that this new type of warfare had a future. So, despite the opposition of the cavalry regiments, the British Army pioneered mechanization until the crimping economies of the 1930s. The Americans, having reluctantly come over to Europe to squash the enemy in 1918 by massive material superiority (as had happened in the Civil War between the North and the South), were not interested in planning for another great war in Europe; they went home and demobilized. The early Red Army under Trotsky, then more professionally under Tukhachevsky, studied this debate and discussed what were called "deep battle" tactics, until Stalin's manic purges in the late 1930s brought everything to a halt. And Japan's planned zones of expansion—chiefly rivers, deltas, and jungles in Asia—were not suitable for tanks, except for very light ones that could manage the narrow roads and wooden bridges.

This left the German military, massively reduced in troop numbers and weapons systems by the Treaty of Versailles, battered by the failing Weimar economy, and marginalized by the corrosive party politics of the time, to wonder how it might recover its old fighting power. With the coming of the Third Reich in 1933, the generals began to see a steady and then dramatic inflow of resources for a military buildup. Most of the senior officers were Prussian conservatives of a traditional bent; their feeling was that the more infantry and artillery divisions, the better. But there was also a sufficient sprinkling of more radically minded officers—Guderian, Manstein, Hasso von Manteuffel, Rommel—to urge the case for fast mobile warfare. The Wehrmacht that opened the Second World War with the vicious assault upon Poland reflected this uneasy balance between old and new, with a small number of mobile motorized and panzer units and a preponderance of slower-moving, often horse-drawn divisions. Still, the faster forces, aided now by their own tactical air forces, had a punch that no one else possessed and the potential to execute that desirable, almost mythical battlefield operation, the *Kesselschlacht*.* They also had a Fuehrer who wanted fast results.

In response, the Wehrmacht and the Luftwaffe produced for their leader a combination of aggressive weapons systems—infantry, panzer, air attacks—that no one else's armed forces had ever encountered before. Was it any surprise that Nazi Germany's more traditionally structured neighbors were blown away, one after the other? They had faced the coordinated force of the three components mentioned above: (1) mobile, highly trained infantry units, wherever possible transported by trucks or rail to save time, working in support of free-ranging panzer battalions; (2) the tactical/operational doctrine of unorthodox, fast attack, designed to push through a chink in the enemy's defenses and range afield; plus (3) an air force, which had in large part been specifically organized for close-in and low-altitude tactical bombardments, to destroy its foe's air force, shatter his communications, and demoralize defending troops who were unprepared for the horrid, whistling sound of Ju 87 Stukas diving at them from the sky.

So how could the Nazi blitzkrieg be checked? First of all, by an opponent who could deploy stronger, tougher, and better-equipped forces (with panzers, bazookas, mines, better tactical aircraft) to counter German operational boldness; and, in turn, being able to launch a heavy counterassault. But this simply could not be done by small or medium-sized nations. The aggressive style of warfare developed by the Wehrmacht and the Luftwaffe could be defeated only by the forces of another great power, or probably two. Practically, this meant the British Empire, the Soviet Union, and the United States. None of these three was ready for large-scale modern combat during the first half of the war, but all possessed great inherent strength in the form of population, raw material resources, and technological expertise. And all three of them

---

*It is hard to extricate this word from the German language: the literal meaning is "cauldron battle," though in practice it was understood to involve an outflanking move, a breakthrough, an encirclement, and, in its highest form, a pincer movement that fully enclosed the enemy's army and compelled its surrender. The ghosts of the ancient Battle of Cannae, and of the recent Schlieffen Plan, haunted this operational dream.

enjoyed a sufficient geographic distance from the Third Reich's forces—America especially, but also Britain (thanks to the RAF and the English Channel) and even the USSR (because of its vast interior, where its military production could be relocated)—to develop their capabilities and allow their inventors, production engineers, and strategic planners to come forward with the instruments necessary to take on the Nazi war machine in great strength.

Second, the two great powers most closely affected by Germany's early successes—that is, Britain and the USSR—could defeat the Wehrmacht's "lightning war" strategy by developing a counterstrategy that took advantage of their own geographic assets. The British counterstrategy, developed in minute detail by the Joint Planning Staff, which worked under Churchill and the War Cabinet, was to defend the country by airpower from the Luftwaffe and develop a vast bombing offensive of their own, to protect the Atlantic sea-lanes, and slowly to advance out of Egypt and the Mediterranean, where German units were badly overstretched and could not hold off forever the accumulating military resources of the British Empire. The Russian counterstrategy, forced upon Stalin and the Stavka (the high command's General Staff) by the success of the Wehrmacht's powerful advances, was to trade space for time until the Germans were so overextended that the counterassault, boosted by enormous military production from factories also safe from Luftwaffe attack, could commence. It was probably not a coincidence that, since Britain and the USSR both needed time to recover, their advances could not commence until late 1942 or early 1943.

The Nazi blitzkrieg was stopped, pummeled, and then converted into a disaster for the attackers in two places: on the barren stones and shifting sands of North Africa and across the great wide plains of western Russia. The first of these twin checks to German expansionism was, as measured by the size of armies and the extent of the casualties, much, much smaller than the second, for it was the war on the Eastern Front that proved by far the greatest check to the self-proclaimed Thousand-Year Reich. Yet the story of how the Wehrmacht was defeated by the Anglo-American armies in North Africa is very much worth including—if only as a comparison and a control—in our analysis of what happened when swift offensive campaigns came up against great powers that possessed the advantages of geography, production, increasingly better weapons, and superior planning staffs. Before we examine the much larger battles that raged across the Eastern Front, Stalingrad, Kursk, and Operation Bagration, it is useful to begin with Egypt and the small coastal town of El Alamein.

Blitzkrieg in the Desert

The first place where this particularly German form of mobile warfare ground to a halt occurred roughly fifty miles west of Cairo, among the ridges and gulleys that stretched directly southward from El Alamein, until one came to the treacherous dunes of the Quattara Depression.[9] In contrast to the massive flat wheat fields of the Ukraine, there was instead a restricted corridor, only 50 miles wide, between the Mediterranean and the Quattara. Apart from the specially equipped units of the British Long Range Desert Group, nothing could operate across the fine shifting sands in the south; Rommel's tanks, trucks, and artillery would be swallowed up as certainly as were the chariots and heavy infantry of the Persian monarch who had tried to march through those same sands six centuries before the birth of Christ. Thus, if the Afrika Korps, urged on by Hitler, was to take Alexandria and then Cairo, it could do so only by bludgeoning its way through that relatively narrow bottleneck south of El Alamein. No one was more aware of that than Rommel himself—and the new commander of the British Eighth Army, General Bernard Montgomery, who had taken up his post in mid-August 1942. Aggression and mobility were pitted against caution and stolidity. But here was a case where geography played heavily to the latter's advantage, provided the British defenses were strong enough.

Over the preceding two years of conflict along the lengthy North African littoral, which had begun in September 1940, the British did not have that strength. Mussolini's large armies in Cyrenaica had made an early but slow advance into Egypt—blitzkrieg this was not. By contrast, the Eighth Army's counterstroke in December, using its faster-moving armored units against a far larger, slower Italian army, seemed to validate the Fuller–Liddell Hart doctrines but then petered out disappointingly the farther it moved from its starting point. There were many later military advances and retreats in this particular theater of war—some coastal towns such as Sidi Barrani changed hands on half a dozen occasions—which suggests that neither side had at this stage the capacity to sustain a prolonged offensive. Thus nothing decisive occurred until the arrival of Rommel's Afrika Korps early in 1941 to bolster the Italian position. Rommel's presence brought a new pace to the advances and retreats. That promising young and aggressive officer at Caporetto in 1917, the brilliant leader of the 7th Panzer Division in the 1940 defeat of France, was not interested in having his troops stay still. Of course, he and his Luftwaffe corps commander needed time to develop bases amd receive their tanks, trucks, and aircraft, but they would soon be ready.

In the meantime, the conflict continued, affected by many considerations such as Churchill's diversion of much of the Eighth Army to the fighting in Greece and Crete during 1941, and then their painful recall. It was also affected always by which side received further reinforcements of infantry, tanks, the critically important trucks, artillery, and aircraft.[10] The fighting was determined overall by the value the two great war leaders placed upon this campaign. To Hitler it was certainly important as a means of keeping Italy in the war and pinning the British down in the Middle East, but it did not compare with the titanic struggle against the Soviet Union; to Churchill at this stage in the war, it was the only theater of land/air warfare in which the battered British Commonwealth forces could pick themselves up, carry the fight forward, and regain their pride.

By the summer of 1942, following these considerations, each side had placed more and more chips on the North African table. Given the global picture of the war at this time, it is surprising that the German high command could continue to furnish Rommel with *any* additional troops and armored units, and impressive, too, that by now certain of the better Italian divisions and regiments, including the Ariete, Bersaglieri, and Folgore, were training and fighting up to Wehrmacht standards.[11] But by this stage Churchill was betting almost everything on a North African victory, even if that implied sending only limited reinforcements to Southeast Asia and leaving the Burma theater essentially dormant. When Allied convoys swung around the Cape of Good Hope and into the Indian Ocean, the troopships and merchant vessels laden with tanks, boxed fighter aircraft, trucks, jeeps, and the rest were much more likely to head toward the Red Sea than to the Bay of Bengal.

The climactic battle for Egypt did not begin on the clear moon-lit night of October 23, 1942, when the Eighth Army's artillery opened fire across the whole line of German and Italian divisions that stood only a few miles to their west. The much-lauded fight that made its way into British legend and collective memory was really the third stage of a protracted encounter that commenced when Rommel, urged on by Hitler, thrust forward in midsummer to crack the British hold upon Cairo and, potentially, much of the rest of the Middle East. Napoleon had tried the same a century and a half earlier, but his venture was far too vulnerable at sea, and thus to Nelson's battle fleet. This time the invaders of Egypt were coming by land. But they, too, were vulnerable.

The preliminary tussle, the First Battle of El Alamein, lasted throughout July 1942 and was remarkable on various counts. German panzers and infantry retained supremacy in the field, even when the odds against them multiplied. A little earlier the Afrika Korps had badly bruised a far larger British force around Tobruk, wheeled back westward to seize the city (to Churchill's immense anger and frustration), pushed the counterassaulting Eighth Army into the desert, then continued its relentless drive to the east, with very small numbers of tanks. This was blitzkrieg warfare at its best, perhaps as impressive as Guderian's breakthroughs in France in May 1940, and it is noticeable that German aircraft were still playing a disruptive role, helping to paralyze the enemy's nervous system and throw his defenses into confusion. Rommel's advance troops actually reached some British defensive positions west of Cairo before many of the exhausted South African, New Zealand, and British battalions themselves crept back to rejoin the Eighth Army. But the German forces were also exhausted, and the cautious British general Claude Auchinleck was content to hold the line and get new forces into position. By July 3–4, 1942, when their dynamic (though temporarily very sick) leader urged them forward again, Rommel's Afrika Korps had only twenty-six tanks fit for operation. That evening (July 4), in one of his most telling letters home to his wife, the field marshal wrote: "Things are, unfortunately, not going as we should like. The resistance is too great, and our strength is exhausted."[12] As ever, fuel was terribly short, cramping his panzers' mobility again and again. To make things worse, the Luftwaffe squadrons were being pulled away to the Eastern Front, a nice example of the folly of trying to fight enemies in three theaters at once.

"The resistance is too great, and our strength is exhausted." Here is the key to this tale. All lightning wars rush out to the skies and seas and the great wastelands beyond, and then they begin to lose their concentration, their density, their power; it is purely a matter of physics. The fact was that the desperately strung-out units of the Afrika Korps were beginning to bump into far too much opposition: ever greater numbers of brigades from Britain itself, from Australia and New Zealand, from the Indian division, from South Africa and the rest; ever more Allied tactical aircraft; ever more of the vital medium and heavy guns and howitzers of the Royal Artillery, which was at last fulfilling its proper capabilities.[13] Very few, if any, of these newer Allied armored infantry units were of the highly mobile and explosively effective character of even a reduced German battalion, but there were simply too many of them, and they were too well equipped, to be blown away. The British Commonwealth divisions would take their nasty defeats and losses, at odds that to this day look deeply embarrassing. But they would not break—which is why the second battle in this trilogy is even more interesting, and more suggestive.

It took place around a long desert ridge called Alam Halfa, to the south and east of El Alamein. The coastal region was so thickly covered by mines, tank traps, and reinforced, well dug-in troops that Rommel felt he could strike only through the southern section of the Allied defenses. Strike he did, but his attacking units were blunted by deep and comprehensive minefields—it often took an attacking commander a while to discover that what looked like open territory was heavily sown with mines—and then by the sheer size of the resistance. The newly appointed Montgomery had the satisfaction of pushing more brigades into the Alam Halfa lines while Rommel watched his much smaller forces come close to running out of fuel, a fate common to so many German and Japanese armies (ironically, their prewar frustration that they lacked enough oil to be truly independent great powers was repeatedly confirmed in their campaigning).* Eventually, in early Sep-tember, he pulled back west. The British, rather to the surprise of some of the German generals, deliberately did not follow. As Liddell Hart observes, this was a battle that "was not only won by the defending side, but decided by pure defence, without any counteroffensive—or even any serious attempt to develop a counteroffensive."[14] That time would come.

There is an important military-technical point to reflect upon here, especially when one considers the fate of German offensive armored warfare for the rest of the war. All the evidence suggests that the slash-ing tank attacks by Rommel's Afrika Korps were blunted not so much by British tanks (Matildas, Grants, Shermans) or regular infantry as by two far less romantic means of war: acres and acres of minefields, and specialist antitank battalions that deployed large numbers of artil-lery and bazookas. The land mine—regarded by many nowadays as one of the most evil weapons of war—had for its small size an extraor-dinary capacity to bring an enemy's fast armored assault to a halt, or at least to force a rethink.[15] Either the attacker undertook to clear a mine-field that could be as much as 5 miles deep or he was funneled into routes without mines but full of deeply entrenched antitank battalions, not only in front of the advancing panzers but, more viciously, on each side, that smashed the caterpillar tracks and penetrated the slimmer sidewalls.

The unglamorous and very cheaply produced land mine became, therefore, a significant determinant of the contours and fate of mobile armored warfare in the Second Battle of Alamein. Not only did the extensively laid minefields further restrict both armies' freedom of ma-neuver, but by having their specialist sapper units bury hundreds of thousands of mines ahead of their own defensive positions, for exam-ple, Auchinleck and Montgomery could even further shorten the op-

---

*Frustrated at Auchinleck's caution even after receiving so many reinforcements, Churchill sent him off to India. His successor, Gott, was killed unexpectedly in an air crash, and Bernard Montgomery was then placed in charge of the Eighth Army. He himself was almost as cautious as Auchinleck, and his chief achievement hitherto had been his careful management of the retreat of his division toward and through Dunkirk in 1940. After Alamein, however, Montgomery became the popular symbol of the restored glory of British generalship.

erational gap between the Quattara Depression and the Mediterranean shore. As the German forces sought to claw a way through the mines— actually, *had* to claw a way through the mines—the defenders had ample warning to strengthen the threatened section of the front. However, this same problem would confront the Eighth Army when it went on the offense as it had the Afrika Korps, for Rommel had a keen appreciation of the value of mines (as we shall see again in his supervision of the Atlantic Wall in 1944), and also had vast numbers laid under the North African sands.*

By comparison with the great battles waged across the Pacific, in northwest Europe, and on the Eastern Front, this campaigning between German and British Empire forces along the North African littoral was small in scale. But it was a superb test bed for the West's innovators to try out newer ways of weakening German military power on land. For example, the two devices most commonly associated with Allied mine clearance efforts—the flail tank, one of the many contraptions in General Percy Hobart's inventory of tricks (see chapter 4), and the handheld acoustic mine detector, devised by the remarkable Pole Jozef Kosacki and given freely to the British Army, were both products of this particular battleground.[16] To later armies across the globe, not having mine detectors in their armory would seem absurd, just as it would seem ridiculous to fight without advanced radar or decryption machines. Like many of the other novel Allied weapons examined in this book, this one also began as a small experiment. In his earlier life Kosacki was an engineer, and after the war he became a professor of technology and engineering. In 1942, however, he was a lieutenant and technical specialist in the British-Polish army, and eagerly desired to say thank you to Britain. It all fitted nicely together, and the mine detector arrived just in time to be used by Montgomery's sappers when the Eighth Army at last took the offensive.

Given the massive reinforcements of tanks, trucks, artillery, and aircraft the Allies were now sending to the Middle East theater, El Alamein was probably Rommel's (and Hitler's) last chance to overcome the

---

*Hence the continuing calls by Libya and Egypt for the British, German, and Italian governments to take steps to have millions of these wartime mines destroyed, since they continue to cause casualties among local civilians and their animals, make oil exploration dangerous, and even deter tourism.

British-led forces, the last chance to seize Cairo and the Suez Canal, both hugely important strategically and symbolically. The pulling back of the Mediterranean Fleet farther east in September 1942 and the sight of confidential files being burned outside Middle East HQ in Cairo (shades of the Quai d'Orsay in May 1940) remind us that the British authorities regarded a German breakthrough as a possibility. But even if there was not a complete German victory, a heavy punch might be sufficiently destructive of Montgomery's forces that Rommel could control the El Alamein–Quattara Depression gap and make it impassable for ages to come. This fight was also Montgomery's (and Churchill's) best opportunity to demonstrate that British Commonwealth forces could beat German panzers and infantry in the field, not defensively as at Alam Halfa but continuously, pushing them back again and again, harassing and sweeping aside attempted blocking points, taking ever more prisoners, and driving the Germans and Italians toward Eisenhower's armies at the other end of the North African shores.

For once, the terrain around El Alamein favored the cautious British armies. Indeed, looking at the story of the land wars between 1940 and mid-1942, a military analyst might argue that this topographic setup was the only one in which British Commonwealth armies had much of a chance of defeating their more experienced German counterparts. The other major card in Montgomery's hand was sheer superiority in numbers—and not just in frontline forces but in fuel and other supplies as well, and in near-total command of the air. By this stage, the Luftwaffe's strength was much reduced, while British and American tactical aircraft squadrons were pouring into Egyptian bases. The ninety-six squadrons available to the air C in C Sir Arthur Tedder (around 1,500 aircraft) across the Middle East not only represented a much larger and more modern air fleet than the mere 350 Italian and German planes that opposed them, but now packed a far bigger punch as well. Medium-range bombers based around Cairo joined Malta-based aircraft in vastly reducing Axis supplies across the Mediterranean; the Royal Navy's famous U-class submarines (also based out of the great caves in Malta's harbors) wrought havoc, too. The Italian merchant ships that got past these blockades were then terribly vulnerable to low-altitude attack as they unloaded in harbor. Axis supply lines along the narrow coastal road were frequently attacked—trucks and staff cars being the favorite target—while Rommel himself displayed an amazing lack of interest in military logistics and supply limitations.

Because of improved shortwave radio communications, Eighth Army units at the front could now call in fighter-bomber squadrons for direct attacks upon enemy panzers, motorized divisions, and troop clusters. If aircraft losses to ground fire were high, replacements were always arriving. Most important of all, RAF tactical airpower, so miserable or nonexistent in the first three years of the war, was coming of age. Air Vice Marshal G. G. Dawson was transforming the air force's hitherto lamentable record in repair and maintenance; above all, the coordinated tactical air doctrines developed by the Western Desert Air Force under Air Marshal Arthur Coningham—that is, with a central command linked by radar and radio both to army headquarters and to the attacking squadrons—proved so obviously effective that it became standard practice for both the Sicily and Normandy operations later.[17]

Under this aerial umbrella, knowing that further reinforcements were flowing up the Suez Canal each week, and having resisted Churchill's demands for action until he was fully ready, Montgomery unleashed a barrage of a thousand-plus guns on October 23, 1942. If, as most experts would concede, artillery was the queen of the battlefield during the slug-it-out fights that characterized the second half of the Second World War, then the Allies' advantage in numbers and firepower was about as great as their aerial superiority (specifically, 2,300 British artillery pieces against 1,350 Axis artillery, of which 850 were very feeble Italian guns). Moreover, the Eighth Army's gunner regiments were at last concentrated in strong groupings rather than being scattered up and down the front. Montgomery's reinforced and varied army (British, Australian, New Zealand, South African, Indian, Polish, and Free French divisions or brigades) could deploy three times more soldiers than Rommel's combined Italian and German troops, and his tank force of more than 1,200 vehicles, including 500 of the more powerful Shermans and Grants, far eclipsed the German-Italian armored regiments in firepower, range, and armor. But the depth of the German minefields slowed things down greatly, and the Eighth Army's painstaking drive to clear them gave Rommel good clues as to where to position his relatively few but terrifyingly effective 88 mm antitank platoons.

This time the sheer weight of Montgomery's pressure meant that the Axis lines just buckled under the relentless pressure; more and more British tanks were knocked out by German mines, bazookas, and the 88 mms, losing at a rate of four to one, but they still kept coming, and they could sustain their losses more easily. The Italian tanks were blown away, the lighter German tanks swiftly destroyed, the artillery barrage continued, the daytime air strikes intensified. By the end of the epic battles of November 2, during which the British had seen almost two hundred of their tanks lost or disabled, they still had another six hundred in hand; Rommel had thirty. Then the German retreat began, often leaving large numbers of less mobile Italian forces behind. The discipline of the Afrika Korps over the next few days in deploying alert and hard-fighting rear guards while the bulk of their transport units filtered through cannot but command admiration, but the blunt fact was that the Germans had lost in the most decisive battle for North Africa.[18]

During the night of November 7, as Rommel used the drenching rains to cover his westward withdrawal from the coastal position of Mersa Matruh, massive Allied invasion forces were arriving off the Moroccan and Algerian coasts to implement Operation Torch. In that larger sense, the German-Italian forces in North Africa were now caught in a gigantic pincer movement, with Eisenhower's forces pushing the defenders to the east and Montgomery's divisions chasing them back to the west until they ended up, surrounded, in Tunisia in early 1943. This was how the German blitzkrieg ended, at least in the Mediterranean theater.

Yet the Wehrmacht still fought on. Part of the reason the British Eighth Army only gingerly followed up—rather than earnestly pursued—the retreat of the Afrika Korps along that very familiar Sidi Barrani–Tobruk–Benghazi route was that Rommel's diminished cohorts would suddenly turn and bite, or would set up prepared positions from which they would give the Commonwealth forces a good drubbing. Then Rommel would retreat a bit more, just as Allied commanders were bringing forward massive aerial and armored forces to assault a force that was no longer there. Perhaps the British and American commanders (or most of them) accorded the Germans too much respect, but the record shows that there was good reason to do so.

The Germans perhaps might have significantly delayed what was later termed the "clearance of Africa" if Hitler had earlier given Rommel the divisions he belatedly ordered into Tunisia on learning of the Torch landings, or if Rommel had been freed from the complications of being formally under Italian command and from the rivalry of his co-equal General Hans-Juergen von Arnim in the final months.[19] To be sure, the massive Allied air, land, and sea superiority really made the outcome in North Africa all but inevitable, but if a German defeat had been much later, there could have been serious knock-on effects for the planned invasions of Sicily and Italy, and possibly a revival of Anglo-American disagreements about when to launch the second front in France. The Fuehrer's erratic interventions (tolerating the Arnim-Rommel rivalry, trickling in reinforcements, sending units to guard against an Allied invasion of Greece) were by now giving the Allies unexpected benefits.

As the Western Allies moved into southern Europe, the German military record on the ground would remain impressive (see chapter 4). The German command chose not to contest the massive Allied landings in southern Sicily, but then held on in the northeastern mountains of that island against repeated assaults by Patton's forces. When the time was ready to go, they smartly slipped away across the Straits of Messina, but they also chose not to stand in the vulnerable "toe" of Italy, where they might be outflanked; instead, they withdrew farther north, the better to resist. And the able Kesselring (after persuading a dubious Hitler that such a strategy was feasible) was preparing successive lines of defense between the Mediterranean and Adriatic coastline and running all the way up that lengthy, rocky, difficult Italian peninsula. When the Anglo-American armies tried to foreshorten this campaign by landing behind the front, the German response (Salerno, Anzio) was ferocious. All such counteroffensives by the Wehrmacht could be contained only by an overwhelming abundance of Allied

ANGLO-AMERICAN ARMIES ADVANCE IN NORTHERN AFRICA AND SOUTHERN ITALY

British Empire and American land forces first begin to push back the Wehrmacht through offensives launched from each end of the lengthy North African coastline. Following those victories, they moved on to Sicily and then into the entire length of Italy.

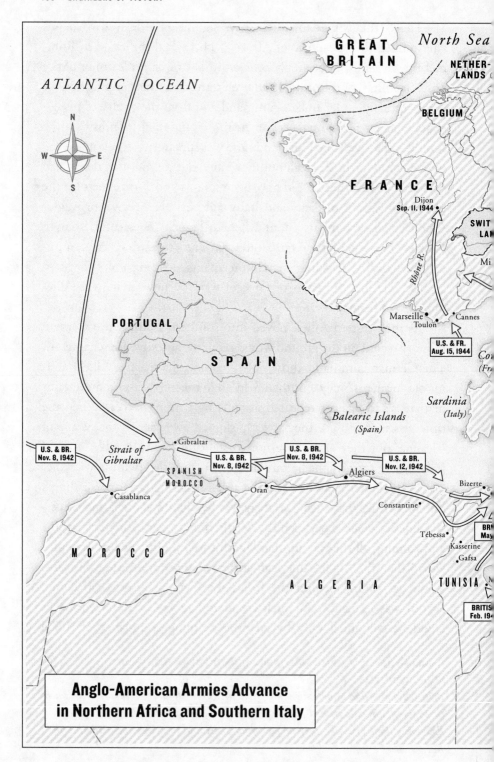

Anglo-American Armies Advance in Northern Africa and Southern Italy

force: total control of the air, the massive use of battleship and cruiser bombardments, and the bringing in of more and more army divisions. More Allied casualties were suffered in the battles in Italy than in any other campaign in the West.[20]

Hindered by meager supply lines, poor intelligence, loss of aerial support, and the constant damage inflicted by Hitler's obsessions and interventions, the Wehrmacht's divisions on the Italian and, later, northwest European fronts showed great resilience right up to the end of the war. The tenacity and operational effectiveness of a seasoned German army regiment or division had no equal in the Second World War. Only superior numbers could beat them. Some years ago the American military expert Trevor Dupuy attempted a systematic analysis of all the main battlefield encounters between German, British, and U.S. army units during the North African, Italian, and northwest Europe campaigns. Division for division, and without much exception, the German units overall had a 20 to 30 percent combat superiority (though even that may be too generous toward the Western armies).[21]

Yet a "superior numbers" argument is not enough. By the end of 1943 the British were introducing into the North African struggle a remarkable number of force improvements: superior radar, superior decryption, a much better orchestration of tactical airpower, much better coordination between the army and the RAF, unorthodox Special Forces units (the Germans and Italians had none), aircraft that were more powerful and more adaptable, the mine-flaying tanks and the acoustic mine detectors, and, sitting above this all, a far better integrated command-and-control system than at the time of the Crete and Tobruk disasters. They had worked it out at last, and Montgomery—and his publicity team—were the beneficiaries of much hard work and deep thought. Even before the epic struggle in the wastelands of Russia, it seemed the German blitzkrieg could possibly be defeated by superior numbers, positioned in depth; but those numbers also needed more advanced weaponry and a superior organization. It had taken the British Army and RAF a long time, and Churchill's relief was enormous.

The Clash of Giants

The 1941–45 war that raged along the broad western regions of the Soviet Union—the Great Patriotic War, or the Ostfeldzug—was, the most widely read encyclopedia entry claims, "the largest theatre of war in history." The struggle was

> characterized by unprecedented ferocity, wholesale destruction, mass deportations, and immense loss of life variously due to combat, starvation, exposure, disease, and massacres. The Eastern Front, as the site of nearly all extermination camps, death marches, ghettos, and the majority of pogroms, was central to the Holocaust. Of the estimated 70 million deaths attributed to World War II, over 30 million, many of them civilians, died on the Eastern Front. The Eastern Front was decisive in determining the outcome of World War II, eventually serving as the main reason for Germany's defeat.[22]

Given this image of colossal contest and gigantic military effort, the reader may well inquire what role there could be here for the problem solvers, the scientists and engineers and organizers, the men in the middle. Their roles were in fact critical once one begins to peel away the layers of explanation as to how the USSR defeated the all-powerful Nazi war machine, and the second half of this chapter will attempt to explain why. Yet there is no denying that it was in the Russo-German War of 1941–45 that brute force showed itself to its extreme and provides its own easy explanation as to why the Wehrmacht lost.

This struggle was unique in its grand combination of mechanized destructive power with Asiatic-horde-like warfare. The existential struggle between Teutons and Slavs was now entwined with an increasingly complicated and ever-changing technological competition. No longer a clash between roughly equal military technologies, such as cavalry, pikemen, and archers, all at the same stages of development, the Ostfeldzug was a struggle between two vastly more complicated sys-

THE RAPID GERMAN EXPANSION IN THE EAST, JULY–DECEMBER 1941
But did the Germans stretch too far?

The Rapid German Expansion in the East, July–December 1941

Territory controlled by Germany and Axis allies on June 22, 1941

to July 9, 1941

to September 1, 1941

to September 9, 1941

to December 5, 1941

tems. Both combatants reached deep into their own very advanced technological and productive resources to bring forth newer or improved weapons that, they hoped, would sweep the foe away. And as each brought his newer weaponry to the front, he brought with it a few million more men to handle those weapons.

Consider the numbers involved in the opening campaign of Operation Barbarossa and as it was waged from June 1941 into the spring and summer of the following year. Not including significant additions from satellite and volunteer military units from countries such as Hungary, Romania, Italy, and Finland, the German high command (OKW) sent around 110 divisions, including 14 fast-moving panzer divisions, eastward in the greatest invasion of all time; additional divisions waited behind for their turn to fight. The successive encirclement operations showed that blitzkrieg could work even across expanding battlegrounds: the defender's air force was smashed on the ground, the forward lines buckled under heavy artillery and infantry attack, and then the panzer armies drove through at two widely separated points, wheeled toward each other, and closed the circle. When Smolensk fell in early August, 310,000 Soviet troops were captured; the seizure of Kiev in mid-September brought in no fewer than 600,000 prisoners, 2,500 tanks, and 1,000 guns; and the so-called Vyazma pocket, holding out to the west of Moscow, collapsed in mid-October when 670,000 surrendered, with 1,000 tanks and 4,000 guns.[23]

What could stop this continued attack, slaughter, and surrender?

First, the weather. In the summer months of 1941 the German assault took its usual form: the Luftwaffe wiped out the unready Red Air Force squadrons on the ground and then turned to help the panzers by strafing and bombing enemy troops; armored columns punched holes through the confused foe, wheeled together for the encirclement (and surrender), then, reinforced by fresh fuel and ammunition, struck farther eastward. But by October and November it had really slowed down. There was simply nothing that the Nazi blitzkrieg could do against the autumn muds, followed by the extraordinarily early onset of extreme frosts and snow—the coldest winter, ironically, since Napoleon's invasion of 1812, and with German troops still clad in summer uniforms. By the end of October, Army Group South was virtually out of fuel, diesel engines would not start, artillery was frozen, and soldiers were crippled by frostbite. By the time spring came, the ground conditions only got worse, because of that notorious Russian weather condition called *rasputitsa*—the surface snow had now melted, but the resulting water could not drain away because the soil below was still deep frozen. Two feet of unfrozen soil quickly turned into a mud bath into which both the German and Soviet armies sank every spring during this great conflict. Some 500 miles from their starting point, with lines of communication difficult to sustain even in the best weather, the Wehrmacht really was literally stuck in the mud.

The second reason the Barbarossa assault faded was simply that the Red Army fought back—not well, not skillfully, not enough to throw back the onslaught, but sufficiently to slow it down. The Russians burned their own barns and bridges, destroyed or poisoned their wells, and pulled back their cattle and horses just as they pulled back their factories; the enemy would get nothing. Third, they would soon throw more troops into the fight, either the hastily assembled new divisions or the experienced cohorts being transferred from the Far East now that it was becoming more certain the Japanese would strike southward and not against Russia's distant eastern provinces. There was probably nothing more frightening to a German soldier in December 1941 than to be lying exhausted and freezing in a trench west of Moscow and then to spot regiments of white-clad ghosts, the hardened troops from the Mongolian Front, advancing through the swirling snow.

In other words, while the Wehrmacht's gains were immense—they had advanced 625 miles by December—they had not produced a breakthrough. From the first fighting until the last, the Germans, so superior in their tactical-operational skills, seem to have persistently underestimated the strength of the opposition. Their intelligence record here can only be described as abysmal, which is perhaps likely to happen when an army thinks of their foe as primitive—even when they faced a general such as Georgy Zhukov, who had badly hurt Japan's Kwantung Army in Soviet-Japanese border fights two years earlier. Signals intelligence doesn't seem to have helped, because the Soviets kept wireless messages to a minimum, and there were no spies to inform them of the many fresh divisions being raised and trained well behind Moscow. Thus, however hard the Wehrmacht hurt the defending forces in those early months, there turned out to be no endpoint, no culmination, no collapse of France. A mid-August 1941 entry in the diary of Franz Halder, the army's chief of staff, nicely captures this dilemma: "We underestimated Russia: we reckoned with 200 divisions, but now we have already identified 360."[24] Even if Soviet divisions were smaller in size than their Wehrmacht equivalents at this time, the numbers are breathtaking.

This was the story all through 1942 and into 1943: the Germans increased their efforts to encircle and then destroy the growing Soviet armies, but without success. All the German military language about breakthrough (*Durchbruck*) or encirclement (*Einkreisung*) suggests that the enemy's lines were thinly held or could be outflanked. But a defensive position 1,100 miles long and 200 miles deep, full of broken bridges, poisoned wells, booby traps, and ruined crops—plus a climate of summer heat, autumn muds, winter snows, and then spring muds— was not a thin, fragile wall at all. The primitiveness of road communications in Russia and the endurance of local inhabitants long used to the harshness of their daily existence all worked against the invaders.

As in the Mediterranean theater, the Wehrmacht's difficulties were compounded by Hitler's repeated and disastrous alterations in the axes of attack after the early failure to take Moscow. Assuming that the siege of Leningrad (from September 1941 on) in the north would continue, even if always stalemated by the city's fanatical resistance, the main strategic choice for the German high command was between a renewed thrust toward Moscow and a new offensive along the southern front toward the oil fields of the Caucasus. The debate among historians today over the merits of a German drive toward the enemy's capital city versus a bold and wider stroke across the Don and the Volga is as fierce as the 1942 disagreements among the German generals; what is not in dispute is that Hitler's follow-up directive that there should be both a Stalingrad offensive *and* a massive move toward Baku—that is, two offensives in the south—was an act of reckless and colossal overstretch. Among other things, how did one keep munitions streaming between, say, the factory in Leipzig and an outlier city in the Caucasus such as Grozny? How did one provide air cover all that way when the Luftwaffe was beginning to lose its critical fight against the RAF both in western Europe and in North Africa and the Mediterranean? Moreover, Germany was now going against a Stavka and its generalissimo who had regained their nerve and were turning the campaigns over to professional generals—many of them much younger than their German equivalents.

The successive and critical campaigns that form the centerpiece of the Russo-German War lie between roughly November 1942 and July 1944: Stalingrad, Kharkov, Kursk, and then the massive westward advance of Operation Bagration. The first was a bloodbath, even more a battle of exhaustion than the fight for Moscow had been. Stalingrad was a deliberately chosen duel between two very aggressive boxers; the hits and the pain were going to be great. So also were the chances of achieving a big win. While various comparisons can be made about the contemporaneous battles of Stalingrad and El Alamein, the most interesting difference may not be so much in the sheer size of the forces involved on each side as in the topographic breadth and thus in the operational opportunities. El Alamein, as we have seen, was fought in a very constricted space, from which neither side could make a flanking attack. By contrast, the choice of each combatant to fight at all costs for Stalingrad in late 1942 was not unlike the choice of France and Germany to endure a great, bloody battle over Verdun in 1917. It didn't actually have to be fought there, and a new opening could be made on either flank—but the sheer commitment by both sides to winning this fight would not allow it. So an epic encounter took place around, across, above, and below Stalingrad.

Ironically, Stalingrad may have been the greatest example of a *Kesselschlacht* in history, except it was achieved against the German army. As the Wehrmacht's forward units struggled to take the broken-down factories and streets on the city's western side, gigantic Soviet armies arrived to strike from the north and the south, entrapping General Paulus's entire Sixth Army and much else besides. This was an intensely personal battle of wills as well. Hitler drove his troops toward the symbolic city; Stalin held that attack off, then moved to encirclement. The Sixth Army had consisted originally of around 300,000 men; on January 31, 1943, to Hitler's fury and contrary to his direct orders, the remaining 90,000 surrendered. Then the Soviets advanced through Kursk and Kharkov, albeit slowly, allowing the Wehrmacht the chance to pull their forces out of the Caucasus and thus to regroup. On the other hand, the Stavka was showing a much more sophisticated appreciation of how to feint against one enemy position, then strike obliquely against the real target. Not for nothing does Liddell Hart have particular praise for these advances, as confirmation of his theory of "the indirect approach."[25]

It remains amazing that in early 1943 the German army could again launch a major strike eastward, taking advantage of the Russians' need to regroup after a big offensive battle as they waited for the newly arrived U.S.-donated trucks to bring up gas, ammunition, tinned food, and spare parts from hundreds of miles in the rear. For a while, Germany enjoyed the advantage of much shorter front lines, which is to say that its extreme overstretch had now been trimmed to merely a serious overextension. And the commanders on the Eastern Front, Manstein especially, were to benefit from Speer's revitalization of the Nazi economy and the arrival of the Tiger and then Panther tanks—in nothing like the numbers of Shermans and T-34s then being produced, but enough to make a serious impact upon the battlefield if employed in strong formations, which they were. By February 1943 the Leibstandarte, Totenkopf, and Das Reich armored divisions, each with several dozen new Tiger tanks, had been released from the central reserve and sent east, along with some fresh infantry divisions from France. All this, and the firm frost-encrusted ground that permitted the panzers to average 20 miles a day, allowed Manstein to unleash his faster forces upon the much-battered city of Kharkov, which fell on March 14, with his SS-Panzercorps eliminating 32,000 Soviet troops. The Red Army was pushed back to the Donets, and then an early spring thaw brought all serious campaigning to a halt. Both sides drew breath for the next round. Eight of the twenty Soviet tank corps had been mauled. Manstein became Hitler's favorite and most trusted commander, at least for a while.[26]

This was not a bad record for a German army now into its fourth year of total war. Yet, as Robert Forczyk points out, while the Russian tank corps may have been mauled, they were not destroyed, and Manstein himself was now mightily impressed by the stubbornness and sophistication of Red Army defenses. What both sides needed was a three- to four-month breathing space, to rebuild, regroup, and wait for the steppes to warm up and dry out. The arguments in Berlin and Moscow about how to conduct the impending summer clash were serious and contentious, with, apparently, both dictators listening hard to their respective commanders. The Red Army's most probable strike would be against the tempting Orel salient in the north, and the Wehrmacht's against the equally tempting Kursk salient farther south. Each was assessing where to place the heavy tanks, where to put the bridging equipment, where to sow the minefields, and how to deploy its rather limited aerial striking capacity. In a way, it was like a massive chess game.

The Germans struck first, on July 5, 1943, against Kursk, with Manstein's heavier forces (including panzer and panzer-grenadier divisions) cutting in from the south, while Hans von Kluge's pincer attempted to drive in from the north. This is probably the culminating battle of the many blitzkrieg campaigns conducted by the Wehrmacht during the war, and the one that best showed its weakness. Immediately the German attackers found themselves encountering enormous strength-in-depth defenses, for Soviet military intelligence knew what was coming and the Stavka had pre-positioned additional armies for the counterattacks. The fighting was most bloody, with the high point being an all-day fight on July 12 between the Soviet 5th Guards Tank Army and the II SS-Panzercorps around the village of Prokhorovka; later the conflict was described by Russian propaganda and subsequent Western authors as the greatest tank battle of all time. Probably it was, for the numbers committed to this single clash, especially from the Soviet side (about 800), were huge. What is more certain are two things: first, that the Russian armored losses were much, much greater; and, second, that nonetheless the Wehrmacht had to abandon the field because it simply could not get through. The "Manstein era" on the Eastern Front was over. Although the casualty figures are, as usual in this campaign, very vague and general, the best guess is that the overall Kursk campaign cost Hitler more than 50,000 troops and 1,600 tanks. Dozens of German army divisions were either destroyed or reduced to shells.[27]

Two days before the tank battle of Prokhorovka, the Western Allies landed in Sicily, which caused Hitler to turn his attention—and to direct many military units—to the Italian front. And on July 12 itself, the Red Army struck at the Orel salient in the north. From here on, one gets the sense of fatigue setting into the Wehrmacht: structures folding, fronts being abandoned (often despite the Fuehrer's manic orders to stand fast), battle-hardened but weary units limping off the field with perhaps only one-quarter of their equipment left intact. And so, despite Manstein's hope that the stupendous Kursk encounter had blunted the Russians' offensive capacities, they kept on coming, this time in the north, in the south, and on the critical central front. Slowly the Wehrmacht's many units were eased out of the Leningrad area, the Crimea, Smolensk, Kiev. The great Operation Bagration of June 1944 was not yet in sight, but the overall trend was clear: Soviet push, German resistance, then German pullback, then Soviet further push.

When, therefore, did the tide turn in the Ostfeldzug? One author has declared that the German failure to capture Moscow during the brutal winter fighting of 1941–42 was "the greatest battle of the war."[28] That conclusion certainly would have surprised Stalin and the Stavka, who were really disturbed by the immensity of the Nazi drive toward the Caucasus, the lower Donets basin, and Stalingrad itself in mid- to late 1942, well after they had held on to Moscow. Most historians today regard the massive defeat of the German armies at Stalingrad as the beginning of the end, the turn of the tide: John Erickson's classic two-volume study of this epic contest is deliberately divided into *The Road to Stalingrad* and *The Road to Berlin*. There, among the shattered warehouses by the Volga and the factories and assembly plants of the once-impressive city of Stalingrad, the Nazi tide began to ebb. R. M. Citino's important article "The Death of the Wehrmacht" assures us that late 1942—El Alamein, the dreadful Caucasus battles, Stalingrad—marked the definite end of German lightning warfare.[29]

And yet, as we have seen, Manstein's reinvigorated panzer armies came back in the spring of 1943 to smash the forward Soviet divisions and capture the entire Kharkov region. Perhaps this was a bid to persuade Stalin to agree to some sort of negotiated peace across east-central Europe, but while it was militarily impressive, it sparked no political response. So in July 1943 Hitler's gigantic military machine launched an even greater assault to the east, Operation Zitadelle, to chop off the impertinent Russian armies lodged in the Kursk salient. It is telling that Erickson, despite the time division chosen for his two massive books, actually titled his chapter on the Kursk battle "Breaking the Equilibrium," and that on Operation Bagration in summer 1944 "Breaking the Back of the Wehrmacht." The full tide, he seems to suggest, turned rather later than the titles of his volumes imply and many other accounts maintain.

Some years ago, the great German expert on this campaign, Bernd Wegner, raised the question of whether it makes sense to search for a turning point at all. Was there in these middle years really any identifiable watershed—Moscow, Stalingrad, Kursk? That is, was there a battle at and after which the course of the war went inexorably in favor of the eventual winners?[30] Perhaps the enterprise was simply doomed from the start. However deeply the Nazi extraction regime mobilized its own economy and plundered its conquered European space, possibly it never could match the fully mobilized resources of the British Empire and the Soviet Union. Add in Germany's declaration of war against the United States on the news of the outbreak of the war in the Pacific, and the odds became far worse.

It is clear that many of the leading German generals slowly came to hold this view, though it is also true that very few of them wished to cease fighting or to seem to contradict the Fuehrer. Proposing a withdrawal almost always risked Hitler's fury. When in November and December 1941, after the first great advances through Russia and the Ukraine, Gerd von Rundstedt suggested that with the coming of winter plus the arrival of unknown numbers of Red Army divisions from the Far East, it would be wise to pull back, shorten the lines, and consolidate, he was dismissed. His name was the first on an increasingly lengthy roll call of wartime German generals fired by their master.*

For their part, divisional commanders were getting uneasy at the massive and growing defects in transport and supply. What was the point of panzer columns streaming through the Ukraine if they kept running out of fuel and ammunition, or if a tank that broke its caterpillar track (as so many did) was a hundred miles ahead of the repair crews? The Wehrmacht's trucks were tied to the few, notoriously poor paved roads, and the vast numbers of horse-drawn wagons were painfully slow in dragging their way through the boggy paths that formed

---

*Ironically, von Rundstedt would be back, as C in C West, during the Normandy battles.

most of the country's land infrastructure. The smarter German generals were also bemused by the smoke-and-mirrors act that Hitler and the OKW kept pulling, increasing the number of German divisions on the Eastern Front simply through reducing their original size and then creating new ones. Thus, in the interval between the fall of France and Operation Barbarossa, the ten original panzer divisions were doubled in number, but each one of them was halved in size, and much the same happened to the motorized infantry divisions, which were trimmed from three regiments each to only two.[31] Perhaps this made those units more maneuverable on the battlefield and gave the army group commanders more flexibility, but the uneasy thought was that the Fuehrer was creating new divisions out of thin air, whereas the Red Army's increases in numbers were real ones. (In total during the war, the Soviets conscripted thirty million men, of which they lost ten million, with still larger numbers of wounded.)

Finally, none of the Wehrmacht generals fighting the Ostfeldzug could be unaware of the fact that by 1942 the Third Reich was now engaged in a global war: they saw so many of their classmates from officer school deployed in Greece, the Balkans, North Africa, France, Norway . . . So Germany was fighting on three fronts, while the Soviet Union battled on only one. Paul Carell notes perceptively that the Allied invasion of North Africa, followed by Hitler's order for the occupation of Vichy France in the crisis month of November 1942, occurred precisely at the time that the Stavka was about to order the Soviet counteroffensive at Stalingrad, yet this news of the Western Allies' push into North Africa and the Mediterranean caused the OKW to hold back, for a while, some of their most effective fighting divisions from the Battle of Stalingrad.[32] Germany could not be strong everywhere, and the Great Patriotic War cannot be described in isolation, even if Soviet writings tend that way. When the Western Allies invaded North Africa, they affected the Stalingrad battles; when they invaded Sicily, they affected Kursk. So a division such as the Leibstandarte, along with various air flotillas of the Luftwaffe, were being shuttled from front to front. This shuttling clearly couldn't and didn't fit into the theory of lightning warfare. And once the Ostfeldzug is placed in the context of a global struggle for power, we can see that it may indeed be misleading to speak of a certain battle, such as Moscow or Stalingrad, as the decisive turning point.[33] It is perfectly feasible to argue that the war on the Eastern Front *was* decisive without having to argue that one particular battle was key to the whole struggle.

This brings us to the final conundrum about the German conduct of the Ostfeldzug, and indeed of the paradox of the "German way of war" after 1941. The whole point of the theory and practice of lightning war was precisely to avoid the static fighting and bloodbaths of the First World War. This newer sort of warfare would be different because fast-moving motorized forces would surprise an enemy's line and drive through it, overrun his rear lines of communication, and force a surrender. But this concept of fast, clinical, and economical campaigning was totally contradicted by Hitler's fanatical belief that the attack upon the USSR had to be a *Vernichtungskrieg* (war of annihilation), followed by the permanent occupation of gigantic tracts of arable land. It was also contradicted by the ghastly and fateful Nazi habit of shooting and starving prisoners of war, ransacking even small villages in the search for Communist Party members and Jews, and foolishly mistreating the millions of Ukrainians who originally turned out to welcome the German troops as liberators. (How might the war had gone with forty million Ukrainians on the Axis side?) Blitzkrieg was supposed to be clever warfare, and so it was in the 1939 Polish and 1940 French campaigns. As it was conducted on the Eastern Front, however, it was really stupid warfare. Thus, as one student of this war has noted, it may be that "the means and ends were in conflict from the beginning."[34]

All this might suggest that the result of the war on the Eastern Front was a foregone conclusion. It was, clearly, *the* campaign of all the major struggles of the 1939–45 war where brute force is most in evidence; sheer numbers, plus weather and distance, are the determinants.[35] Yet the problem solvers do have a vital role here. Some field commanders, support managers, scientists, and engineers responded to the stress of mass warfare in a cleverer way, and in the pages that follow we will provide examples. And the Red Army's organizers had, in confronting the greatest blitzkrieg attack ever, figured out things better than those pursuing that particular form of warfare. The USSR did not win this epic contest solely because it poured more human beings into the fight than did its enemy. It won because it slowly developed means of stemming German armored assaults, reducing their core forces to shreds, then moving forward to regain the conquered lands. The Soviet Union—not unlike, say, the Americans in the Pacific—won its critical campaign because it was putting an assembly of weapons, tactics, and command structures into an overall tool kit that could not be broken. Understanding how the German form of lightning warfare came to a halt on the banks of the Volga and the wheat fields of Kursk requires further analysis.

The Curious Case of the T-34 Tank

The weapons, organizations, and techniques that helped the Red Army turn the tide on the Eastern Front were many, but in an order of significance that is much different from our generally held notions of which Soviet weapons brought the blitzkrieg to its defeat. The present author, like many military historians, has long assumed that the T-34 tank and its later improved variants such as the T-34-85 were by far the most effective weapons deployed in the Soviet counterattack. The assumption is understandable. The accolades for this armored behemoth are limitless. "The greatest tank of all time," "the most versatile tank of the Second World War," and "the weapon that shocked the Germans" are among the more common descriptions. As early as July 1941 OKW chief Alfred Jodl noted in his war diary the surprise at this new and thus unknown *wunder*-armament being unleashed against the German assault divisions.[36] Paul Carell, in his postwar work "Hitler's War on Russia," admiringly recounts a startling confrontation between a lone T-34 and an advance group of Wehrmacht panzers and infantry at the battle around Senno on July 8, 1941:

> T-34! Now it was the turn of the Central Front to experience that wonder-weapon. . . . "Direct Hit!" Sergeant Sarge called out. But the Russian did not even seem to feel the shell. He simply drove on. Two, three, and then four tanks were weaving around the Russian at 800–1000 yards' distance, firing. Nothing happened. Then he stopped. His turret swung around. . . . This one was making for the anti-tank guns. The gunners fired furiously. Now it was on top of them.

When three of these mastodons were unexpectedly stuck in a swamp and abandoned by their crews, Guderian himself came to examine them. It is reported that he walked away silently, thoughtful and long-faced.[37] Until the much heavier German tanks joined the war, the Wehrmacht's Mark III and Mark IVs were at a distinct disadvantage—at least on a one-to-one basis.

The impression, and the consequent legend, of the T-34 as a stunning battlefield conqueror was reinforced by the testimony of defeated German generals after the war. Panzer leaders such as Friedrich von Mellenthin said, "We had nothing like it," and Ewald von Kleist declared it to be "the finest tank in the world." Guderian somberly noted: "Up to this time we had panzer superiority, but from now on the situation was reversed. . . . Very worrying." The author of the study that records those statements goes on to add, "Who could be better judges?" This view was also very strongly (and perhaps more convincingly) reinforced by the testimonials of the Red Army tank crews themselves. Finally, and for obvious propaganda purposes, the Soviet Union in later years could point to the T-34's success as a symbol of socialist ingenuity and national industrial strength. This tank's acclaimed power certainly convinced a large number of Western historians, not to mention a long list of foreign defense ministries that purchased it well after the Second World War was over—as late as 1996, variants of the T-34 were in service in at least twenty-seven nations.[38]

It is likely that captured German generals would have had a good reputational reason to praise a weapon they themselves never possessed, and that post-1945 Soviet propaganda might have had an even better reason to extol the Red Army's great instrument of war. Even if we may discount those sources, however, there remains the indubitable fact that whole tank armies of T-34s confronted the Wehrmacht's powerful panzer groups at Kursk in 1943, and that eventually it was the depleted German divisions that limped away from the field. And it was also the case that the powerful forward thrusts of the Red Army from summer 1944 onward, culminating in the eventual seizure of Berlin, were spearheaded by fast-moving and enormous columns of Soviet armor. But none of that later history means that the T-34 was a wonder weapon, something in a class of its own as soon as it appeared on the battlefield in July 1941. The blunt fact was that the early T-34s had masses of design flaws, were highly unreliable in battle, and suffered from the "on-off" attitude of Stalin and the Stavka toward building main battle tanks in the late 1930s. The development of this imperfect yet potentially great fighting machine thus has many features of the parallel history of the B-29 Superfortress bomber and the P-51 Mustang fighter (see chapters 5 and 2, respectively).

But this case was, arguably, even worse, given the length of the time the tank had been in development. For example, even the most enthusiastic of its Western admirers admit to its poor turret design and to the scarcity of radios.[39] Moreover, when Operation Barbarossa commenced, the Red Army possessed only about a thousand of these newer tanks, the battalion crews had not trained in them, and even when they got to the fighting they were scattered up and down the front individually and thus lacked the concentrated punch of a German panzer column. They also did not work with Soviet infantry, thus having no cover on their flanks, and they quickly ran out of fuel.[40] Finally, all the production and updated design schemes were badly disrupted by the German invasion and the desperate need to shift entire factories from around Leningrad and Kharkov to the distant Urals (the Voroshilov Tank Factory No. 174, for example, went from Leningrad to Omsk, more than a thousand miles away). By September 1942, even the famous production lines at the Stalingrad Tractor Works had to cease working as the German divisions closed in. Perhaps these were just early teething problems, sorted out by the fighting at Stalingrad in 1942 or at the latest by the 1943 Kursk battles. But the catalog of the T-34's defects was much longer—and longer-lasting—and its actual battle performance in the middle years of the war was uneven. While it did eventually become a great weapon on the battlefield, at the very least the time frame of its road to success needs to be shifted.

The provenance of the T-34 is rather murky. Most military planners of the interwar years had studied the British Army's use of tanks on the Western Front in 1917–18 and subsequently tried to keep abreast of both the technical advances and the futuristic literature on this weapon. Many developments were proposed, despite the global fiscal crisis and strong disagreements inside militaries as to what exactly a tank's role in future conflicts might be. It was in these confusing times that the legendary American inventor and manufacturer J. Walter Christie produced his new M1928 design. Since the U.S. Army had different specifications in mind and the unorthodox Christie refused to change things, he turned instead to foreign clients, such as Poland, Britain, and the USSR. Thus it was that two of his new tanks (without their turrets, and documented as "farm tractors") were sold to the fledgling Red Army. This was not as unusual as it may seem, since the Russians were also buying British Vickers tanks, and the British themselves purchased a Christie design and turned it into their own Cruiser Mk III tank.[41]

Several aspects of Christie's invention intrigued the Red Army. It was the first to use sloping armor (40 percent incline) on the front of the vehicle, which vastly increased its resistance to enemy shells. Second, Christie was a genius in designing more sophisticated suspension systems—vital for cross-country ops—and he gave his M1928/M1931 vehicles a variable or coil-spring suspension. Despite delays caused by Stalin's purges and disagreements among the Red Army generals about the purpose of tanks, Soviet designers began to combine Christie's ideas with the Vickers designs, formulating their own hybrid models, the A-20, the A-32, and then the T-34.

By the late 1930s the lead design team was housed at the Kharkiv Komintern Locomotive Plant (KhPZ), under the leadership of engineer Mikhail Koshkin, an unusually gifted inventor and manager who also gained the ear of Stalin in the internal Red Army fights about light cavalry tanks versus heavier battle tanks. Despite all these disputes, modifications were made to the nascent T-34. The gasoline engine was replaced by a new V-12 diesel; whether it really was less flammable than a gasoline engine and thus less dangerous if a tank was hit remains open to question, but from postwar testimonies the alteration certainly seemed to make the Red Army crews more confident. Second, sloped armor was provided all around, and in retrospect one can see the shape of the modern battle tank emerging. Finally, the main gun was now bigger (76 mm). Very long-range testing drives of the prototype T-34s in the winter of 1939–40 (during which Koshkin caught pneumonia and died shortly afterward) persuaded the Stavka to increase production.[42]

That decision did not give the Red Army any kind of wonder

weapon during the first two years of the war. There were still far too few T-34s, and they were badly deployed (scattered rather than concentrated). The great bulk of the crews were raw novices, pulled from the farm and the factory, forced through the most basic military training, then pushed into a weird contraption that was hard to operate. They were often ordered to attack difficult targets, and they came under ambush from all sides. It is hardly surprising in these circumstances that the T-34's advantages over contemporary German tanks (such as better armor and better maneuverability in mud and snow) were neutralized. Yet in that early part of the war, almost every other armed service in the world was also grappling with teething problems with their own tanks, torpedoes, high-altitude bombers, and long-range fighters. It was precisely the task of the middlemen in the Second World War to solve these many problems.

Remedying the technical defects of the T-34 took a very long time. One might think that the problem was over by mid-1943, the time of Kursk and eight months after Stalingrad, but it was not. Had things been fine, why would the people's commissar for the tank industry, V. A. Malyshev, have visited Tank Factory No. 112 just a little while *after* the battle for Kursk to rail at the poor performance of the T-34s against the German armor and bewail the disproportionate loss of Russian tanks and their vital crews whenever they were required to take out a single Tiger or Panther?[43] Clearly, in his view at least, the T-34, even those arriving from the factories in mid-1943, was still not good enough.

The problem lay with the tank's design and operational weaknesses, which for obvious reasons were kept secret at the time. Recently the Russian historian A. Isaev has summarized in English a set of postwar recollections and interviews with former tank drivers, commanders, and other crew members of the T-34s.[44] All these men were clearly fond of their armored steed but were also quite candid about the vehicle's defects: its stiff controls, its tendency to leak water, its cramped spaces. Their testimonies are confirmed by a very different source: a Soviet summary of a detailed report by the designers, engineers, and tank men at the famous U.S. Army Proving Ground in Aberdeen, Maryland. The Russians had provided Aberdeen with a T-34 at the very end of 1942, presumably to get feedback. The Americans were all engineers, and their summary is clear, clinical, and balanced. The Main Intelligence Department of the Red Army, which translated and summarized the findings for the Stavka, had no time to quibble—they were, after all, in the middle of the Battle of Stalingrad and needed all the useful information they could get.*[45] The American testers admired the T-34's sloping silhouette (one wonders if they knew that their predecessors had turned down Christie's proposal fourteen years earlier), which was "better than that of any American tank." They loved the "good and light" diesel engine, and bemoaned the fact that the U.S. Navy monopolized access to diesel factories in America. The T-34's gun was simple, dependable, and easy to service, and the aiming/backsight device was "the best in the world." It could climb an incline much more swiftly than British and American tanks, thanks to the wider tracks. But then comes the Main Intelligence Department's summary of the much larger list of defects spotted by the Aberdeen analysts.

The air filters were lamentable, so a large amount of dirt very quickly got into the engine, which overheated and then damaged the pistons. The armor was mainly soft steel and could be better strengthened with alloys that included zinc. Because of poor welding, the T-34s leaked badly in heavy rains and water crossings, thus disabling the electrical equipment and even the ammunition. The American testers in the repair factories also pointed to the weakness of the track (the holdings and pins were too thin) and the poor performance of the engine air cleaners—leading to more clogging, probably because the exhaust tubes blew directly downward and created massive clouds of dust for the tanks following. The transmission was terrible and probably disabled many more T-34s than did the enemy. The radios worked poorly until a British design was copied en masse, and there was no internal means for communication among the crew, apart from the commander tapping his foot on the driver's shoulder. Finally, the poor commander's compartment itself was a horror, for he had far too many jobs to accomplish when a battle erupted, in a cockpit that was too cramped and

---

* It was probably the general major of the tank armies, Khlopov, who signed off on the report, though the signature is incomplete. One has the sense that his office was very pleased to use a U.S. "neutral" report to push the blame for all these weaknesses back to the designers and manufacturers.

with levers and pedals so stiff that altering them often required a smart blow from the supplied sledgehammer.[46]

The list was taken seriously, for the Russian summary concluded with nine recommendations, every one following the Aberdeen assessment. The dilemma was that to make these changes, production would have to cease at major tank factories at a critical point in the Great Patriotic War. The newer versions might also bring out new problems, and certainly the crews would have to be retrained. So should the Russians seriously reduce current output to gain a longer-term advantage, or should they just continue with the existing flawed machine, since any and all weapons were needed for the current fight? Between early 1942 and late 1943, the Red Army felt it had no choice. It just had to keep sending T-34s to the front, even with disproportionate losses, in order to stem the Nazi tide and to begin bleeding the panzer armies. Design improvements could come only when the spring and autumn muds reduced the level of fighting. It is therefore not surprising that Russian tanks were still losing heavily at Kursk and that Malyshev was bewailing the kill ratios—but what were the production managers to do when they had not been allowed to push through the necessary design changes right after Stalingrad? And the Kursk battles revealed a fresh problem: even the T-34's newer 76 mm gun was inadequate against the Tigers and Panthers. A three-man turret was really needed, along with torsion-bar suspension for rough terrain. But, realistically, all that the T-34 factories could do over the winter of 1943–44 was to keep on producing lots of leaky tanks.

The real successes of the T-34s came only in early to mid-1944—in curious parallel to the introduction of the P-51 Mustangs into the air war over western Europe. The newer Russian tank models had much better, tighter construction, a far heavier 85 mm gun, a roomy three-man turret with a radio in it, better air filters and overall air flow, a vastly improved periscope for all-round view, and stronger and broader tracks. Moreover, the support systems, the greater efficiency of the recovery crews, and the faster truck-transported fuel supplies all added to the Red Army's effectiveness. Here again, as with other Second World War weapons systems of this middle period, the changes were all incremental and could be realized only when circumstances allowed it and when the harsh experiences of the earlier fighting had been learned. In this case, the desired improvements could be made in the repair shops and factories after the German retreat from Kursk. Output of the older T-34s slowed in the fall of 1943 but was not stopped entirely, while production steadily shifted to the impressive newer types of armored vehicles, in preparation for the great advances of the year ahead.

The second major improvement between 1943 and 1944 was not technical but tactical. The T-34 units had not been well used against determined German defenders. Even the more powerful T-34-85 versions were likely to lose a one-on-one battle against the Tigers and Panthers, and despite the overwhelming Soviet superiority in numbers it was no fun for commanders or crews to watch four of their own tanks destroyed to kill off a single enemy. In addition, by late 1943 the Wehrmacht had the extremely fearful *Panzerfausts* (bazookas), which were more effective than any Soviet portable antitank weapon. And there were always mines. Grinding into German-held cities, especially as the Nazi lines shrank, significantly reduced a T-34's mobility. So why not set it free, or at the very least create some free-roaming brigades that could run around the enemy and strike in unexpected places? Why not take a leaf from Guderian's book and give the newer, faster T-34-85 units their head? One might then break a hole in a weaker part of the enemy's lines and let "deep battle" unfold—a term with special meaning in the Red Army. During the last full year of fighting on the Eastern Front, as armored divisions pushed through Bessarabia, Romania, southern Poland, and Hungary and on toward Austria, this was what was to happen.

By the time improved versions reached the battle-hardened Soviet tank armies in early 1944 and a much more experienced and confident Stavka orchestrated the great sweep toward Berlin, the T-34-85 could justifiably be called the most all-round battle tank of the Second World War. It still could never outgun a Tiger in a one-to-one shoot-out, but it had a better mix of maneuverability, range, and gun power. By 1944, in fact, everybody's tanks—American, German, and Russian—were more lethal and effective than they had been a mere two years earlier. But then, so were everybody's aircraft.

So the historian seeking to explain the Red Army's growing effectiveness against the Nazi blitzkrieg between 1942 and 1944 will need to look elsewhere, *not* rejecting the important role played by the T-34 tanks but acknowledging that this much-improved tank was only one of a mix of Soviet weapons systems on this critical and hellish battle-field.

The Tank-Killers: PaKs and Mines

When Montgomery and Zhukov met in Berlin in early May 1945, amid their celebrations and mutual award-givings, it is not too fanciful to suggest that they might well have chatted about their respective campaigns against their common Nazi foe. The campaigns the two field marshals had directed had much in common, first in their containment of a ferocious German ground attack and then, when the time was right, in how they struck back. Both had placed immense faith in situational warfare, which meant employing the "small warfare" weaponry of antitank missiles, minefields, ditches, and other field obstacles to blunt the assault of the panzers, and only then moving forward.* El Alamein and Kursk were different in time, space, and numbers, but they were not so different in essentials. In both cases, the defending side stopped and irreparably damaged the German attackers before starting a steady advance that would never be thrown back.

The attention Zhukov paid to massive minefield investments equaled and possibly surpassed the care devoted to the weapon by Rommel and Montgomery (no American general, so far as I can tell, bothered with mine warfare). Mines were cheap and easy to produce in vast numbers, easily transportable to the battlefield, capable of being readily hidden in the ground and covered up, and able to be arrayed in all sorts of checkerboard formations. They did not need to be powerful enough to destroy an oncoming tank or jeep; just damaging a wheel or breaking a track was enough. And then there were the antipersonnel mines, with a lighter detonation but still sufficient to kill or wound the

---

*I use the term "small warfare," *Kleinkrieg*, here in the German navy's sense of smaller weapons that could hurt or sink far bigger craft. Thus in naval warfare this includes torpedoes, submarines, and mines, all of which could destroy battleships.

German infantrymen running alongside the tanks. Finally, as the battles around El Alamein demonstrated, minefields—or, rather, the attempt by the attacker to break through them—gave the defender valuable time to prepare his own moves.

This was demonstrated once again, and in a much larger form, in the Battle of Kursk, which can lay claim to being the greatest minefield battle of the war. Like Montgomery eight months earlier, Zhukov and the Stavka supporting him were well prepared. The layout of the battle lines of the Eastern Front in June 1943 suggested that a renewed and massive Wehrmacht offensive, which everyone knew was coming, would attempt to envelop the Kursk salient and offer another example of a German *Kesselschlacht*. The task of the well-trained and ferocious Wehrmacht divisions was to execute the pincer movement. The task of the hardened Red Army was to frustrate that maneuver. And nothing frustrated swift panzer attacks like deeply sown minefields.

During the spring months, Mark Healy records, the Red Army laid tens of thousands of mines across the entire salient, all of them soon to be hidden by the growing summer wheat. "The density of the minefields, particularly between the strongpoints, was remarkably high, with antitank mines averaging 2,400 per mile and antipersonnel mines 2,700 per mile." The Soviet defensive minefield belts, carefully laid by the invaluable sapper units, were, one scholar notes, between 16 and 25 miles deep. They thus completely blocked the prospect of lightning armored strikes.[47]

When and where the static mines did not disrupt the panzer thrusts, the Red Army's antitank weaponry did so. The Russians, like the British, traditionally placed enormous reliance upon field artillery, the "gunners": that is, upon a combination of mortars/howitzers, regular heavy field guns, and screaming Katyushas (multiple-warhead rockets loaded and launched on a Red Army truck) to beat off enemy attacks, assault enemy lines, and blow the enemy to pieces before the tanks and infantry had even started moving. At Kursk, the Red Army deployed more than 20,000 guns and mortars, a number far larger than the Wehrmacht possessed at this stage in the war. The 85 mm dual-purpose guns were especially effective, because the Red Army had at last learned to group them into very large clusters and to site the individual weapons in ditches, in concrete buildings, and in thousands of prefabricated and even portable concrete pillboxes. Almost all of this weaponry, offensive and defensive, had actually been around in embryonic form in 1941. But, like the early T-34 tanks, they had always been scattered—"in driblets," to use John Erickson's favorite term—and there were never enough of them to stabilize that wide, collapsing front.

In addition, however, two weapons were specifically designed to disable enemy tanks. The first was a primitive yet effective antitank rifle (the Degtyaryov PTRD-41) that a simple infantry platoon could use to blow off a panzer's track or any other exposed part, including the driver. Although a crude weapon compared with later American bazookas and German *Panzerfausts,* it was effective because the Soviets employed more of them by far than any other army, and expected their platoons to get right in close to the German tanks. The Red Army often won the Ostfeldzug by still being primitive in many ways (thus the early T-34s, while easily knocked out of action by superior German armor, could be cannibalized for spare parts) and also by being willing to expend more of its troops than anybody else. To the Stavka, losing half a platoon in order to cripple two panzers was worth it.[48]

The second was the much more powerful antitank gun, at first a long-barreled 45 mm weapon with a mixed record of success—fast-moving panzer columns blew many of them away—followed by the much more powerful 57 mm ZiS-2. Production of the latter, which was designed by V. G. Grabin and built at the vast Artillery Factory No. 92 in Gorki, was begun in 1941, inexplicably halted through 1942, then resumed in earnest when the Red Army witnessed the sheer defensive strength of the new Panther and Tiger tanks and realized that their own 45 mm guns were not powerful enough. The 57 mm guns slid off the production line from June 1943 onward, only a short while before the July 5 German offensive against Kursk. Once again, one is struck by how late in the day the tide turned in this and in other campaigns, as improved weaponry finally reached the battlefronts. About this particular clash, John Keegan suggests that "Kursk may be regarded as the first battle in which the anti-tank gun . . . actually performed the role intended for it—to deflect and if possible destroy attacking enemy tanks without recourse to supporting armor."[49]

Kathy Barbier's description of how the Red Army put the pieces together before the Kursk onslaughts cannot be bettered. By July 1943, she explains,

the Soviets [had] created a series of strong points with heavy concentrations of guns in the areas which were most likely to be used by the enemy. They then combined individual strong points into anti-tank areas. In any given anti-tank area, there were at least three company strong points. Each held 4–6 anti-tank guns, 15–20 anti-tank rifles, several tanks and self-propelled guns, and an engineer platoon capable of attacking tanks with mines and grenades. The Soviets used all types of artillery, including anti-aircraft guns, in the anti-tank defences, which were from 30 km (18.5 miles) to 35 km (22 miles) deep.[50]

The point, again, was that neither the antitank rifles nor the PaK bazookas—nor, for that matter, the silent mines—needed to do anything more than disable an oncoming tank. If its caterpillar tread was damaged, the behemoth was immobilized, its capacity for attack destroyed. Additionally, as we have noted, the Wehrmacht was not very effective in bringing up its repair teams, and even at this stage in the war it did not appear to grasp that smart logistics are the foundations of victory.

That strategic polemicist Fuller, despite all his enthusiasm for rapid, long-range military strikes, often remarked that while the panzer's purpose was inherently offensive, the infantry's role was inherently defensive. The problem was that Fuller and his ilk assumed that a swift armored blow would always work, and thus the offensive would prevail. But that simply could not happen through the *eight* successive lines around Kursk, nor the many lines around El Alamein when Rommel launched his assault. And behind those lines were groups of enemy tanks waiting to counterattack when the assaulting panzers had lost half or more of their initial strength.

Thus the record suggests that the practical problems facing the Wehrmacht by this stage in the war on the Eastern Front included (1) running out of ammunition and fuel, (2) those insidious minefields, and (3) the effective Red Army antitank platoons. The geographic problem was that there were no longer any weak places that would permit a Guderian-like exploitation. It was therefore no wonder that the official war diaries of the German armies that struck both from the north and from the south of the Kursk salient in July 1943 kept reporting that they were trying a change of direction or a switch to another flank, always in the hope of finding a weak spot. There were none.

Finally, from mid-1943 onward, all those Russian weapons systems had one further advantage: the mobility brought about by the continual stream of Studebaker trucks and the ubiquitous jeeps. Mutual Cold War chauvinisms later produced a silly debate about how much or how little American Lend-Lease aid actually "helped" the USSR during the war, and it is quite true that the majority of Red Army vehicles (58 percent of its 665,000 trucks by war's end) was produced in the country itself. Yet it is also true that the American trucks and jeeps were significantly more robust and reliable, that the frontline Soviet commanders insisted on having them, and that they were exclusively used to carry guns and ammunition for combat units, while the Russian trucks were employed to bring up follow-on supplies and carry back the wounded. (A nice symbiosis is observable here: American trucks, brought over in British naval convoys, helped Zhukov's frontline mobility.) By 1944, ironically, a completely motorized Russian antitank regiment could probably move around faster than a tank regiment itself. No other army managed that.[51]

The Red Army's successful defense at Kursk and then its steady advance westward over the next year were helped by three other advantages it held over the Wehrmacht: bridge-building capacities, deception techniques, and the assistance of a vast network of partisans. The first two of these offer fine examples of midlevel organizations providing solutions to acute military problems. How, for example, did one get three, four, or five Soviet armies and their heavy equipment across big rivers in the steady reconquest of Eurasia after the summer of 1943? The next chapter will examine the story of the Anglo-American efforts, often at first unsuccessful, to cross great stretches of water and land on an enemy-held shoreline. But this was also a massive logistical challenge for any Soviet army group (consisting as it did of tanks, self-propelled guns, heavy artillery, Katyusha rocket launchers, trucks, headquarters equipment, and all the rest) because of the physical obstacles posed by the extremely wide and reed-bed-surrounded rivers so familiar in the history of Russia—the Volga, the Don, the Dnieper, the Dniester, the Vistula.

This was a logistical problem that the Russians, because of their geography and their history, had had to deal with over the centuries—in 1799, for example, the great General Alexander Suvorov took an entire army to the Alps, obviously crossing a large number of rivers in between. While pontoon construction does not occupy a high place in the planning documents about Soviet rearmament plans in the 1930s, the fact was that by January 1, 1942, as the battle for Moscow was drawing to its close, the Red Army already possessed 82 engineer battalions and a further 46 separate pontoon bridge battalions, which turned out to be of great importance three years later in advancing upon Berlin. By January 1944, the Red Army had no fewer than 184 engineer battalions and 68 pontoon bridge battalions.[52] One has the sense that the Soviet system accorded much more importance to engineers in general than did the German high command.

The logistical problem was worked out by the design bureaus of the People's Commissariat of Defense (NKO)—though we still lack many details of who these designers were because of the unavailability of the archives—in the emergency created by the launch of Operation Barbarossa. At first they couldn't do much, for it was difficult to design and build pontoon bridges when the invader was crashing through Minsk, Kiev, Smolensk, and on to Moscow. Yet by the spring of 1942 the NKO was sending forward its first kits, fairly primitive all-wooden ferries (rafts, really) that could transport troops and light equipment across a river, and which could be strung together to form the underpinnings of an improvised and very rapidly made bridge, then dismantled to be used elsewhere. These "pontoon-bridge parks" relied upon simple, sturdy, and mass-produced equipment; for that reason, the factories never built the equivalent of the American LCTs (landing craft, tanks). Yet by September 1942 the first of the new-model TMP pontoon bridges were reaching the front, just in time for Stalingrad. Their professional assembly crews needed only three to five hours to erect across the Volga—which was a mile wide at points—a bridge with a total load capacity of 80 tons. This was a remarkable feat.[53]

At the same time, another group of NKO designers and production teams was working on specialized equipment to be used in one of the Red Army's most favored tactics—*maskirovka,* or deception ploys. The idea was as old as fighting itself—Scipio, Frederick, Napoleon, and Sherman all used deception techniques, and the Western Allies were to do the same, very successfully, before and during the D-Day landings. The Stavka regarded deception as an essential part in Russia's advance westward, for two obvious reasons: first, they shared the universal respect for the German army's capacity to react swiftly and fiercely to an assault on any front, and second, attacking across a wide and swift-flowing river was about as precarious an undertaking as landing across coral reefs, because of vulnerability to enemy fire from the shore. It was important to strike where the enemy wasn't, or where he was only holding a weak line. The NKO therefore threw itself into the production of an inventory of goods, from camouflage netting and artificial terrain-masking materials to mock-up and dummy tanks, artillery, and other large weapons. False trenches, fake airfields, and promising-looking tank parks were also constructed. These were all handled in the field by special *maskirovka* companies assigned to each Red Army front for the twin task of hiding where the army was strong and falsely pointing to locations from which an attack would not be made.[54]

The successful deployment of these new techniques was immensely assisted by the rapid growth of Soviet partisan groups within the German-held territories from about mid-1942 onward. Here the folly of the cruel Nazi treatment of the Ukrainians and other ethnic groups within Stalin's loathed empire came back to haunt the German high command. The harsher their reaction to partisan activities (mass shootings, hangings from trees, the burning of churches), the more they drove the people—men, women, and children—into resistance. Apparently the difficulties they encountered with their repressions in Crete, Greece, and Yugoslavia conveyed no lessons to the Wehrmacht. Thus from Brittany to Belarus, from Norway to Rhodes, the German armed forces faced the double task of handling the assaults of the Grand Alliance and responding to sporadic attacks from within their conquered lands. The Soviet special-operations coordinators, like their British equivalents, worked assiduously to supply the partisans with small arms, dynamite, radios, and explosives experts. As soon as the Kursk

THE PLANNING BEGINS

*Franklin D. Roosevelt and Winston Churchill seated in the garden of the villa
where the Casablanca Conference was held in January 1943. Grouped behind
them are the British and American chiefs of staff.*

EARLY ALLIED REVERSALS

*Despite being in close convoy, a British merchant ship sinks quickly after being tor-
pedoed in the North Atlantic, with an Allied destroyer nearby.*

A pall of smoke hanging over the harbor in Suda Bay where two Royal Navy ships, hit by German bombers, are burning on June 25, 1941. British Naval and military forces were smashed by the German blitzkrieg assault on Crete.

German Panzers advancing unopposed in Ukraine, September 1941.

An American B-17 Flying Fortress bomber crashing out of the sky during a daytime raid over Germany—a typical fate for many such aircraft in 1943 and 1944.

This scene on the beach after the Dieppe raid in August 1942 shows the many dead Canadians and their broken vehicles.

The waterfront at Betio in Tarawa Atoll, Gilbert Islands, in 1947, showing how the sea wall foiled the advance of American tracked amphibious vehicles.

An artist's impression of the battle of Midway, June 1942.

An iconic image of British infantry advancing through the dust and smoke of the battle of El Alamein, September 1942.

The SS Ohio entering the Grand Harbor at Valletta in Malta on August 15, 1942, following Operation Pedestal; one of the hardest-fought convoy operations of the war. Hit by an Axis bomb, the U.S. oil tanker could only be saved by two Royal Navy destroyers "linking arms" and towing her slowly for the rest of the journey. She was so badly damaged that she had to be scuttled in order to offload her absolutely vital cargo. Malta's fuel stocks were assured and were never again in such threat.

Newly trained American troops resting after their unopposed landings in Casablanca, Morocco, November 1942.

THE TOOLS OF VICTORY

The Cavity Magnetron: This miniaturized radar device gave a huge advantage to British and American forces because it could be placed in aircrafts and smaller warships.

By mid-1943, an aircraft like this Vickers Wellington bomber had been transformed into an acoustical machine with enormous weaponry for the destruction of U-boats.

The sloop HMS Mermaid, *which possessed all of its high-angle armament and detection equipment by 1944.*

The Hedgehog, a 24-barreled mortar, was mounted on the forecastle of the British escort vessel HMS Westcott.

USS Bogue: *This cheaply produced warship provided additional air cover not only for the critical North Atlantic battles of 1943, but also for escorting the massive U.S. troop flows to North Africa and the Mediterranean. Many* Bogue-*class ships were transferred to the Royal Navy under the provisions of the Lend-Lease program.*

Captain. F. J. Walker was the most successful anti–U-boat commander in the battle of the Atlantic, seen here directing a sister warship in its attack on a German submarine.

The Leigh Light, used for exposing U-boats on the water's surface at night. Here it is fitted onto an RAF Coastal Command Liberator aircraft.

The role of long-range American, British, and Canadian patrol aircraft over the Atlantic waters was absolutely critical to the Allied victory in the West. Allied aircraft actually sank more U-boats than did warships. The PBY Catalina above is returning to its base in Gibraltar, crossing Europa Point on the north shore of Morrocco.

USS Wasp *loading Spitfires in the Clyde, April 1942.*

In April 1942 the U.S. Navy loaned their carrier, USS Wasp, *to assist the British in their desperate need to fly Spitfires to the beleagured island of Malta. In early 1943, the British Admiralty loaned the new fast carrier HMS* Victorious *to Admiral Halsey's Southwest Pacific Fleet as a partner to the USS* Saratoga. *At that time, the* Saratoga *was the only American carrier operating in the Pacific. This illustration shows the* Victorious *joining its American partner in Noumea Bay, New Caledonia, Southwest Pacific, May 1943.*

USS Essex *leaving Norfolk, Virginia, laden with aircraft for the Pacific War, May 1943.*

HMS Anson: *Another brand-new British battleship steams out to join the fray. By 1943, only the American and British Navies were launching heavy warships.*

THE EASTERN FRONT

The spring thaw made the going tough for both sides on the Eastern front, in this case for a German motorcycle unit.

The armored struggles during this conflict were the largest and most confused of the entire war. Here German Tiger tanks are getting ready to strike at Kursk. The Wehrmacht still relied, however, on a vast number of horse-drawn wagons, which were painfully slow in dragging their way through the boggy paths and dirt roads of the country.

The early T-34 tank had extremely confined space for its three-man crew. The tank commander not only had to shout course changes to the driver, and aim and fire the gun, but also assist in loading the shells.

The advanced T-34/85 tank had a bigger turret and a more powerful gun, much transformed from the earlier prototype.

German two-engine bomber: Heinkel 111s

Japanese two-engine bomber: Mitsubishi G4M known as the Betty

RAF Lancaster bomber: This massive four-engine bomber, like its American equivalents, the B-17s, B-24s, and B-29s, carried a much larger bomb-load than eight to ten Axis medium bombers.

Once the pride of the German fleet, the battleship Tirpitz *lies silhouetted against the snow-capped mountains of Norway after being destroyed in November 1944. In the foreground is a crater left by one of the six-ton Tallboy bombs dropped by a Lancaster during the same raid in the Sanne Strait.*

B-17s flying toward Germany with long-range fighter cover overhead, at last.

The American B-29 Superfortress in squadron flight toward Japanese cities. Their main base in the Pacific theater was the island of Tinian in the Marianas, captured in June–July 1944.

ENGINEERS OF VICTORY AND THEIR PRODUCTS

Ronnie Harker

It was Harker, the British test pilot, who suggested that putting the new Rolls-Royce Packard Merlin engine into the American-built P-51 fighter would transform its range, speed, and overall performance. Here, a Merlin engine is lowered into the legendary Mustang chassis.

Captain Pete Ellis

American marines landing from barges on a beach at Guadalcanal, beginning an attack on the Japanese on September 7, 1942. It was Captain Pete Ellis who first conceived, as early as 1919, the strategy of amphibious landings on small islands across the Pacific.

Major General Sir Percy Hobart was responsible for many of the specialized armored vehicles, known as Hobart's Funnies, that were deployed in North Africa, the invasion of Normandy, and later actions.

One of Hobart's "flail tanks" in action. The turret is temporarily reversed; the flails are driving forward.

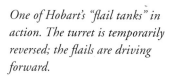

Admiral Ben Moreell (1892–1978) was a brilliant engineer and a distinguished naval officer who persuaded Franklin D. Roosevelt to allow him to create the new Construction Battalions (CBs or SeaBees) shortly after the attack on Pearl Harbor.

The Mulberry harbor at Arromanches, with lorries rolling ashore loaded with supplies in June 1944. The Mulberry harbors were built by the SeaBees in Pembroke Harbor, towed to the Normandy beaches in sections, then reassembled to aid the invasion armies.

Germany also possessed brilliant wartime engineers. The above image shows the Messerschmitt Me-262, the first jet fighter of the war, which was far faster than the most advanced Spitfires or Mustangs.

The modified version, however, with bombs attached underneath, seriously impaired the plane's speed and stability in flight. Hitler's obsessiveness about bombing thus overcame Willy Messerschmitt's engineering genius.

Field Marshal Bernard Montgomery leaves the Brandenburg Gate after a ceremony to decorate Soviet Generals in Berlin, July 12, 1945. With him are marshals of the Soviet Union: Georgy Zhukov (second from left), Vasily Sokolovsky (center right, background) and Konstantin Rokossovski (right foreground). Montgomery and Zhukov were perhaps the two most successful "positional generals" of the war.

battle began, an estimated 100,000 guerrillas began sabotage and snip-ing. During June 1943, partisan groups destroyed 298 locomotives, 1,222 railroad carriages, and 44 bridges in the area directly behind Army Group Center—a worrying sign to Manstein and the other gen-erals as they prepared for the assault upon Kursk. A year later, just be-fore the Red Army's next big push (Operation Bagration, to be described later), Belarusian guerrillas were detonating tens of thousands of charges against German rail lines and bridges, and waiting for the order to at-tack straggling German army platoons after the big knockout blows were delivered. Still, while the partisans could assist the Red Army, they were not themselves capable of turning the tide on the Eastern Front. As one scholar has noted, their greatest impact occurred when the main Soviet armies were already heading toward Berlin.[55]

The partisans also supplied lots of information about German troop movements. It was, admittedly, often difficult to make much sense of this, because by 1943–44 reports were coming in from dozens if not hundreds of guerrilla groups, and the Stavka faced a problem very like the one that had confronted American planners before Pearl Harbor—too much "noise." The Soviet high command's intelligence challenge was that it needed specific data: exact times, exact numbers, exact line of first attack, exact route of the *Kesselschlacht* move, and so on. Pretty well everyone—the Lucy Spy Ring in Switzerland, British Enigma decrypts, Die Rote Kapelle, members of the Cambridge Five, captured prisoners—was reporting in an urgent way that a large Ger-man offensive would come in June or July 1943, but what did that tell Zhukov that he didn't know already? The Stavka had already issued three separate alerts in the month of May alone. By early July, the sus-pense at the front was barely tolerable. Then, at 2:00 a.m. on July 5, a forward Russian platoon captured a German engineer clearing mine-fields. Zhukov and Konstantin Rokossovski at once decided to begin massive spoiling fire (the contrast with the forced inactivity on June 21–22, 1941, is striking). The struggle for the salient was now on. But what role intelligence played in it still seems unclear, and the de-bate among scholars over Kursk and Russian intelligence continues un-abated.[56]

Another source of important intelligence data to the Stavka was the frequent reports of the aerial intelligence service, which had improved enormously since the battle for Stalingrad. Aerial sightings of enemy positions below and of units on the move (and, even better, aerial photographs) were extremely valuable for many obvious reasons: they were more or less immediate, and they could be verified far more efficiently than could a radio message from the Belarusian forests. They thus could give the Russian commanders an earlier chance to alter the balance of their forces, and the Stavka the chance to move around its reserve armies. Finally, they were useful to the Soviet Air Force in targeting its tactical air missions.

Soviet Airpower?

As with the other elements mentioned above, the Red Air Force also grew immensely in numbers and capacity during the years after 1942. Assessing its real effectiveness, however, is a different matter. How important was the contribution of Soviet airpower to the campaigns along the Eastern Front during the war? The historian has to peel back layers of contemporary and then Cold War propaganda that obscure the real story. Additionally, many of the archives on the Russian side are still inaccessible, while the memoir literature is often unconvincing. This in its turn leads to a very heavy reliance upon German sources. Yet some things are clear.

The first is that the roles played by the Red Air Force were much more specific than the far more numerous tasks that had to be carried out by the American and British air forces. There was, obviously, no air war at sea, so no carrier aircraft, and therefore no carriers nor their necessary escorts. There were no massive deployments of Coastal Command–type airpower—not just planes but crews, ground staff, air bases, intelligence and training personnel, production lines, fuel and ammo supplies—needed to carry out the grinding, titanic fight against the U-boats. Nor was there any real effort at strategic air warfare, employing massively expensive long-range four-engine bombers, which were key elements in American and British grand strategy. The Soviet air contribution was restricted to two elements: fighter forces, to defend against Luftwaffe attacks on the homeland and to gain aerial control over the battlefields, and tactical air forces, to bomb, strafe, and otherwise interdict the Wehrmacht's efforts in the field as the Russian forces advanced.

Even within these more limited fields, it still remains difficult to assess the effectiveness of the Red Air Force, for two reasons. The first is that we lack really sustained, high-quality comparisons of their fighting capacities. It is quite easy to accept that the performance of all Russian planes steadily improved, especially from 1943 onward, but that was the story everywhere. What is harder to establish are the claims made near and after the end of the war that, for example, the Yakovlev-3 (Yak-3) was as good as the latest Spitfires, for the simple reason that the two aircraft never fought each other, let alone engaged in a sustained aerial tussle like the Battle of Britain in summer 1940 or the massive operations of Allied fighter forces across western Europe to destroy the Luftwaffe squadrons from February through April 1944. Where were the battles like the Great Marianas Turkey Shoot (see chapter 5), where the Hellcats' superiority over the Zeros was complete? If what we have learned about the T-34 is any guide, then mere figures regarding Soviet fighters' respective speed and engine horsepower are simply not enough. Was the Yak-3 better than the enemy in real aerial combat? We will never know.

The second difficulty for scholars in this field is even greater: how is one to assess the Soviet Union's gradual ascendancy in the air in the final eighteen months of the Great Patriotic War when most of the Luftwaffe's fighter groups, and all of their best aces, had been pulled away to fight across western Europe? Once Barbarossa unfolded, the German air force was fighting on three fronts (eastern, western, and southern), yet until the end of 1942, 50 percent of its air forces were campaigning in the East. Then came the British and American bombing of the cities of the Third Reich, especially the RAF's devastation of Hamburg in July and August 1943, which in particular seems to have shaken up Speer, Goering, Milch, Galland, and the others a great deal. All of them felt, in consequence, that many more resources had to be devoted to the defense of Germany's industrial core, even if they needed to be withdrawn from elsewhere. As is detailed in chapter 2, the American and British bombing efforts over Germany were blunted, though in part by a relocation of Luftwaffe squadrons from the east.

This relocation intensified with the Allies' massive aerial campaigns over western Europe in early 1943. The impact upon the Ostfeldzug was clearly great. By April 1944, the German air force had denuded the Eastern Front, leaving only five hundred aircraft of all types to confront more than thirteen thousand Soviet planes.[57] At the same time, of course, Germany had to concentrate the vast bulk of its dual purpose antiaircraft/antitank batteries (more than ten thousand guns and half a million men) on the defense of its own cities, rather than trying to shoot the Red Air Force out of the sky and bring Zhukov's advances on land to a grinding halt. Stalin always complained to Churchill that the West was not doing enough to bring Germany down. He never acknowledged that the air war over Russia might have been won, albeit indirectly, by the air war over the Ruhr.

So, was the achievement of the Soviet air force, the VVS, really that of a first-class service, or was it pushing against a weakened German air force increasingly manned by novice fliers? The VVS may have become ascendant only when the war was essentially won. More particularly, it is hard to find an occasion (better, repeated occasions) when a squadron of eighteen Fw 190s fought it out with a squadron of eighteen Yak-3s and the latter won decisively. The respected Russian fighter pilot Kozhemyako records in his memoirs a single occasion when his Yaks spotted four Fw 190 fighters, but the latter simply increased speed and flew away (they could not have outrun a Spitfire or a Mustang). Actually, one wonders how many Fw 190s were left on the Eastern Front by late 1944, apart from those squadrons being reluctantly converted to fighter-bomber status to replace the ailing Stukas and engage in the land war. For Operation Bagration, the Red Air Force–to–Luftwaffe ratio on fighter aircraft may have been as lopsided as fifty-eight to one, which if true makes nonsense of any real comparative assessment. When the fabled "Red Phoenix" rose, was it because there was no opposition and the remaining Luftwaffe groups were supplying garrisons and bombing pontoon bridges, not fighting for control of the skies?[58]

This particular aspect of air conflict in the Second World War therefore remains unclear, despite some impressive scholarly studies.[59] Clearly, the Red Air Force was a nonentity early in the war. On the first day of Barbarossa operations alone, the Russians lost around twelve hundred aircraft (almost all probably hit on the ground) to the German loss of a mere thirty-five planes. Then came the gradual Russian come-back. By the closing stages of the Battle of Stalingrad (when the street fighting was so dense that Stukas could not be used in close-support operations), the soldiers of the Red Army could see a few planes above them bearing a red star. Simultaneously, the inter-Allied pressures played their part. The Luftwaffe now faced a collapsing North African/ Mediterranean front, the arrival of the American bombers in England, and Harris's tearing apart of cities such as Cologne, all of which forced the high command to pull the better pilots and swifter fighters back to the Reich. And the VVS was at last getting better planes, the Yak-9 fighter and the greatly improved Ilyushin Sturmovik, an exceptionally tough, rather slow, low-flying tank-buster of an aircraft that was de-signed to assist the Red Army's advance to the west.[60] By 1944 the Yak-3 was in service and playing havoc with German transport aircraft, light bombers, and those awkward Focke-Wulf 190 conversions. The Red Air Force was also acquiring its own aces (including female fliers), although even as late as Operation Bagration it continued to lose more planes than did the Luftwaffe. Overall, though the real significance of Soviet airpower during the Great Patriotic War remains shrouded in mystery, it probably played far less of a role in the turning of the tide on the Eastern Front than its early propagandists proclaimed.

The Beginning of the End: Bagration to Berlin

On June 22, 1944, as the Western Allies were slowly advancing beyond sight of the Normandy beaches, and as American troops fought yard for yard across Saipan, the Red Army launched Operation Bagration, a massive assault upon the German-held central front that involved sev-eral times as many ground-based forces as the combined totals of all those involved in the Marianas and D-Day attacks. On the Eastern Front, nothing was small in scale, but the size of the Red Army and Air

RED ARMY ADVANCES DURING OPERATION BAGRATION, JUNE–AUGUST 1944
Although less well known than the battles of Stalingrad and Kursk, this Soviet vic-tory over the Wehrmacht was the largest during the entire war on the Eastern Front. The detail in the map on the next page illustrates the sheer size of this assault, involv-ing as it did 1.7 million Soviet troops. It coincided with the campaign in Normandy, the capture of Rome, and the seizure of the Mariana Islands.

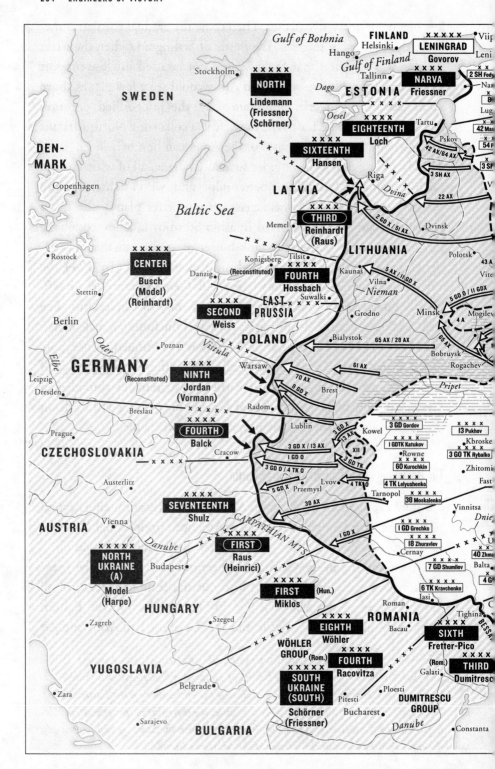

Red Army Advances During Operation Bagration, June–August 1944

Red Army advance
June 22–August 19, 1944

Force's commitment to Bagration exceeded anything that had gone before, whether Moscow, Stalingrad, or Kursk. Appropriately, the operation was named by Stalin after imperial Russia's most aggressive general, Piotr Bagration, who died of his wounds while blocking Napoleon at the great Battle of Borodino. It may not have been a coincidence that Bagration, like Stalin, came from Georgia, nor that the operation named after him was launched exactly three years to the day after Operation Barbarossa. It was the Soviet Union's massive return blow.[61]

The numbers of troops, tanks, and guns involved in this enormous Soviet surge from the Ukraine to Poland are difficult to comprehend from the perspective of post-1945 limited war, and impossible to count with accuracy. Should we include all the Soviet forces facing their German counterparts from north to south, or just those involved in the punch into the center? Should we include the vast follow-on divisions initially held in reserve, or just those used in the initial Bagration assaults? Probably it makes no difference; though the varied and conflicting figures simply baffle the mind, all arrive at a massive total. According to one account, "At the commencement of the offensive, Stavka had committed approximately 1,700,000 combat and support troops, approximately 24,000 artillery pieces and mortars, 4,080 tanks and assault guns and 6,334 aircraft. German strength at the outset was approximately 800,000 combat and support troops, 9,500 artillery pieces, but only 553 tanks and assault guns and 839 aircraft."[62] At the Second Battle of El Alamein, Rommel had a mere 27,000 German troops; the comparatively large British Empire force of 230,000 men was far smaller than any one of the five Russian army fronts (that is, full army groups) pushing westward in June 1944.

But Operation Bagration was not just a matter of brute force. At long last, the Red Army had reached its full potential. It was still feeling the effects of Stalin's purges of the 1930s when the Germans attacked, and then had to suffer greatly in all the early battles, with disproportionate losses. The service had had many failures, and from top to bottom it had been slow to learn important lessons from its humiliations or even from its bruising victories. Yet the high command had eventually learned to put all the pieces together in the most impressive battlefield recovery of the entire war. By the time of Bagration, Stalin was willing to entrust two members of the Stavka, Aleksandr Vasilevsky and Zhukov, with coordination of the key central section of those five gigantic fronts that stretched from the Baltic to the Ukraine—provided, of course, Zhukov gave his suspicious boss daily updates.

In this vast battle, the Red Army encountered the reverse problem of the "expanding torrent": the defender's lines were contracting and becoming more concentrated in the troops-to-land ratio. Yet even in June 1944 the length of the German Eastern Front was still enormous, and the sheer size of the Soviet ground forces was overwhelming. Ironically, the more the Wehrmacht retreated, despite Hitler's raging counterorders, the more the Ostfeldzug came to resemble the Western Front of 1914–18. But this time the attacking side had a force advantage of at least three or four to one; in many individual encounters, it was more like ten to one.

Every piece of the elaborate machinery that was needed in a major land operation worked for the Red Army. The *maskirovka* was superb. While the OKW was anticipating a further large Soviet offensive during the summer months, it had no idea where the main assault would be launched—their best guess was that it would come in the south or perhaps on the Baltic Front. The OKW greatly underestimated the size of the force that would be deployed against Minsk in the center. The T-34-85, now in vast numbers and with its many defects corrected, was virtually unstoppable. On a one-against-one basis, a single Tiger or Panther tank was still more powerful, but by this stage Soviet armor tactics were much better developed, with the faster T-34s nimbly moving around the slower Nazi tanks or leading them into an antitank ambush. The river-crossing engineer battalions were at their peak, and much helped by intelligence from partisans on the other side as to where, and where not, to attempt a landing across the Berezina and the Dnieper.

Moreover, possessing much better trucks and supply services than before, the Red Army's capacity for tactics of deep penetration and exploitation were at last feasible. (Proposals for such operations had been hammered out as early as 1935, only to be discarded in the purges.) And, ironically, while the Russian armies moved from stupefying Stalin-driven rigidity to greater flexibility, the German land forces abandoned their historical *Bewegungskrieg* (fluid warfare) for a nonsensical clinging-on to every redoubt, because their Fuehrer had ordered no retreat—thus making the Soviet encirclements easier.[63] By now, also, the Red Air Force's ground attack squadrons were in control of the airspace above the battlefield. Above all, the Russian infantry—the Russian common soldier and his sergeant, really—had great experience of how to fight, house by house, street by street, hedgerow by hedgerow. Just as the Red Army had more and more tough, experienced divisions, the Germans had fewer and fewer.

While the Stavka ensured that all five army fronts would be constantly probing, to keep the enemy unsure, it then ordered the attacks, always in variegated timing (the Polotsk offensive on July 1, the Vilnius offensive on July 7, etc.), with the chief punch to be in the center, to capture Minsk and then to head westward to the Polish border. The fighting here was more ferocious than anything since Stalingrad, and the result was the same—an entire German army captured and another mauled. In the twelve days between June 22 and July 4, 1944, Army Group Centre lost twenty-five divisions and 300,000 men, in what Zaloga calls "the most calamitous defeat of all the German armed forces in World War II." The overall German losses in Operation Bagration were a numbing 670,000 killed, wounded, or captured, though the Russian losses were even higher (178,000 killed and 587,000 wounded). Truly, the road from Moscow to Berlin ran with blood. The difference was that the USSR still had millions of men to draw upon, and for a single strategic purpose. And it was not fighting alone, because at long last the Anglo-American-Canadian armies were fully engaged in battle. By the time the Normandy campaign in the West was over, in August 1944, the Wehrmacht in France had lost another 450,000 men. Little wonder that Allied forward divisions were reporting that most of the German soldiers captured in the final year of the war were either older men or Hitler Youth, recruited in desperation to replace the depleted able-bodied battalions.

Now the ring was closing. When Bagration began, Soviet armies were about 750 miles from Berlin. At the same time, Patton's armies in western Normandy were approximately 650 miles from Berlin. Two months later Paris was liberated, and Red Army units, which had moved a far greater distance, were looking at Warsaw from the other side of the Vistula. Meanwhile, Allied armies in Italy were bumping up against the Gothic Line, well north of Rome. All of them were tired, slowing down, and needing replenishment from the vast supplies coming up from the rear. Yet all were consoled by the fact that in June 1944 the Third Reich had suffered colossal and irreversible defeats in the field. It was, as one historian puts it in the title of his recent book, "blitzkrieg no longer."[64] It had not really been lightning warfare for a while, but now it was truly over.

Thus, when Bagration ended in August 1944, with the Red Army's hard-fighting frontline troops exhausted and adequate supplies of fuel, ammunition, and spare parts once again some way behind, everyone knew who had won this campaign. The Finns nimbly extricated themselves from Nazi thrall and signed a peace deal with Moscow in September, while in the south Romania and Bulgaria also parted company with the Thousand-Year Reich. Now it was on to Berlin, a story of astounding German resistance in the face of overwhelming odds.[65] It was of course a hopeless resistance, like that in the West, but one still fueled by ideological passions that gripped Wehrmacht units as well as the soldiers of the Waffen-SS, and by a pathological fear that surrender would mean a death not unlike those the Germans had inflicted upon their own defeated foes on the Eastern Front in the preceding years. It was also gripped by revolting, and rising, atrocities against any domestic foes of the doomed Reich.[66]

The overwhelming significance of this titanic campaign stands out. If one couldn't defeat the German army in the field in Russia, as well as weakening it in North Africa, Italy, and France, then the war would not be won. Victory at sea and victory in the air, as critical and absolutely necessary as they were, simply were not enough. One of them gave protection to the Allied lines across the seas. The second gave aerial protection for the vital counterattack points of Britain and North Africa and allowed the buildup of the thousands of Allied bombers that would cripple the Third Reich's industries and infrastructure. If Hitler refused to surrender, though, somebody's army had to crush Germany's formidable ground forces, march into the Fuehrer's bunker, and end the war. In the main, it was the Red Army that did it. As a consequence, 85 percent of all Wehrmacht losses occurred on the Eastern Front.

Clearly, the Germans had never anticipated that Russian soldiers would be so tough, and that Russian designers and engineers would supply those soldiers with increasingly effective weaponry. And, with a few exceptions mentioned below, they also ignored the geographic folly of what they were attempting. One of those exceptions was Halder. Before he was dismissed as army chief of staff, he had repeatedly voiced concern about this overstretch. His famous diary entry (see page 176), on having reckoned with around 200 Soviet Army divisions at the onset of war and now counting at least 360, is followed by this lament: "Time . . . favors them, as they are near their own resources, while we are moving further and further away from ours. And so our troops, sprawled over an immense frontline, without any depth, are subjected to the incessant attacks of the enemy. Sometimes these are successful, because too many gaps must be left open in these enormous spaces." A month later, Halder was retired, von Rundstedt had been dismissed for suggesting a winter pullback, and the fatal advances toward Stalingrad and the Caucasus were going ahead. By 1943 and 1944, many more generals had come to realize that they had occupied too much space, but if they spoke out or retreated from their fronts, they too would be dismissed.

Halder's replacement as army chief of staff from 1942 until 1944 was Kurt Zeitzler. Long after the war was over, Zeitzler sat down and made some geographic calculations in an intriguing essay called "Men and Space in War." When Barbarossa began, the Ostfront was approximately 438 miles long. By Christmas 1941 the Wehrmacht had advanced 625 miles and, with its allies, held a front of about 1,125 miles, with significantly smaller manpower totals. By the summer of 1942 they spanned 1,250 miles, by which time Zeitzler was pleading for Hitler to give up space—but in vain. By the end of 1942 the Eastern Front was roughly 1,800 miles long, while at the same time German troops had to garrison an Atlantic Wall that encompassed 1,250 miles, and Danish-Norwegian coasts stretching to about 1,500 miles. That still left the North African boundaries, the Balkan coasts, Crete, Rhodes, and other islands. (It is perhaps no coincidence that the hyperenergetic Zeitzler had a nervous breakdown in July 1944, and could only write about this fatal overstretch long afterward.) In theory, all the fronts of the Nazi empire were defensible, provided the Wehrmacht had the sort of technological and tactical superiority over its foes that for generations the Roman army possessed against the barbarians outside its gates. But that was never true for Germany in the air and at sea during the Second World War, and by 1943–44 operational and logistical superiority on land was also distinctly moving to the Allied side.[67]

The Eastern Front and the North Atlantic

Strange though the comparison might initially seem, the Ostfeldzug resembled another epic, long-lasting campaign of the Second World War: the Battle of the Atlantic. They took place in roughly the same latitudes, one in the western reaches of Germany's military endeavor, the other on its eastern extremities. Hitler's twin enemies here, Britain and the Soviet Union—and their respective warlords, Churchill and Stalin—regarded their struggle as existential. By continuing to fight on despite early disasters, each of them drained German resources (armaments, steel, ball bearings, trained personnel, detection equipment, air cover) to the west or the east. Then the RAF bombing of German cities and factories exacerbated the Third Reich's resources dilemma, even before the Americans joined in. Russia's land effort was always connected, distantly but definitely, to Britain's naval and aerial actions, because each forced Berlin into allocation decisions.

Both the Battle of the Atlantic and the great struggle across the Eastern Front were battles of attrition, grinding on for year after year, swaying backward and forward, with each side pulling in more and more reinforcements, introducing newer weapons, employing the latest intelligence, fighting at the northern and southern fringes but always needing to get control of the central area. Across the early years of 1941 and 1942, these great naval and military struggles for Europe unfolded with extraordinarily heavy losses—though never enough to push any combatant out of the war, as had happened earlier to countries such as Poland, France, Norway, and others. The deeper resources of the great powers made for a larger and longer war. Only in mid- to late 1943, however, did the conclusion become clear: Doenitz's U-boats could not win in the Atlantic, nor could Manstein's panzers win in Russia. The defeated German forces, land, air, and naval, would still remain dangerous and would still fight very well, but their days of going forward were over. Gone, too, were any hopes of negotiating some sort of 1918-style compromise peace. The "unconditional surrender" demanded at Casablanca meant just that.

Second, in neither the Atlantic nor the Ostfront campaign did any single wonder weapon transform the course of the fighting. The Battle of the Atlantic was not won *particularly* by Enigma, nor by the Hedgehog and the homing torpedo/depth charge, nor by miniaturized radar, nor by long-range aircraft and escort carriers, nor by operational research, nor by Liberty-ship construction output, nor by the hunter-killer groups, nor by the Leigh Light, nor by the bombing of the U-boat pens and shipyards. But by the middle of that decisive year of 1943, they all came together to help turn the tide; the historian who overemphasizes any one of these aspects distorts the larger, holistic account. In much the same way, the Great Patriotic War was not won chiefly or overwhelmingly by superior Red Army intelligence, or by increasing aerial supremacy, or by the T-34-85 tanks, or by the PaK units and the minefields and the river-crossing battalions, or by better logistical support, or by better fighting morale. Victory required all these elements, and they needed to be organized. But it took time to assemble the various pieces, just as it takes time to bring together an orchestra and train it to deliver a fine performance.

By 1943, the British Admiralty, and especially its Western Approaches Command, had at last achieved that satisfactory level. By around the same time, perhaps a little later, the Stavka and the forward operating army groups under Zhukov and Vasilevsky had done the same. Given the sheer number of armies, divisions, regiments, air groups, partisan cells, engineer battalions, railway and bridging teams, and behind them transport managers, production planners, and factory leaders, the overall organizational feat of this badly damaged Russian state was simply astounding, and is still not fully recognized in the West even today. By contrast, while the Germans fought extraordinarily well in all theaters of war, added a special racially driven ferocity to their fighting on the Eastern Front, and produced some first-class weapons for that grim campaign, the full and efficient orchestration of all the pieces in the Third Reich's Ostfeldzug never occurred. Those despised, Jew-ridden, bolshevized, peasant-heavy, and backward Slavs, those fragile serfs of the inept Communist regime, actually managed to organize the way to victory better than the famously efficient Prussians and the fanatical Nazi *Supermenschen*.

One final thought: it may be many years before historians are allowed to mine the archival sources that will disclose the full story of the middlemen and organizations that contributed to the turning of the tide in the Great Patriotic War. We know how the great land battles in the east unfolded, and we have good biographical details on a figure as eye-catching as Zhukov.[68] And we have a better idea nowadays about how Stalin and the Stavka ran the show. But what about the lesser-known contributors to the Soviet victory? Who were the problem solvers in that part of the story, the equivalents to the innumerable players on the Anglo-American side whose tales are so readily accessible? Clearly they existed and made enormous contributions in the three years between Barbarossa and Bagration, creating ever-smoother feedback loops between the top and bottom levels of the Soviet war machine and producing, eventually and after many setbacks, the instruments for the Red Army's smashing victory. Their achievements are manifest, but their own histories are not yet known to the world.

CHAPTER FOUR

HOW TO SEIZE AN ENEMY-HELD SHORE

Fair Stood the Wind for France
When we our sails advance,
Nor now to prove our chance
Longer will tarry.

—MICHAEL DRAYTON (1563–1631), "AGINCOURT"

The Earl of Chatham, with his sword drawn,
Stood waiting for Sir Richard Strachan;
Sir Richard, longing to be at 'em,
Stood waiting for the Earl of Chatham.

—CONTEMPORARY DOGGEREL VERSE ON THE FAILURE OF
THE WALCHEREN EXPEDITION, 1809

The special feature of amphibians is that they can and do inhabit two worlds, the sea and the land. Generally they are happier and safer in the watery element, and creatures such as turtles and seals advance onto the shore with increasing vulnerability, though no one would regard a fully grown Nile crocodile as easy prey. The very act of moving from sea to land is full of risk—unless, of course, that step is unopposed and takes place in favorable weather and topographic conditions. In wartime, such circumstances are not often at hand.

This chapter is about the evolution of amphibious warfare during the Second World War until the great Normandy invasion of June 1944. By its very nature, it is not about peaceful and uncontested landings of men from ships to shore, such as William III's crossing to southern England in 1688, which brought an end to the Stuarts' reign. The focus here is upon military operations against a shoreline held by defenders determined to frustrate the intended invasion. It covers failed assaults from the sea, but it is essentially a study of how certain organizations found solutions to one of the most difficult challenges facing any army: how to land on an enemy-held shore under counterattack.

In that sense, like the other chapters of this book, it addresses the central question of how to gain the advantage over the enemy and thus contribute to the winning of the conflict. It is intimately linked to three of the other four chapters. The Pacific War of 1941–45 (chapter 5) is about battles in a theater where landing on a distant and often hostile shore was at the heart of strategic success or failure. Thus, while devoted to Allied amphibious warfare in Europe, this chapter cannot be separated from the virtually simultaneous campaigns waged in Guadalcanal, the Gilberts, and the Marianas, because of so many similarities. But it also cannot be divided from those in the first two chapters. In the saga of the West's defeat of Nazi hegemony of Europe, three successive steps were intended to fit smoothly together, like a tightly glued triangle. The first was control of the Atlantic seas and the defeat of the U-boats, with sea power greatly assisted by airpower; the second was the domination of the skies, with the Allied air forces in the United Kingdom and North Africa reliant upon continuous seaborne munitions, fuel, and parts; and the third was the invasion of the enemy's shores, with the vulnerable armies protected both by sea and air. The colossal Normandy operation was thus to be the spectacular fusion of maritime, aerial, and land power, the apotheosis of combined arms.

Sea Landings in History

The history of amphibious warfare goes back well before the modern term itself. The massive landing by the Persians at Marathon, the ill-fated Athenian expedition to Sicily in 415 BCE, Caesar's invasion of Britain in 55 BCE, and some of the Crusades are invoked as examples of assault upon the land from the sea.[1]

Looking back to those earlier ventures can help to clarify the enduring, historic features of this special form of warfare. It is not about raids upon an enemy's shore, such as Sir Francis Drake's attack on Cadiz and other Spanish ports in the 1580s. Those were strikes from the sea, but a permanent lodgement on the beachhead followed by an advance upon the rest of the mainland was not intended. Operations such as the assault upon Cadiz usually had a smaller, more specific purpose, such as throwing the enemy's intentions into disarray (Drake's assault was a preemptive disruption of the Armada) or hurting his offensive capacities (like the Zeebrugge Raid of April 1918, where the British planned to block egress by U-boats from the German-occupied port), or were simply persistent, small-scale attacks to stretch out and, it was hoped, wear down the defenders. Royal Marine commando units carried out many of that sort of raid throughout much of the Second World War, compelling Hitler to order the stationing of vast numbers of Wehrmacht troops along Europe's western shores, from northern Norway to France's border with Spain. In late December 1941, for example, a commando raid successfully destroyed the German power station, factories, and other installations at Vaagso, halfway up the Norwegian coast, and in February 1942 another famous raid attacked and seized vital radar equipment from the Bruneval station, near Le Havre.

But these were not invasions; at Bruneval, the commandos actually parachuted in, seized the machinery, and *left* by the sea.[2] Some of them had specific utility, such as the acquisition of the radar equipment, or the later midget submarine raids on enemy merchant ships in the Gironde (the "Cockleshell Heroes"). Sometimes, perhaps, the merits were psychological; they certainly were to Churchill, who almost immediately after the fall of France—and well before the Battle of Britain— ordered the Chiefs of Staff to propose "measures for a vigorous, enterprising and ceaseless offensive against the whole German-occupied coastline."[3] Finally, even the smallest raid, whether a success like Vaagso or a failure like Guernsey (July 1940), produced lessons: about training, command and control, land-sea communications, weapons used, vessels used, accuracy of prior intelligence collection, and so on.

It is the lessons of larger and more purposeful amphibious operations that claim attention here. The first was that specialized troops and specialized equipment were needed to carry out a successful invasion against a determined land-based enemy. Sometimes, perhaps, a hastily flung-together unit, if it possessed the element of surprise, could pull off an operational miracle, but when launched against a foe who had prepared its defenses well, such attacks were usually a recipe for disaster.

It is therefore not surprising that historians call our attention to two innovations by the army of Philip II, since that service was one of the driving forces behind the "military revolution" of the sixteenth and seventeenth centuries. The first was the creation by Madrid of specially trained troops assigned to their various armadas and experienced in moving from ship to land; the Royal Spanish Marines were born in 1560s operations to recover Malta, and other powers followed by establishing their own such units. The second was the establishment of specific weapons platforms and the implementation of suitable tactics for their success in battle. Thus, in the May 1583 Spanish operation to recover the Azores from an Anglo-French-Portuguese garrison, "special barges were arranged to unload horses and 700 artillery pieces on the beach; special row boats were equipped with small cannons to support the landing boats; special supplies were readied to be unloaded and support the 11,000 men landing force strength."[4] The attackers also practiced deception, a partial force landing on a distant beach and distracting the garrison while two waves of marines got onshore at the main point.

The third, equally important general lesson was that those who ordered an amphibious operation, whether it be the king of Spain in the 1580s or Churchill, Roosevelt, and the Combined Chiefs of Staff in 1942–43, had to eliminate interservice rivalry and create some form of integrated command. Rivalry among allies is one thing (Wellington often claimed that having enemies was nothing like as bad as having allies), but rivalry between the armed services of one's own nation is altogether more serious. In many cases, operational failure was due to a lack of appreciation of what the other service could or could not do, or even how the other service thought. The doggerel at the beginning of this chapter about the Earl of Chatham and Sir Richard Strachan was not chosen merely as an example of puckish Regency satire. The Walcheren invasion of 1809 was a disaster. The place was badly chosen, being a low-lying island ridden with malaria; there were no serious preparations (tools, barges, intelligence) for an advance from the island into the Netherlands; Chatham did little with his 44,000 troops, and Strachan and his ships stood offshore. There was no planning staff and no integrated command structure. It was a total mess, neither the first nor the last of its kind.

The final lesson was the oldest of all: that no matter how sophisticated and integrated the armed forces involved in a landing were, they were always going to be subjected to the constraints of distance, topography, accessibility, and the weather conditions of the moment. The internal combustion engine conquered much of time and space. Against the blunt force of a gale, it was greatly hindered and reduced in its power (as we saw from the physical difficulties that Churchill had in simply getting to Casablanca). Given that the tides changed daily—in the Atlantic, there were very large vertical rises and drops—and that a storm could come up swiftly, there was always great unease at the idea that forces would be landing upon an open shore, even a lee shore.

Wherever possible, then, invasion planners, thinking also of the follow-on troops and supplies, desired a safe, functioning harbor in which their ships could rest securely and through which reinforcements could flow. The problem, of course, was that any good harbor worth its name was going to be heavily defended by cannon, bastions, outerworks, innerworks, and possibly mines and hidden obstacles, while the invading troops and their transports would be offshore, churning away in collective seasickness and the ebb and flow of the tides before the bloody assault was made. The history of amphibious warfare is thus also replete with examples of attacks that were repulsed—in 1741 the British put 24,000 men, 2,000 guns, and 186 ships against Cartagena de Indias (Colombia), yet still were driven off by a much smaller Spanish garrison holding a massive fortress. Trying to seize an enemy harbor naturally provoked an enormous defensive reaction and most probably would be fatal; landing on beaches, whether nearby or farther away, exposed the troops to the watery elements and also forced them to bring their own communications systems (bridging equipment, repair units, spares) until they reached the enemy's roads. But deciding against any amphibious attack and staying with a land campaign (as the Allies did in Italy between 1943 and 1945, apart from Anzio) meant that one could not take advantage of the opportunities of maritime flexibility and would instead be forced to grind on. One of these operational options might be a winner, but it was impossible to say in advance which one it was.

In sum, assaults from the sea were a gambler's throw; perhaps only airborne attacks could be riskier. It was not just about ships dropping off soldiers and equipment and then sailing away; it was about integrated combined warfare in the face of hostile fire and often in extremely difficult physical circumstances. It called for an almost impossible construct: a smoothly functioning joint staff under a single commander, with everyone knowing his place and role due to systematic preinvasion training. It relied upon superb communications in the face of enemy efforts to disrupt them, and it required the right weaponry. After that, it might just be feasible.

With all these lessons of history available (and some earlier campaigns were studied at nineteenth-century staff colleges), one might have thought that pre-1914 armed services would have been better prepared than they were for flexible, carefully prepared strikes from the sea when the Great War finally came. This should have been particularly true of policy makers and senior strategists in London, reared as they were in the "British way in warfare."[5] But much less attention was given by those strategists to the lessons arising from the Crimean campaign (clumsy, but actually successful in forcing Russia to ask for terms) than to the rapier-like strikes of the Prussian army against Denmark, then Austria, then France, in the 1860s. If future European wars were to be decided so swiftly, in the first summer and autumn of campaigning on the main battlefields, what was the point of peripheral raids? It was a question that advocates of amphibious warfare found hard to answer. There was another reason so little amphibious warfare was practiced during the First World War: the larger strategic situation. This war was overwhelmingly a European *land* war and thus a generals' war. The mass armies of the Central Powers were contesting for terrain against the mass armies of France, Britain, and (later) the United States in the west, that of Russia in the east, and Italy's in the south. Since the Anglo-American armies were already deeply inside France by 1917–18, there was no need for a massive amphibious landing on French shores. Mines, torpedoes, and coastal artillery prevented any Allied thrusts into the Baltic; seaborne operations that did occur there were German-Russian strikes in a secondary theater. All significant nations of the Mediterranean were either Allied (France, Italy, and their colonies, plus Egypt) or neutral (Spain, Greece), which only left Turkey and the Levant as possible target areas. Britain's Japanese ally controlled the Far East and easily gobbled up the exposed German colonies there.

Thus, for all the pre-1914 talk by Admiral Jacky Fisher and others about the army being a "projectile" fired onshore by the navy, it wasn't clear where that missile could be fired, even if the British generals agreed to be so dispatched (which, once settled in France, they didn't). Taking over Germany's colonies in Africa and the Southwest Pacific was relatively uncontested, except for a disastrous amphibious operation in November 1914 by British-Indian forces against the Tanganyikan port of Tanga, which should have been a salutary lesson in how poor training, communications, equipment, and leadership can turn an imaginative strike into a fiasco.[6] But lessons are salutary only if they are learned.

Alas, the lessons of Tanga were not, as was most readily demonstrated in the greatest example of a failed amphibious invasion of the twentieth century: the 1915–16 Gallipoli campaign, as notable a conflict as the Athenian assault upon Sicily, and just as disastrous. Even today, Gallipoli receives much attention, not just on account of its historical resonances (as witnessed at every ANZAC Day commemoration in Australia and New Zealand, or in the Turks' celebration of Mustapha Kemal, later known as Ataturk) but also because of our fascination at the spectacular gap between its grand strategic purpose and its disastrous execution. Perhaps no operation other than this one better illustrates the feedback loop—in this case, a wholly unfavorable one—between what happens on the ground and at sea, and how the general course of the war can be affected by tactical mishap.[7] By the single stroke of pushing a force through the Dardanelles, its principal advocate (Churchill) maintained, a tottering Russia would have its sea-lanes to the West restored and thus be kept in the war; on the other side, the supposedly fragile Turkish power (it had joined Germany in November 1914) might be pushed into collapse, and the Balkan states of Greece, Bulgaria, and Romania might be tempted out of their neutrality.

While the strategic reasoning was attractive, the operation itself was a catastrophe. It began with a purely naval attempt in March 1915 to rush the Straits; by the time the Allied fleet escaped from the Turkish-laid minefield, it had lost four capital ships (three British and one French), with a further three badly damaged—an outcome worse than the Royal Navy's losses at Jutland a year later. After that, infantry units were assembled from various sources—French regiments in the Mediterranean, British units from Egypt, India, and the home country, brand-new Australian and New Zealand divisions en route to the Western Front. In late April 1915, having given the Turks plenty of time to bring up reinforcements, they began to land on the craggy, ravined, thorn-covered hills of the Dardanelles Peninsula. Try as they might, the Allied forces could never get control of the higher ground and suffered appalling losses. Each side threw in more and more divisions, but the situation did not change. In December and January, in swift nighttime moves that surprised the Turks, the Allies pulled away from the beaches, admitting defeat, and sailed for home. They had lost 44,000 men and had another 97,000 wounded (more than all U.S. losses in the Korean War). Turkey's casualties were even higher, but they had won.[8]

The Western nations had proved to be much better at getting off a Dardanelles beach than landing on one, let alone moving on from their early lodgement to their chief inland destination. In retrospect, the reasons for this defeat became clear. The weather in the Straits was always extremely fickle, ranging from the intense heat of the summer months (without adequate water supplies, an army withers like a bush, and the sickness rate soars) to the intense storms and blizzards that poured out of the Bosphorus as winter advanced. The topography is intimidating, with steep slopes, sudden crevasses, and thornbushes everywhere. The landing areas, especially where the Australian and New Zealand units came ashore, were inhospitable and virtually impossible to move out from. Allied intelligence about what to expect was weak, the forces had not been trained for this kind of operation, and fire support from the offshore vessels was incomplete, in part because it was hard to see where the Turks were, in part because the bombarding squadrons were steadily forced away by enemy mines and submarines (three further capital ships were sunk within the next month). The landing craft that brought the men to the shore were, apart from a few prototypes, not landing craft at all. Finally, both the weaponry and the tactics of the raw units ordered to advance up this craggy terrain were simply inadequate for the job. Supervising this unfolding fiasco was a command structure that brought back memories of Sir Richard Strachan and the Earl of Chatham—except that this time the casualties and the immensity of the failure were far, far greater. In consequence, the line to Russia could not be opened, Turkey stayed in the war and fought to the end, Bul-

garia joined the Central Powers, and the other Balkan states stayed neutral. Slightly over a year later, imperial Russia began its collapse.

After Gallipoli, British interest in amphibious operations waned,
not surprisingly. More and more resources were needed for the colossal
struggles along the Western Front, and in consequence exotic and difficult landings from the sea were now frowned upon. At French urging,
an Allied army did establish a beachhead in Salonika later in 1915, but
it never really got very far from the shore for the next three years—the
battalions there were aptly named the "Gardeners of Salonika." By the
next spring the French were fighting for survival in Champagne and
Flanders, and therefore opposed all eastern adventures. If the British
were much more tempted to campaign for the territories of the Otto-
man Empire after 1915–16, it was by large-scale *land* assaults, eastward
from Egypt, northward from Basra. The army leadership simply wasn't
interested in its divisions being dropped off on hostile shores; the navy
was concentrating upon bottling up the High Seas Fleet in the North
Sea and trying to avoid losing the Atlantic convoys' battle against the
U-boats. The Zeebrugge Raid of 1918, however well executed, was just
a raid, nothing more. Nor did the American entry into the war change
attitudes; millions of doughboys sailed safely into Le Havre and were
marched overland to the front. During 1917–18 the U.S. Marine
Corps was located far inland, fighting along the Aisne and the Meuse
rivers.

In sum, the First World War discredited the notion of amphibious
warfare. And when the dust of war had settled and the new global strategic landscape revealed its contours—roughly by 1923—there were
obvious reasons this type of operation had few followers. To be sure, in
a badly defeated and much-reduced Germany, in a badly damaged and
scarcely victorious France and Italy, and in an infant Soviet Union,
there were many thoughts of war, but none of them involved the pro-
jection of force across the oceans. Japan was in a liberal phase, and the
military had not yet exerted its muscle—even when it moved to take
Manchuria in 1931, that was a land operation that had nothing to do
with attacking beaches or seizing ports. By the late 1930s things would
be different, with large Japanese merchant ships carrying landing craft
and vehicles during their attack upon the lower Yangtze. During this

post-1919 period, then, only two of the seven great powers gave any thought to amphibious warfare.

One of those two powers was Britain, although economic stringency and the embarrassment of Gallipoli (refought in many a wartime memoir) pushed combined operations into a dark and dusty corner; the result was the occasional small-scale training exercise, a theoretical training manual, and three prototype motor landing craft. Only the 1937 Japanese invasion of mainland China and then the 1938 crisis over Czechoslovakia would force a resumption of planning and organization. On paper, things began to improve. The Inter-Service Training and Development Centre (ISTCD) was set up, specialized landing craft and their larger carrier ships were designed, and the manual for amphibious assaults was updated. But this was all *theory*. The midlevel officers worked well together and had fine, advanced ideas, but they still lacked the tools. A large-scale exercise off Slapton Sands, Devon, in July 1938 was badly affected by near-gale conditions and ended in chaos. This galvanized the ISTCD into further serious planning, and it is to their credit that they anticipated virtually all of the practical difficulties that amphibious operations would throw up during the Second World War itself. Yet at the outbreak of that conflict, remarkably, this truly interservice unit was disbanded. The army was off to France, the air force was bombing Germany, and the navy was awaiting high-seas battle with the Kriegsmarine—so where on earth would one carry out combined operations? And who was interested? All but one of the ISTCD officers returned to their fighting units in September 1939.[9]

The other country interested in amphibious warfare was the United States, because of its lengthy shores, multiple harbors, and flat beaches; because of its cherished memories of the War of 1812; and because it had possessed, since the founding of the Republic, its own Marine Corps with special campaign memories ("From the halls of Montezuma to the shores of Tripoli"). But the story of the U.S. Marine Corps before and during the 1941–45 war in the Pacific is more properly kept for chapter 5 of this book, where both the similarities and differences with amphibious warfare in Europe will be apparent. This chapter will focus on the European theater alone.

Amphibious Operations from 1940 to 1942

The chief feature of amphibious warfare during the first three years of the Second World War was *not* the West's implementation of their interwar plans but the extraordinary success and shock of the Axis victories across so much territory, very often coming by sea.[10] Thus, for example, by all physical and technical measures, the Wehrmacht's successful conquest of Norway in a couple of months of 1940 against a potentially vastly superior Anglo-French opposition was (and remains) one of the greatest surprise strikes in military history. Just glancing at the map—with the harbors and airfields of the entire eastern Scottish shoreline from Rosyth to Scapa Flow having been pre-positioned for decades to block a German drive to the north—makes one gasp at the Allied failure to respond to Hitler's strike at this strategically critical possession. It is true that in the naval battles offshore, and in the Narvik fjord, Admiral Raeder's fleet received a battering from which, perhaps, it never fully recovered but the Royal Navy simply could not deal with Luftwaffe attacks farther south, and the Anglo-French military units put ashore usually found themselves against better-positioned and better-trained Alpine troops.

Above all, then, soldiers from the sea found themselves for the first time up against hostile airpower. This was not understood by the sometimes overly historically minded Churchill, whose daily urgings for action in Norway were accompanied by a dire lack of appreciation of the confusion on the ground and the intimidating effect of German medium bombers and dive-bombers.[11] Perhaps the only good thing that came out of the Norwegian fiasco was the parliamentary vote against Chamberlain's government and the prime minister's replacement by Churchill himself in the crowded days of May 1940. Yet even if the new, hyperactive, and imaginative war leader of the British Empire was only a few months later to watch how the Germans' failure to gain aerial control over the Channel rendered completely impossible *their* amphibious operation, Sealion, he had not yet come to understand that the same restriction applied to the Royal Navy itself. Without secure air cover, even the most powerful warships could not operate securely off an enemy-held shore. What the Turkish minefields had done off Gallipoli, Nazi dive-bombers could now do along all of western Europe's coastal waters.

This lesson was rubbed in during 1941 and 1942, the years when the Empire suffered defeat after defeat. The large-scale British attempts to assist in the defense of Greece and then to hold Crete ended in calamity. Perhaps no army in the world could have stopped the horde of mechanized and armored divisions that an enraged Hitler had ordered into the Balkans in April and May 1941, bursting their way through Yugoslavia to southern Greece in an amazingly fast time. But the British, having completely intimidated the Italian fleet earlier, did have command of the sea, and Crete was an island. Without airpower, however, that mattered little. The German landing of 3,000 paratroops around Maleme on May 20 unhinged the defending battalions, and the Luftwaffe heavily punished the Royal Navy's efforts to either reinforce the garrison or, just a couple of weeks later, pull off the exhausted troops. This time there were no Hurricanes and Spitfires present, as there had been over the skies of Dunkirk, and thus none of Churchill's exhortations to stand firm could prevent a 250-pound bomb going through the deck of the destroyer HMS *Kashmir*, which sank in two minutes (five other destroyers and three cruisers suffered the same fate). The Royal Navy had no point defense and forward-picket warships, as the U.S. Navy was to deploy off Okinawa to blunt kamikaze attacks in the summer of 1945. A mere six months later (December 9, 1941), bombers of the elite 22nd Air Flotilla of the Japanese Naval Air Arm ripped apart the new British battleship HMS *Prince of Wales* and the battle cruiser HMS *Repulse* off the coast of Malaya in under three hours. The newer warship had been designed to handle Billy Mitchell–type claims that planes could sink battleships, and possessed 175 antiaircraft guns that could pump out 60,000 shells a minute.[12] But even that was ineffective against well-executed aerial attacks.

What worked was aerial control over the shoreline and over the approaching waters. Airpower in the Second World War created winners and losers; either they had it or they didn't. For the Norway operation the Luftwaffe deployed 800 combat aircraft and 250 transport planes, while the British had far fewer of both and the French none. For the invasion of Crete, the German air force had 500 transport planes (and 100 gliders), 280 bombers, 150 dive-bombers, 180 fighters, and 40 reconnaissance aircraft. All these planes created a logistical nightmare on the cluttered southern airfields of Greece. No doubt the British would have welcomed such a problem.[13] The Japanese deployed 34 high-level bombers and 51 torpedo bombers in one single strike against the *Prince of Wales* and *Repulse;* there was no RAF land-based cover, and the Royal Navy aircraft carrier intended for the squadron, HMS *Indomitable,* had been damaged by grounding on trials. At exactly the same time as the sinking of the British capital ships, Japanese planes from Formosa knocked out almost all the modern U.S. aircraft at Luzon, thus allowing a virtually uninterrupted amphibious assault on the Philippines and farther afield.

After early 1942, therefore, one thing was clear to both American and British planners involved in the West's counterattack: no assault from the sea could take place without absolute control of the air, and not just above the landing beaches themselves. Air supremacy had also to be established over the maritime approaches to the invasion theater against enemy aerial, surface, and submarine forces, and against any air-based and land-based counterattacks that might be launched upon the beachhead.

Even if they had command of the air, and even if they were not encountering German veteran units, they also still had to figure out how to land on the shore. The first Allied amphibious attack against an enemy stronghold was the ill-named Operation Menace of September 1940, an action taken by the Royal Navy and the Royal Marines to assist General Charles de Gaulle's forces in capturing the Vichy French colony of Senegal, with its key harbor and naval base of Dakar. The results were embarrassingly bad in almost all aspects, and after a few days the Anglo–Free French forces had been forced to abandon the operation without even getting ashore. Command and control was terrible. The British commander General Noel Irwin had his HQ in the cruiser HMS *Devonshire,* which at one point had to race northward to drive off some Vichy French destroyers; he was then transferred, with staff, to a transport, and then to the battleship HMS *Barham,* which became engaged in a close-in slugging match with the guns of the Dakar fortress and those of the formidable new battleship *Richelieu.* The offshore vessels came out of it the worse for wear, and the single Vichy French submarine in the area badly damaged the second British battleship, HMS *Resolution*. The limited air strikes from the carrier HMS *Ark Royal* made no impression at all. Perhaps it was a good job the troops did not get to fight on land, as in Evelyn Waugh's bitingly satirical novel of this operation, *Men at Arms*. It must have been difficult for the extremely frustrated Irwin to control his language as he composed his final report on the way back to Gilbraltar, but he did manage to get in two compelling points: (1) without a special combined operations HQ ship, this type of operation would never work, and (2) ships going into a shooting match against forts was a losing game—a truth that Churchill continued to contest.[14] Nelson, the boldest of naval commanders, often stated that putting warships against well-defended ports was not a wise move. This was still true a century and a half later.

The British had another try in April and May 1942 with Operation Ironclad, against the important Vichy French island possession of Madagascar, or at least against its strategic northern port of Diego Suarez, to forestall any Japanese assault. This time things went better. The French resistance in and around the main target was slight and was willing to surrender once some army and marine units got around them (the garrisons in the south held on in the jungles for a while longer). The land forces included the borrowing and deployment of an entire British Army division en route to the India/Burma theater. French air defenses were minimal, while cover from the fighter squadrons of the two British fleet carriers was ample. And there was at last a headquarters ship, converted from a liner, which was separate from the bombardment group (which only needed to fire for ten minutes), the assault forces, and the distant coverage of the Eastern Fleet.[15] Consequently, this operation was much easier than the Dakar embarrassment, and at Combined Headquarters Command there was much rejoicing at having gotten something right at last.

There was one further positive aspect to Operation Ironclad that deserves more attention than it generally gets: the landing force had sailed from the river Clyde (Glasgow and its lower ports), which meant that British sea power was projecting military units a staggering 9,000 miles to the northern tip of Madagascar. Now, it is true that by this stage of the war the rapidly expanding British Army was sending regular troop convoys with heavy warship protection from the Clyde around the Cape to the Middle East and India, which made the overall logistics more familiar to Admiralty planners; all that was needed was to employ one of those divisions in transit (in this case, the 5th Infantry Division) to land in considerable support of the commandos, and to add a far larger than usual number of protective warships from Force H and the Home Fleet. Even so, the planning had involved orchestrating whole groups of troopships and escorts traveling at different speeds and from different ports in order to launch an ultra-long-range strike from the sea, a considerable logistical feat. (Just planning for meals at sea for 40,000 men was a massive exercise in itself.)

Combined Operations Command (COC) had been set up at Churchill's order in June 1940 to carry out raids up and down the coasts of occupied Europe. Ironically, perhaps the fall of France created an opportunity for the revival of combined operations! With the prime minister's encouragement, the range of targets became more ambitious and the sober-minded Chiefs of Staff had to keep reining them in. Dakar was COC's first trial by fire, with the results described above, and some critics felt justified in their doubts. Yet, prodded on by an impatient Churchill, increasingly aware that their new American allies would insist upon large amphibious operations in Europe as soon as was possible, and impressed by the arrival in October 1941 of the youthful, vigorous, and massively overpromoted Vice Admiral Lord Louis Mountbatten to head the COC and become a colleague on their committee, the British Chiefs of Staff had to swallow the fact that increased resources had to go to amphibious operations.

And so they did, though such a move placed enormous demands upon an already greatly overstrained British shipbuilding industry. Fast passenger ships were converted into very large landing ships for infantry (LS-I), a couple of Dutch ferries were adapted as troop carriers, and hundreds of smaller landing craft were laid down, many in the United States. The first, critically important landing craft for tanks went to trials. More and more personnel—naval as well as military—were recruited. Combined Operations was no longer a dead end; could one name another role that was going to be as important as the "beach master" in a landing on a hostile shore, directing the incoming troops, trucks, and tanks, moving them and their supplies farther inland, and swiftly eliminating the bottlenecks? In the next year the number of

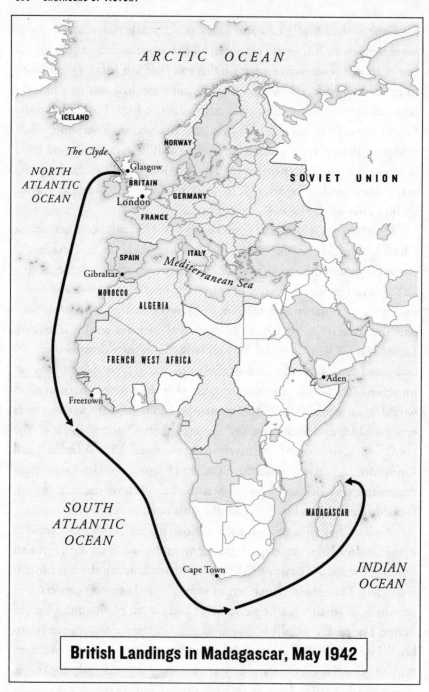

British Landings in Madagascar, May 1942

BRITISH LANDINGS IN MADAGASCAR, MAY 1942
Although Operation Ironclad was relatively unknown, it demonstrated the growing Allied capacity for very long-range amphibious operations. It was the first major British counterattack on the Axis and its satellites.

training camps and practice beaches increased, almost all along the chilly waters of Scotland and far from Luftwaffe reconnaissance (and if one wanted to practice landings in choppy seas, that was the place). The first properly designed tri-service headquarters ships were being created and would soon be available. Mountbatten's drive was infectious, but it really was important to have this Madagascar success.[*16]

It was therefore all the more disappointing that these signs of promise were soon to be followed by another badly botched amphibious operation, at Dieppe. Fortunately, it was followed only three months later by the largest—and most successful—amphibious operation the Allies had attempted so far, in North Africa.

Dieppe and North Africa: Warning and Encouragement

Why pay much attention in our story to a small-scale raid on a French port that lasted only one day (August 18–19, 1942) before the attackers were kicked back into the sea? By all measures of size and intention, the Dieppe Raid was nothing like as significant as Walcheren, Gallipoli, or Crete in the history of amphibious warfare; even if it had proven to be successful, it was not intended to go anywhere. Yet while it was certainly not successful, it pointed to many a lesson that British Empire planners still had to learn. It was a badly arranged operation that led to a disproportionate number of casualties, chiefly Canadian. In that setback, ironically, lay its significance: it had the same perverse utility as had the losses in Convoys HX 229 and SC 122 in March 1943 and the appalling USAAF bomber losses during the October 1943 Schweinfurt-Regensburg raids. All three heavy blows dealt by the Germans at the tactical-operational level—convoy escort, strategic air, and amphibious landing—compelled the Western Allies to rethink seriously their previous assumptions and to search energetically for new weapons, tactics, training techniques, and organization.

But Dieppe was somewhat special as compared to the convoys and the bomber raids, which were inherently strategic to begin with. It was always thought of as a *test,* a trial run against the Atlantic Wall, which

*The operation commenced on the same day as the Battle of the Coral Sea: an Allied comeback in two oceans.

the Wehrmacht had been constructing ever since the decision to abandon Operation Sealion in late 1940. This operation was intended to produce lessons that might help preparations for later, greater actions. Many historians have used this utility argument—"the lessons of Dieppe"—as its ultimate justification, while Canadians have been outraged at the idea that the first deployment of their troops in the European theater was as a form of guinea pig in an experiment, and so badly bungled.[17]

The planning and execution of the raid was urged on by Mountbatten's Combined Operations. They would seize and hold a major enemy-defended port for a short time, gather information, and gain a chance to measure the German reaction. All this made sense militarily, and it also reflected political realities: both Stalin and Roosevelt were pressing for an early opening of a second front in France, the British public was restive at the setbacks in North Africa and the Far East, and the Canadian public was wondering whether their troops would ever be launched into battle. The original operation, code-named Rutter, was actually planned for early July, although, ominously, it was disrupted by a nighttime Luftwaffe attack upon the vessels assembling in the harbor. Renamed Operation Jubilee, it was dispatched across the Channel five weeks later, as the largest combined operation in the region hitherto. Some 6,100 troops were involved, the bulk of them in two Canadian infantry brigades. Most of the obvious features for an amphibious assault were incorporated. There would be aerial protection, a naval bombardment, and two flanking attacks as well as the main assault. The landing craft would also be carrying tanks. The convoy of 250 vessels would be preceded by minesweepers and escorted by destroyers. Specialist forces such as the commandos (plus fifty U.S. Rangers, the first American land troops to fight in Europe, and the first to die) would join in, and the central forces would hit the beaches just before dawn. On paper it looked good. Fair stood the wind for France.

In the middle of the night, the northern flank group bumped into a small German convoy going along the coast and got into a fight with enemy torpedo boats, coming off worse; the German land defenses were thus alerted. Only a small contingent from No. 3 Commando managed to get ashore, climb the cliffs, and snipe at the shore batteries. In the south, No. 2 Commando had a much easier task and carried out its demolitions. In the center, once landed, the main force simply could not get off the beaches, the steep shingle being as much an obstacle as the iron and concrete barriers; the tanks ground to a halt in the pebbles.

The destroyers offshore made little impression upon the German defenses and could not communicate with the troops onshore. The large number of Spitfire squadrons were operating at their extreme range—the opposite of the Battle of Britain conditions—and although part of the plan was to inflict a lot of damage on the Luftwaffe, in fact the opposite happened: the RAF lost 119 aircraft, the German air force only 46. German bombs and shells sank a destroyer and thirty-three landing craft. The Canadians suffered terribly as they tried to scramble back to the sea. Of the 6,100 men dispatched, more than 1,000 were killed and 2,300 captured; many of the survivors came back seriously wounded. About 1,000 troops never had the chance to get ashore. By midmorning it was over.

The literature about the Dieppe Raid has split into two separate categories. The first consists of multiple expressions of Canadian outrage (and not just in books but in movies, songs, and poems) against British military incompetence—not unlike the ANZAC criticisms about Gallipoli, even if the casualties at Dieppe were far fewer. The second argues for the benefits of the tactical-operational analyses of the existing weaknesses in the raiding plan and thus the longer-term benefits of trying it out. Churchill certainly believed it had been worthwhile when he explained this operation in his history of the Second World War. And that unrepentant buccaneer Bernard Fergusson, of Black Watch and the Chindits, whose last amphibious operation was to be the Suez debacle in 1956, concludes his account with these words: "Out of the fire and smoke and carnage on the beaches of Dieppe emerged principles whereby many lives were to be saved, and victory to be won."[18] A critic might observe that even without Dieppe much was going to be learned from other amphibious operations that would help the D-Day planners—after all, the North African landings were only three months away.

Still, many lessons *were* learned from the debacle at Dieppe. Intelligence preparations were inadequate, and it seems that Mountbatten and his staff had pushed ahead with their intentions without the Joint Intelligence Committee knowing about them. How could the planners not know that a German coastal convoy would be in the same waters that night, or that German forces along the coasts had been put on high alert, with additional machine gun battalions recently brought in to Dieppe itself? The strength of the German defenses was not properly measured, nor the nature of the terrain appreciated—how exactly would one get a heavy Churchill tank up a steep pebble beach, and even if the vehicles managed to crest such a rise, how would one get them past solid antitank walls without special equipment? Ship-to-shore movements were clumsy, and few if any of the Canadian commanders had amphibious warfare experience. Landings were late, sometimes in the wrong place. There was no control from offshore, because there was no command ship. Daytime aerial support was inadequate because the RAF did not have full command of the air. There was no preceding heavy bombing by Bomber Command, nor provision made for tactical air strikes. The strength of the naval bombardment was completely inadequate; if 15-inch battleship guns could not make much impact at Gallipoli, why should 4-inch destroyer guns do so off Dieppe? Above all, there was the chief blunder: making the main attack against a heavily held harbor rather than on some less protected part of the coast. If the Allied planners wanted to test the possibilities of seizing a defended port, at Dieppe they got their answer.

But there was something else about the value of the Dieppe experiment that was larger and rather more nuanced, more of a psychological lesson. The second front, whenever it came, was definitely going to take place along the Atlantic-swept waters of France *and* against well-trained German troops. That was a combination of challenges that really had to be tested. If the results of the raid confirmed all of Alanbrooke's worries—and supplied him and Churchill with the ammunition to argue for postponing Operation Overlord through 1943 and into 1944—it also presented the Anglo-American planners with a new and much higher benchmark. When, eventually, they did come ashore in France to pursue a full invasion, they were going to have to be very, very good.

That was so, of course, for one further worrying reason. Although well emplaced, the German garrison in and around Dieppe was actually not very large. The 571st German Infantry Division had around 1,500 men in the area, but only 150 of them were there to pour fire onto the main beaches; yet that, with the defensive works, turned out to be enough (again, one thinks of Cartagena de Indias in 1741). The Atlantic Wall would never again be so minimally held, and the next two years would see more and more German divisions moved to that front and fantastically more obstacles, pillboxes, and minefields put in place. In sum, each side could learn much from what had really been a small-scale operation at Dieppe.

That term certainly could not describe the Operation Torch landings in North Africa in November 1942, although the chronological proximity of the two operations is useful in allowing us to pinpoint the differences between them. Operation Torch was an amphibious invasion, not a raid, and it was very large. Although initial Allied planning had called for even more extensive operations, the troops landed with the purpose of capturing Casablanca, Oran, and Algiers totaled close to 75,000 in the first stage, with many more to come. It was also an Anglo-American venture, and the first of its kind in the war. It was not only combined, then (as between three armed services), but joint (as between the two Western Allies). Here was another reason it had to work: could two rather different military cultures, in the heat of battle, avoid the almost inevitable frictions of alliance politics and different organizational and training systems, different weaponry, and different control systems?

That it did so was because Churchill, Roosevelt, and the Combined Chiefs of Staff had been hammering out the principles of Allied jointness since at least their August 1941 meeting off Argentia Bay, Newfoundland, and the prime minister's rushed visit to Washington that same Christmas. Some generals and admirals accepted the sinking of differences and sharing of roles better than others—Eisenhower and Tedder were quite superb in this, both having heroic amounts of patience—but mutual necessity forced the pace. Both partners could see that in places where they would fight a campaign together (Southeast Asia, the Mediterranean, northwest Europe), they should accept the principle of the theater commander in chief coming from one combatant, the deputy commander from another. Therefore, given that the overwhelming bulk of the Operation Torch invasion troops would be American (partly because the British Army was just completing the El Alamein battle at the other end of North Africa, partly to avoid antagonizing the Vichy French forces bitter at earlier British attacks), the overall command went to Eisenhower, with Admiral Andrew Cunningham becoming deputy commander as well as being named the overall Allied naval commander. The naval and air force jointness was something of a formality here, though, since the American units operated along the Atlantic landing areas, and the British inside the Mediterranean. Headquarters for Eisenhower was, appropriately enough, Gibraltar, which had fulfilled that key strategic role for well over two hundred years.[19]

Another main difference with Dieppe was the extreme distances that the Allied invasion forces had to cover. The U.S. forces for the landings in Morocco sailed from a variety of East Coast ports directly across the central Atlantic in time for the November 8 coordinated landings. The British and American troops headed for Oran and Algiers sailed a few days later from the Clyde.* The heavy covering squadron Force H, whose task it was to neutralize any attacks by either the Vichy French or the Italian fleets, had left earlier from Scapa Flow, so as to pose a massive presence in the western Mediterranean. Fast squadrons, slow squadrons, refueling squadrons, merchant ships, landing ships, tugs, colliers, advance submarine patrols—all had to be in the right place at the right time. No wonder Eisenhower was nervous.

There was much, much more to be accomplished in the logistics of a large amphibious operation than simply sending five hundred different ships to sea, but the most important thing to note is that these modern invading armies not only required naval and merchant navy support for their original landing but also needed a constant and expanding supply of seaborne provender as the land campaigns unfolded. Given the crisis in Allied shipping by the beginning of 1943, the farther the Anglo-American forces in North Africa advanced—and the more that the stubborn German defense under Rommel and von Arnim held out—the greater became the need to divert merchantmen to the Medi-

---

* Both as a point of receipt for oceanborne supplies and millions of U.S. troops and as the takeoff point for so many maritime routes from Britain to the wider world, the Clyde (Glasgow) and the Mersey (Liverpool) played an extraordinarily significant strategic role throughout the Second World War. They were a long way from German bomber bases in France, they had very large port infrastructures (berths, docks, dry docks, warehouses, shipyards, connections to a dense railway system), and they had easy access to the Atlantic and beyond.

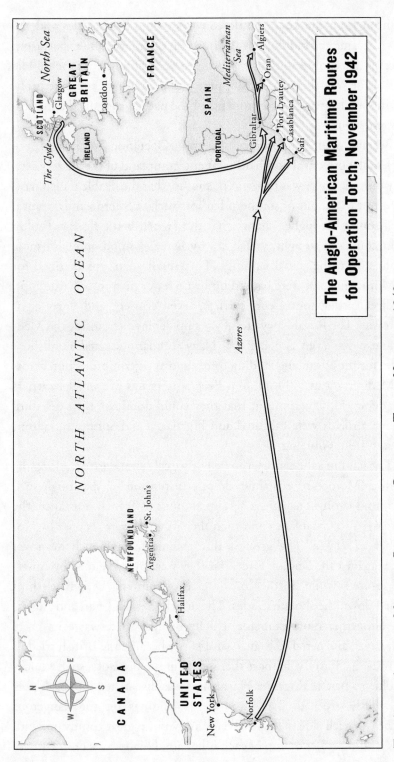

The Anglo-American Maritime Routes for Operation Torch, November 1942

THE ANGLO-AMERICAN MARITIME ROUTES FOR OPERATION TORCH, NOVEMBER 1942
Once again, as on p. 230, the extreme distances that could be covered by the Allied command of the sea are vividly shown here.

terranean from the key Atlantic convoys to the British Isles and the slower became the buildup for the invasion of France, quite apart from the critical task of feeding the British people.[20] Throughout the middle years of the war, shortages of shipping and of landing craft were probably the single greatest determinants of the pace of the Allied amphibious campaigns.

Because of the importance of getting Operation Torch right, the Allies took both naval bombardment and command of the air very seriously indeed. There was the faint possibility that the sizable Italian navy might put to sea out of southern harbors such as Salerno and Taranto, and a somewhat higher chance that the French battle fleet at Toulon might come out to fight, which is why Force H stood on guard some distance from the actual landings. The American troops destined for the Moroccan coast were backed up by no fewer than three battleships, one fleet carrier, four escort carriers, seven cruisers, and thirty-eight destroyers. The Royal Navy's total warship deployment inside the Mediterranean was even larger. Royal Navy submarines acted as offshore guides for the advancing landing forces and as distant protection across the Mediterranean. All told, four fleet carriers and five smaller carriers were deployed, which meant that they could dominate the skies until the first airfields were captured and Hurricane and Spitfire squadrons flown in from Gibraltar.

The landings themselves can be fairly easily summarized. Along the Atlantic (Moroccan) coastline, despite disruption by the rough surf, beachhead confusions, and a certain amount of French resistance, the approaches to Casablanca and then the city itself were secured by November 12, 1942. The greatest threat came from French destroyer strikes against the landing force, but they were overwhelmed by American cruisers while, further offshore, fire from the U.S. battleships pinned down the shore batteries. The attacks against Oran and Algiers were altogether more dramatic, chiefly because there was even more land, naval, and aerial resistance, and partly because the British tried to run warships into the harbors themselves in the hope of a swift knockout blow; a pair of Royal Navy warships was dispatched into each port and swiftly crushed. The amphibious landings on the dangerous beaches on each side of Algiers were thrown into great confusion, with units landing either on top of each other amid high waves or in entirely the wrong place—errors that, Barnett observes, "would have been ruthlessly punished had this been a coast defended by German troops."[21] Then French destroyers came out to fight but were pummeled by the fast, 6-inch-gunned British cruisers. Coastal batteries shot it out with HMS *Rodney*, and Dewoitine fighters tackled the Spitfires landing on the first captured airfields. In the background to this resistance the armistice negotiations came to a successful conclusion, and within three days Operation Torch had achieved its objective—follow-on reinforcements could now flow into Casablanca, Algiers, and Oran.

Some of the lessons learned from all this were at the critical lower level of combat, such as that the distant lobbing of battleships' shells was pretty hopeless but close-in volleys against coastal batteries could be devastating; equally clearly, grabbing an airfield onshore as soon as possible produced a massive gain. The absolute necessity of a command ship was again confirmed. The Oran operation was smoothly supervised from the HQ ship *Largs* by a Royal Navy commodore, Thomas Hope Troubridge; the head of the U.S. Army forces, Major General Lloyd Fredendall; and the combined air forces commander, the irrepressible Jimmy Doolittle of the USAAF. Off Algiers, in the state-of-the-art HQ vessel *Bulolo*, Rear Admiral Harold M. Burrough of the Royal Navy, Major General Charles W. Ryder of the U.S. Army, and Air Commodore Vyvyan Evelegh of the RAF supervised their respective forces. Here, in two medium-sized vessels, was successful interservice and inter-Allied cooperation at the next level down from that of the Combined Chiefs of Staff. This was not the case off Casablanca, where no such independent HQ vessel existed for the American forces, which at one stage led to General Patton and his military staff being carried away from the landing area when the cruiser USS *Aurora* (in which they were located) turned to neutralize French destroyer attacks.

There were two larger lessons. The first was further strong confirmation that assaults from the sea upon or into enemy-held ports, in the age of quick-firing guns, mines, and torpedoes, were just a folly. On the other hand, the beach landings had all shown how difficult it was to place a large force of men on an open stretch of coastline against powerful tides, strong winds, poor visibility, and natural obstacles such as sandbanks. Unbreachable ports, or unlandable shingle and reefs? Yet if the waters and beaches of Oran were difficult enough, what would it be like attempting a much greater operation onto the shores of northwest France, where the Atlantic rollers had a clear 3,000 miles of ocean to build up their power and funnel it through the Channel? The second major lesson—a continually nagging one for Allied planners—was that this had not been a defeat of German troops. Some Vichy units had battled valiantly, but the majority were clearly relieved when Admiral François Darlan decided upon a cease-fire, then a surrender. Hitler's answer to Darlan's defection was to order the occupation of Vichy France and then to pour German (and some Italian) divisions into Tunisia, where from November until May 1943 Rommel and his successor generals held off vastly greater Allied armies pressing them from west and east in a most impressive display of both defensive and counteroffensive warfare (see chapter 3). Despite this tough German resistance, Operation Torch had been relatively easy for the Allies, and a great relief.

After Torch

The small Dieppe Raid and the large North African invasions each had the same array of practical and overlapping issues: the length of the landing zones and how far apart from one another should they be; the timing and length of the naval bombardment; the management of the actual touchdown on the beaches; preparations for topographic and enemy-made obstacles. The questions were endless. When did naval control cease and army commanders take over? How did one manage stage two, when one might be pushing further forces ashore while pulling casualties and empty fuel trucks back to the ships? How did one neutralize aerial counterattacks? Yet every question was serious and had to be addressed.

The experiences from these two operations were also the only recent ones Allied military strategists and other officials could draw upon from the European/Mediterranean theater when Churchill and Roosevelt met at Casablanca and demanded amphibious results in the year ahead. A thoughtful planner might have put it to the two great men, and to the Combined Chiefs of Staff, this way: "Sirs, we have some evidence of the problems ahead simply from a small-scale raid on the coast of France that was bashed apart by the German defenses. And we have additional evidence from our much larger and successful invasion of North Africa of the challenges of putting major armies onshore, despite distance, the tides, the storms, and the many logistical and communications problems. We do not know what will happen when we combine those two things: (1) we land a lot of troops on a hostile shore, whatever the weather, and (2) that shore is occupied, and easily reinforced, by a large number of extraordinarily tough German divisions. Not Italians, not Vichy French, but Germans who are exceptionally able defensive fighters. We have no way of calculating what will happen then."

There is no record that anything like that was said aloud. If it had been, it would not have stopped the political leadership from issuing the Casablanca directives. It certainly would not have halted Roosevelt, so enigmatic and aloof about all specific policies and methods, yet so confident in the capacity of the United States to win any foreseeable conflict; to him, one guesses, all the problems were about time and place, about waiting for massive resources to come out of America's factories, but never about the final victory. And although Churchill occupied a more beleaguered strategic position, scared about weaknesses that Roosevelt didn't seem to appreciate (the U-boats) and worried about his beloved empire, the Briton could also be confident at Casablanca in calling for future landings in the defeat of the Axis. This was the moment he had waited for and had a multitude of ideas about: the turning of the tide.

The unfolding of the Allied offensive during the next sixteen months tended to confirm that imaginary strategist's sober predictions. After Dieppe, no significant military activity occurred along the shores of France or in the rest of western Europe. What would be the point? Until one had won command of the sea and then command of the air, and until the U.S. Army had enough trained divisions and sufficient landing craft, the Western Allies simply could not open the second front, much though Stalin and Roosevelt pressed for it. Thus, unless the Americans in sheer frustration shifted the bulk of their manpower and weapons to the Pacific (Admiral King's preference), the pressure upon the Axis had to be placed in the Mediterranean, as Churchill and the British Chiefs had argued all along. And by this time they had very considerable land, naval, and aerial forces in North Africa that surely should not lie idle for a year. It is important to note that Eisenhower also argued for further operations, perhaps against Sardinia, perhaps Italy. There is no place here to argue about the wisdom or folly of the overall Mediterranean strategy, or to get bogged down in debating Churchill's deeper motives for being so aggressively inclined in this theater of war.[22] The fact is that this major offensive drive took place, and it took place in three amphibious hops—from North Africa to Sicily (July 1943), then the crossing of the Straits of Messina together with the double landings at Salerno and Taranto (September 1943), and then the final effort to hasten the advance upon Rome by the amphibious strike at Anzio (January 1944).

How did these three very large-scale amphibious invasions in the six months between July 1943 and January 1944 better prepare the Western Allies for that greatest assault of all, Operation Overlord? Everyone, even the more cautious Churchill and Alanbrooke, knew that a gigantic attack upon German-held Europe had to come by early 1944 at the latest. But could their staffs and their armed forces get it right at the operational and tactical levels, so as to produce a strategic breakthrough and not a humiliating retreat to the sea? Herein lay the significance of the Mediterranean operations, because for the first time the Allies would be seeking to invade and capture German-held coastlands.

Operations Husky (Sicily), Avalanche (Salerno), and Shingle (Anzio) were all successful, but none of them was perfect, far from it— at a certain critical stage, it looked as if the Salerno and Anzio forces would be forced to pull out, so serious was the opposition. All of them gave the Allied staffs fresh opportunities for learning, for rethinking, for reorganizing, for technical and tactical modifications, and for turning to newer, better weapons. It has been claimed frequently that the Overlord landings were the greatest and best-organized military-naval operation in modern history. If that was so, it must also be said that never had a venture as grand been so much rehearsed by the operations preceding it.

The Anglo-American-Canadian invasion and takeover of Sicily (July and August 1943) was also the Allies' first entry into Europe, more than three turbulent years since Dunkirk, more than two years after the British Empire forces had been swept out of Greece and Crete, more than two years after Hitler attacked the USSR. It was a major blow to the Axis; it led to Mussolini's overthrow and to Italy's surrender. It was an important testing ground for newer Allied divisions, especially the American ones, and a significant further exercise in the goal of perfecting a very large, combined-service, amphibious operation.[23]

Yet even when the last German and Italian troops retreated across the Straits of Messina into mainland Italy on August 17, 1943, it could not be said that the Sicilian campaign had been a stunning or decisive success. Allied planning and preparations for this invasion had been held back by the amazingly determined Axis defense of Tunisia until almost mid-May. The Sicilian garrison was small—of course: eight German divisions had been captured in North Africa, and many others were deployed on Hitler's orders to forestall a possible western invasion of Greece, Sardinia, or even southern France. Ultra decrypts confirmed that the Axis had only two scratch (i.e., rebuilt) German divisions and four Italian divisions on Sicily, plus ill-equipped coastal defense units. By contrast, the Allies planned an enormous simultaneous landing force of eight divisions—considerably larger than later invasions of Normandy or the Marianas, larger too than the Philippines, Iwo Jima, and Okinawa operations.* Some 150,000 troops were landed on the first day alone, and a total of 478,000 overall, along a pair of 40-mile stretches of coast. On the beaches themselves, the invaders were mostly unopposed, although the first American units faced some Italian fire. The 4,000-plus Allied planes should have had virtually complete command of the air, yet the Luftwaffe managed to get in some blows. Vast fleets of landing craft and warships lay off the beaches, occasionally interrupted by enemy fire or planes; naval support was massive.

This was therefore not a real test. Probably only two lessons for the future could be drawn from the Sicily operation. The first, a negative one, was that when the two Wehrmacht divisions entered the fray, reinforced by paratroops and then more ground forces, both the British and American land advances slowed down, even when calling in additional units from Tunisia. Moreover, while the German troops gave way to Patton in the west of the island, they found it easy to hold their eastern

---

*While the number of Allied troops involved in the initial Sicily landings was greater than those landed on June 6, 1944, on the Normandy beaches, the latter operation involved millions more follow-on troops.

mountain line even when they were fighting merely to delay the Allies until they had gotten most of their own trucks, tanks, guns, and supplies across the Straits of Messina—which they did almost uninterrupted. Field Marshal Harold Alexander's report to Churchill that on August 17, 1943, "the last German soldier was flung out of Sicily" may well have given Allied planners in London a false impression.[24] The second lesson, much more encouraging in nature, was that despite strong offshore seas and winds that blew the American and British airborne troops all over the southern parts of the island, the two amphibious commands had another excellent opportunity to iron out many of the stage-by-stage problems of overcoming coastal obstacles, pushing troops and supplies quickly off the beaches, coordinating with aerial and naval support forces, and communicating with the all-important HQ ships and the admirals in charge. The latter, moreover, were the two best amphibious naval admirals that would emerge in the European theater: Sir Bertram Ramsay of the Royal Navy and H. Kent Hewitt of the U.S. Navy. Ramsay had enormous experience, but Hewitt learned very fast, and he easily understood that separate though parallel landings made for the greatest clarity.

The next step was the invasion of Italy itself. Even taking this further advance had been a matter of dispute between the American and British chiefs, though American worries about new delays to Overlord were weakened by the Italian government's clear eagerness to surrender and be brought under the Allied umbrella. This important top-level debate did result in an agreement to invade southern Italy, but that action was delayed until September 3, when Montgomery's Eighth Army crossed the Straits of Messina, followed on the ninth by a flanking British attack on Taranto and a far larger Anglo-American assault upon the Salerno beaches, farther north, under Mark Clark's command. The landings were unopposed, there were no mines or barbed wire, and the 3,000 Italian troops who surrendered were in fact willing to help unload the British ships—anything to be distant from the Germans. (One staff report states that the only resistance was put up by a puma that had escaped from the Reggio zoo.) Cautiously the British and Canadian forces pushed inland, slowed only by German demolitions. This caution was even more in evidence during the "assault" on Taranto, a hastily assembled job carried out, again unopposed, by troops of the British 1st Airborne Division, temporarily without their air transport but also without any land vehicles when they got ashore. Within the next two weeks, the ports of Brindisi and Bari were taken in a similar uncontested fashion. Then the British 5th Corps simply sat and waited for further orders. By an appropriate coincidence, Allied planners had previously code-named the Taranto campaign Operation Slapstick.

The landings around Salerno were a different story. The attack was apparently known beforehand to every barber in Malta and Tripoli, and certainly to the Germans, who put their troops on alert hours before the landings (despite which Clark, to Hewitt's dismay, forbade naval covering fire, at least for the American divisions under his command). While not as large as the Sicilian campaign, Operation Avalanche was certainly a grand affair, involving four divisions, plus additional commando and Ranger units—55,000 troops initially, with another 115,000 to come—carried in and supported by around 700 landing craft and warships. Given its size, and the tactic of dispatching the landing craft to the beaches, about 8 miles away, around three o'clock in the morning (of September 9), it is not surprising that there were confusions and congestions. Nor is it surprising that the American landings near Paestum took a battering from German artillery until destroyers were at last released to go inshore and silence that fire.

What *was* remarkable was the capacity of the lonely 16th Panzer Division, about half the size of an Allied division and with only 80 tanks (Clark in his memoirs claims there were probably around 600), to contain the British advance, then switch units across the Sele River to keep the American troops in check, and after that go back to work along the British lines. On the second day the fresh American 45th Division got almost 10 miles inland and then was hustled back by a counterattack from a single motor infantry battalion and eight German tanks. Nor could the carrier-borne aircraft prevent frequent Luftwaffe attacks upon the crowded beachheads. By the end of the third day, when they had hoped to be in Naples, the Allies were still in two narrow and separate lodgements—and the Wehrmacht's reinforcements were arriving.

Even now, it remains hard to judge how close the Allies came to a

Gallipoli-sized disaster on the fourth and fifth days of this invasion, when the Germans simultaneously sealed the mountain passes, put a powerful force (the 29th Panzer-Grenadier Division) between the two beachheads, and drove the British out of Battipaglia and the Americans out of Persano, sending the latter reeling back to the sea. At one point the Germans were within half a mile of the beaches, and late on September 13 Clark was making plans to reembark his Fifth Army HQ and to evacuate the 6th Corps, possibly to the British beachhead.

The chaos that such a move would have caused appalled Clark's own naval and military officers, and Eisenhower and Alexander in turn firmly dismissed the possibility. But the shock of retreat prompted them not only to dismiss divisional commanders from their posts but also to call in all and any Allied forces in the entire theater. Large numbers of battleships, cruisers, and destroyers were deployed to pummel the German positions—with spotter aircraft, the *Warspite*'s 15-inch shells were crashing into targets 10 miles inland. All Allied aircraft in the Mediterranean, whether strategic or tactical, were made available: a convoy of tank landing ships headed for India was diverted to Salerno, Matthew Ridgway's 82nd Airborne Division was rushed across from Tunisia with only light equipment and dropped into the American sector, while the 7th Armoured Division (the Desert Rats) reinforced the northern beachhead. Rather more slowly, units of Montgomery's Eighth Army were coming in from the south, distracting German attention if not contributing to the fighting.[25]

After one last attack upon the British sector on September 16 was frustrated by enormous Allied naval and aerial bombardments and large numbers of tanks, Kesselring pulled his diminished four divisions back to a newer defensive line south of Naples. As a parting shot, the new radio-controlled gliding bombs badly damaged HMS *Warspite* and sank the Italian battleship *Roma,* which was en route to surrender; British and American heavy cruisers, also badly hurt, limped or were towed back to Malta. It is hard, really, to term Operation Avalanche a success at all, since the best performances were from the outnumbered German defenders. In that same month of September 1943, the Wehrmacht was falling back hundreds of miles along the entire Eastern Front under the batterings of Ivan Konev, Nikolai Fyodorovich Vatutin, and Konstantin

Rokossovski's armies. The contrast was embarrassing, putting further moral pressure upon the Western Allies to launch the repeatedly postponed invasion of France even if Alanbrooke and many experienced planners were doubtful of the outcome.

Looking back a full twelvemonth after Casablanca, then, it could hardly be said that the American or British record in putting troops ashore in the face of enemy resistance was impressive; in truth, it was pretty terrible. There was, moreover, another humiliation to come, along the Anzio beaches in January 1944. In some ways, the story of the Allies' difficulty in implementing their scheme to dislodge Germany's control over central Italy was the most embarrassing one of all. The statistical odds on each side—measured in divisions, armor, air control, special forces, naval support, or anything else—seemed to offer the prospect of a grand Allied invasion not far from Rome, thus cutting off Kesselring's stubborn defense of the lower Gustav Line. Unfortunately, the Americans and British once again bumped into those most uncomfortable creatures, namely, German combat divisions. Many of the more experienced Allied units—and the better commanders—had been pulled back to the United Kingdom in anticipation of Overlord. Even so, they had enough material force to carry out this flanking amphibious operation and thus quicken the hitherto laborious advance up the rugged Italian peninsula. What they did not have was the imagination to anticipate, and then handle, the German counteroffensive. In this case the traditional British Army wariness about advancing too fast against Germans was compounded by the unwillingness of the American land commander, Major General John Lucas, to do much more than consolidate the beachhead.[26] Instead of the spring of the panther, Churchill famously noted later, there were the floppings of a beached whale.

The swiftness of Kesselring's reaction to the Anzio landing (he had previously been preparing for the Allies to invade *north* of Rome) was astounding. Within a week, urged on by Hitler, he had reorganized the entire German army command system in central Italy, shifted troops from the north and the south to the threatened area, and placed elements of no fewer than eight battle-hardened divisions around Anzio under the formidable General Eberhard von Mackensen's direct command.* When Lucas ordered the first advance—more than a week after the landings, which had taken place without opposition on January 22—the Wehrmacht was completely in place. By this time the overall Allied land commander in Italy, Mark Clark, was ordering attacks against the well-held Gustav Line farther south in order to relieve the hemmed-in amphibious forces—the exact reverse of the original plan. These gestures included the fabulously stupid bombing of the greatest Benedictine monastery in Europe, Monte Cassino, inhabited only by monks and refugees. Meanwhile, German artillery was harassing the Anzio beachhead, the Luftwaffe was bombing the ships offshore, and Kesselring was attacking on all fronts. By mid- to late February there was an acute danger of the Allied armies being pushed back into the sea, in a replay of Dieppe. Instead, it was a replay of Salerno. Reinforced by fresh units, and with new commanders on the ground, the American and British divisions just managed to hold their own, and when the persistent winter clouds at last cleared, in early March, Allied bombers and fighter-bombers compelled von Mackensen to cease his offensive. But with a mere five divisions he could still keep the Allied lodgement in check, some way from Rome, while his other units rested.

These narratives suggest that the chances of a successful Allied invasion of western France in 1943–44, which Hitler and the Wehrmacht high command were bound to regard as much more threatening than the campaign for Rome, were very uncertain, not just because of the much more difficult tidal conditions in the Atlantic but also because of who would be there to oppose it. If Kesselring could get eight divisions to the Anzio beachhead within a week, how many more could the equally bold Rommel get to Normandy or Pas-de-Calais in that same amount of time?

Let us pause for reflection here. What the above campaign analyses suggest is that the argument for an invasion of France in the year 1943

---

*General Eberhard von Mackensen had been chief of staff of one of the German armies invading Poland in 1939 and of another army invading France in 1940. By the next year he was with Army Group South on the Eastern Front. By 1943 he was in charge of the First Panzer Army at the critical Battle of Kharkov. He was then moved to command the German Fourteenth Army in Italy. In other words, he probably had more battlefield experience than *all* of the Allied divisional commanders at Anzio put together.

was utterly wrong. Without control of the Atlantic, with the Allied bombing campaign being beaten back, with so many raw troops (and inexperienced commanders), and without the experiences of large-scale amphibious landings, an attack upon Normandy would have been, well, silly, not to mention tragic. It is terribly important to re-cover the many doubts among Allied planners during 1943 and 1944 about the chances of a successful move into France and then a march to Berlin. If this interpretation be true, then it tilts against the overwhelming "inevitablism" of much of the literature on the Second World War, an inevitablism so eloquently captured in the titles of Churchill's own multivolume history: *The Grand Alliance* (volume 3), *The Hinge of Fate* (volume 4), *Closing the Ring* (volume 5), and *Triumph and Tragedy* (volume 6).[27] It is worth recalling that, shortly after giving the final order to go on June 5 and only a few hours before the entire operation started, a lonely Eisenhower sat down and wrote one of the most astonishing letters ever written by a military commander. In this contingency communique he acknowledged that the Normandy invasion had failed, that he had withdrawn the Allied forces, and that he himself accepted full and undivided responsibility.[28] Of course, this letter was totally secret; had even an inkling of the text gotten out over the next few days, the shock waves among the nervous Allied forces would have been devastating. It is just one more confirmation—perhaps the greatest of all—of Ike's character, his generosity of mind, his deep awareness that mankind does not know what the next day will bring, and his instinctive understanding of Clausewitz's warning that as soon as the battle begins, all the greatest prewar planning lies in the hands of the gods.

It is thus another sobering reminder, and from an impeccable source, that the struggle between the Axis and the Grand Alliance was not preordained to be over within another year. It is also a reminder of what the string of previous Allied landings in North Africa, Sicily, Salerno, and Anzio both had and had not revealed to the team of increasingly experienced military leaders and their planners. It had taught them a million lessons: about aerial reconnaissance and support, bombardment, special landing craft, beach-clearing devices, offshore command ships, logistical flow, interservice cooperation. All of these, they now recognized, just had to work well, and work together. By June 5, then, the Allied planners knew many, many things, just like the fox in

Archilochus's poem.* But there was one big thing they didn't know and would not know until the end of the next day: whether the formidable German army, with much more intimidating beachhead defenses and a far larger number of divisions than they had faced during any of the Mediterrranean landings, would drive them back into the sea.

Doing It Right: The Normandy Beaches and Breakout, June 1944

The minesweepers did it first, of course. On the evening beforehand, dozens of them set out from Spithead and the Solent to cut ten paths through Rommel's outer line of maritime explosives. Behind them came 2,700 vessels (not counting the 1,900 landing craft in their mother ships), carrying in this first wave alone 130,000 soldiers, 2,000 tanks, and another 12,000 vehicles. As was now standard practice, Royal Navy submarines were used to mark flight paths for the Allied aircraft and to guide the leading assault ships to the beaches. Essentially uninterrupted by German naval or aerial attacks, this great force emerged from the sea as dawn broke on June 6, and thus began the recovery of western Europe. The Royal Navy, conscious of its heritage, named the operation Neptune, not Overlord.[29]

Four out of the five amphibious landings worked well; the fifth, at Omaha, stumbled badly, though only briefly. At the end of that historic day, all those men (plus the airborne divisions ahead of them) had been landed, and many more were crossing the Channel behind them. They had not been pushed back into the sea or held for long, if at all, on the beaches. Like the waters of a breached dam, the Allied armies poured inland, meeting resistance in some places and swirling past those obstacles in other places. Covered by airpower, the tanks, mobile artillery, and motorized infantry surged through gaps and places of least resistance. By the end of June, American and British units were 20 miles inland and heading south, while the whole Cotentin Peninsula had been liberated. A month later, U.S. troops were close to the Loire Valley and rapidly wheeling east. On August 25, 1944, Free French and American troops entered Paris.

How did they do it?

---

* Or in Isaiah Berlin's classic essay "The Hedgehog and the Fox" (1953).

Of all the features that contributed to the Allied victory in Normandy, the key one, surely, was that of command and control. Without it, all else, even superb performance at the tactical level, would fail. The massive, complex orchestration of the Anglo-American-Canadian invasion of western France needed a superbly talented control organization. Fortunately for the invaders, they had an Allied Combined Operations Command under Admiral Bertram Ramsay that had been brought, through earlier experience, to remarkable standards of coordination of planning. It was no coincidence that the commanders who had shown themselves most adept in joint, interservice operations in the Mediterranean theater—Eisenhower, Tedder, Ramsay—were brought back to London in late 1943 to run the show, as were field commanders such as Montgomery, Miles Dempsey, Omar Bradley, and Patton, who could focus on the land battles. Upon their return, they found detailed provisional invasion plans that could be modified and improved upon, such as by their decision to land on five beaches, not the original three. They also found the early drafts of logistical schemes of the most staggering, yet necessary, detail, which could then be further refined.* Every ship's captain, every divisional commander, would have his own briefing book. Nothing would be left out.[30]

The genius who masterminded all this was a rather modest British naval officer. Bertram Ramsay (1883–1945) had entered the Royal Navy at the age of fifteen, served in the Dover Patrol during the First World War, and after peacetime service retired in 1938. A year later, Churchill bullied him into reentering the service. (How, really, did Winston recognize and promote such people—Ramsay, Hobart, William Slim, Tedder, and even Alanbrooke himself?) Ramsay's war record is extraordinary. Appointed as C in C, Dover Command, the most ancient of all in the Royal Navy, he ran with great efficiency the Dunkirk withdrawal in June 1940, then controlled the Narrow Seas until he was appointed the chief amphibious planner for Torch and Sicily. He was then summoned home, at sixty-one years of age, to become a full admiral and the naval commander in chief of the Allied Naval Expeditionary Force for the Normandy landings. Tragically, Ramsay was killed in

---

*Allied military units, for example, were barracked in 1,108 camps across England and Wales. How did one get them all to the southern invasion ports in good order?

France in an aircraft crash in January 1945 while heading to a conference with Montgomery about the future pursuit of the war. But in
many ways his job was done, and much of what follows is his story.[31]

There were three other positive prerequisites for the success of Overlord: command of the air, command of the sea, and well-handled deception and intelligence. Take any one away, and the story could have been
different. There were also two negative or "might-have-happened" factors that could have been decisive in ruining the enterprise—the weather
in the English Channel and the nature of the German military position-
ing and response.

Command of the air over western Europe called for a number of
things. The first was the broader isolation of France, in particular to the
west of Paris, from the sort of armored counteroffensives that Kesselring and von Mackensen had orchestrated so successfully in Italy; this
time the fabulous Wehrmacht capacity to recover and strike back fast
had to be blunted. The second was the direct aerial coverage of the
beachheads and landing waters, so that the troops would indeed know
that if there were planes overhead, they were Allied fighters. The third
was provision of tactical bombing support, both against the immediate
German coastal defenses and, as the breakout unfolded (if all went
well), against further opposition in the field.

The wrecking of German communications to and from the Atlantic shores—the western part of the so-called Transportation Plan to
cripple the defender's capacity to reinforce—was overwhelmingly successful. The scheme itself had been controversial among Allied commanders for some months beforehand (see chapter 2), but at the end of
the day the bombers really did come and bomb. A gigantic armada of
Allied medium and heavy bombers, diverted temporarily from their
strategic bombing campaign against Germany, plastered enemy-held
railway lines, roads, bridges, and marshalling yards, inflicting paralysis
upon almost all networks. While the heavy bombers worked better farther east, generally disrupting everything from fuel dumps to railway
stations, the smaller aircraft carried out remarkable close-action attacks;
on May 7, for example, eight P-47 Thunderbolts completely took out
the critical railway bridge across the Seine at Vernon, and many more
bridges and crossings suffered the same fate.[32] For successful German
panzer leaders, so used to freedom of movement across the North Afri-

can sands or the Russian steppes, the fact that northwest Europe was riddled with medium-sized rivers and vulnerable bridges turned out to be a dreadful restriction upon their assumptions about tank mobility. And, as they would find out, if an armored column was stuck down a long lane waiting to cross a river somewhere in France, they would very soon be at the mercy of the RAF's rocket-firing Typhoons.

There was also no repeat on D-Day itself of the Luftwaffe's deadly attacks off Norway, Crete, Dieppe, or even Salerno. The odds were impossibly distorted. On June 6 itself, close to 12,000 Allied planes were in the air, from well before dawn right into the night. By contrast, only 170 aircraft fit enough to fly were available to the Luftwaffe's Third Fleet, covering all of western Europe, so it was virtually impossible for them to get to the Normandy beaches against 5,600 Allied fighters. Both high-altitude and low-altitude attacks were anticipated in the aerial-defense plan. One aviation expert noted that "beachhead air-cover consisted of a continuous overlapping patrol of Spitfires at low level, with four P-47 Thunderbolt squadrons above them, and a squadron of P-38 Lightnings acting as cover for the ships off the five beachheads."[33]

Along with command of the air came control of the sea. Here again the Allied forces were overwhelmingly strong, and the invasion was supported by 7 battleships, 23 cruisers, more than 100 destroyers, and upward of 1,000 other fighting ships. All the American and British battleships were of the older World War I and Washington Treaty types (USS *Nevada*, HMS *Rodney*, etc.); they were now judged incapable of keeping up with the newer *Iowa*-class and *King George V*–class fast battleships on the high seas but were eminently suitable for—and adapted very well to—the task of onshore bombardment. HMS *Warspite* (with one turret and one boiler still not working after the glider-bomb attack at Salerno), blasting away with great effect at Le Havre, then Cherbourg, then Walcheren, was in her final campaign of a career that had earned her more battle honors than any other ship in the Royal Navy. Old-fashioned coastal monitors with 15-inch guns came into their own

THE D-DAY INVASIONS, JUNE 6, 1944
Five enormous and simultaneous amphibious operations are orchestrated by a brilliant single command structure.

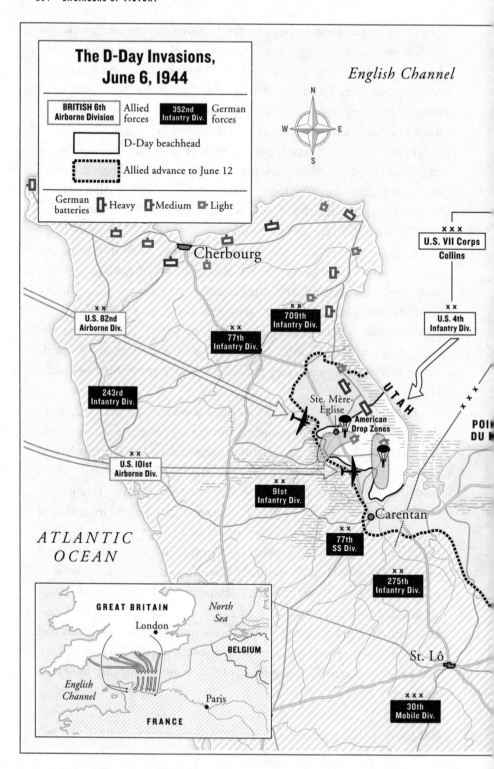

The D-Day Invasions, June 6, 1944

| BRITISH 6th Airborne Division | Allied forces | 352nd Infantry Div. | German forces |

D-Day beachhead

Allied advance to June 12

German batteries   Heavy   Medium   Light

English Channel

N W E S

Cherbourg

U.S. VII Corps Collins

U.S. 82nd Airborne Div.

709th Infantry Div.

U.S. 4th Infantry Div.

77th Infantry Div.

243rd Infantry Div.

Ste. Mère-Eglise

American Drop Zones

UTAH

POI DU

U.S. 101st Airborne Div.

91st Infantry Div.

Carentan

ATLANTIC OCEAN

77th SS Div.

275th Infantry Div.

GREAT BRITAIN

North Sea

London

BELGIUM

English Channel

Paris

St. Lô

FRANCE

30th Mobile Div.

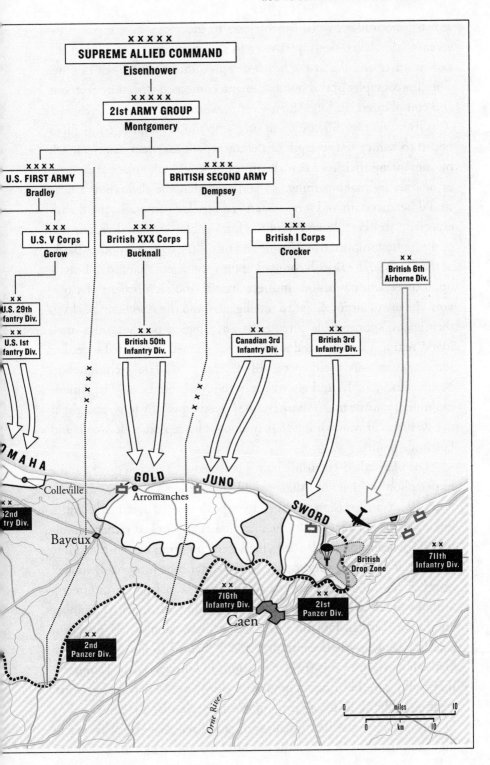

xxxxx
SUPREME ALLIED COMMAND
Eisenhower

xxxxx
21st ARMY GROUP
Montgomery

xxxx
U.S. FIRST ARMY
Bradley

xxxx
BRITISH SECOND ARMY
Dempsey

xxx
U.S. V Corps
Gerow

xxx
British XXX Corps
Bucknall

xxx
British I Corps
Crocker

xx
British 6th Airborne Div.

xx
U.S. 29th Infantry Div.

xx
U.S. 1st Infantry Div.

xx
British 50th Infantry Div.

xx
Canadian 3rd Infantry Div.

xx
British 3rd Infantry Div.

OMAHA

Colleville

GOLD
Arromanches

JUNO

SWORD

Bayeux

xx
62nd Infantry Div.

British Drop Zone

xx
711th Infantry Div.

xx
716th Infantry Div.

Caen

xx
21st Panzer Div.

xx
2nd Panzer Div.

Orne River

miles   0   10
km   0   10

as fortification destroyers. Much closer to the shore, as many as fifty-seven of the Allied destroyers were to site themselves just behind the assault waves and fire at the beach defenses until the very last minute. The choreography of this bombardment, combined with the enormous and complicated landings themselves, was amazing.

Given all the difficulties of any amphibious landing, the Allies hoped to reduce the strength of the immediate German counterattack by convincing their foes that the invasion would take place elsewhere, or at least by making things so confused that the defending armies would be uncertain and split up across a very long front. To the British especially, strategic deception of the formidable German divisions was of the highest importance. Unable and unwilling to fight another Battle of the Somme, the British counted upon a stratagem that rested heavily upon diversion, confusion, indirect attacks, the recruitment of partisans, the use of airpower, faked intelligence, and the search for cracks in their enemy's formidable defenses. It was a logical position for a small island nation to adopt, and a policy that seemed confirmed—despite some failures—in their successful deception warfare techniques in North Africa, Sicily, and elsewhere in the Mediterranean. The American military, while much more inclined psychologically to go straight at the Wehrmacht, was willing to play along. Hence Fortitude North and Fortitude South.

The plan called Fortitude North reflected a Churchill dream and a Hitler phobia—that the Allies would come upon Germany from the north, with an invasion of Norway, a push toward Denmark, and a linking with the Soviets. On the map, it actually looked rather appealing. Logistically, as the British chiefs kept pointing out to the prime minister, it would be very tough. To the Americans it probably looked as much of an "indirect approach" to Berlin as a landing in Greece. Still, it was left as a possibility, and commando raids and RAF bomber attacks on German lookout posts and air bases along the Norwegian coastline kept up the deception. It greatly helped to pin down lots of German garrison troops (twelve divisions in Norway, six in Denmark) that could have been better employed elsewhere, so it was also worth creating a phony invasion army in northeast Scotland for a while. As D-Day got closer, Allied intelligence could tell that the Wehrmacht high command was increasingly doubtful that any major attack (or any attack at all) would occur in Scandinavia, so the deception techniques for Fortitude South were correspondingly increased.

The story of the complex Allied efforts to persuade Hitler and the Wehrmacht high command that the chief assault would come in Pas-de-Calais is entangled in myths, realities, and testy counterfactual arguments. What is undoubted is that Fortitude South was elaborate and cunning, and assumed many forms. The few German agents in Britain who were turned (the "double-cross" agents) fed suitable information back to Berlin. Bombing raids, resistance activity, beach reconnaissance teams, and BBC transmissions, along with another whole panoply of tricks, were deployed. American and British theatrical backroom staff, disguise artists, and set designers joined the war effort. Tens of thousands of inflatable full-sized tanks and trucks were openly arrayed across fields near Kentish ports, even while a stupendously effective and total blackout was imposed across the counties of Hampshire, Dorset, Devon, and farther west, where the real invasion armies lay. German aerial reconnaissance of the western ports and camps was effectively blocked. A not very happy General Patton, with a recognizable radio sign, was temporarily dispatched to take command of this mythic army. Allied intelligence would always want to know Rommel's location, so it was sensible to assume that the Germans would desire to know where the hyperaggressive Patton had been stationed.

But how great a contribution to the D-Day victory did these deception measures make, as compared with the many other factors discussed here? It was considerable. Overestimating the number of Allied divisions bivouacked in Kent, the Wehrmacht kept no fewer than nineteen of its own divisions in the Pas-de-Calais region (four behind Dunkirk alone); by contrast, there were only eighteen divisions between the Seine and the Loire. Days *after* the Normandy landings, von Rundstedt and many other experienced generals still thought that those invasion forces, though growing in size each day, were a feint—the fact that they had occurred west of Caen was actually *proof* to them that the big onslaught would be near Calais. While releasing some German units to move forward to the beaches, the army high command held many more back; messages from double agents that the landings were a decoy caused Hitler and von Rundstedt to cancel an earlier order and send two divisions back to Pas-de-Calais as late as June 10. Before, during, and after the real landings RAF Lancasters flew up and down the Channel between Dover and Calais, dropping continuous sheets of "window," while a fleet of small ships steamed back and forth below; what could *that* portend?

The deception continued. As late as July 3, Hitler's OKW chief Jodl told the Japanese naval attaché in Berlin that "Army Group Patton" was soon to lead eighteen infantry divisions, six armored divisions, and five airborne divisions across the Channel; this was, Jodl opined, "obvious," an astounding misinterpretation. When the high command at last recognized (for some generals, it took until mid-July) that nothing was going to happen in the Narrow Seas, it was impossible to swing those critical divisions southward, partly because many of them were static units but chiefly because of the paralyzing effect of Allied tactical bombing of roads, railways, and bridges.[34]

But the history of intelligence is rarely straightforward. In June 1944, it certainly was not the case of the blind defender against the all-seeing invader. The Wehrmacht had detailed knowledge of Allied landing techniques, the role of specialist beach-clearing teams, the strong emphasis upon airborne forces, the relocation of RAF and USAAF squadrons, and the southward rumble of more and more divisions from the Clyde landing-wharfs toward the Channel. In addition, having carefully studied the pattern of their enemies' naval bombardments and initial infantry landings in Mediterranean operations, Rommel and his staff made sure that most of their fortified bunkers and pillboxes were sited obliquely to the beaches. Thus obscured from Allied naval bombardment, they were still able to cover a large expanse of the shore as enemy troops staggered out of the water. Yet what the Germans did not know was the most important intelligence piece of all: *where* would the Allies land, and *when*? They had reports of landing craft assembling in the Essex ports, but also in the harbors of Devon; they detected new American divisions setting up camp behind Portsmouth, yet others arriving (next to the dummy units) close to Folkestone. Although the key Wehrmacht intelligence officers sometimes exaggerated the Allied numbers, it is not difficult to feel for them when they were being repeatedly pressed to answer the critical question—Pas-de-Calais or Normandy?—especially as it was known that Rommel and von Rundstedt held contrary views.

Allied intelligence about their German opponents was a mirror image of German intelligence about the Allies. To be sure, London wanted to know every detail regarding the Wehrmacht's defenses, location of units, special obstacles, numbers of tanks, and so on. Given their absolute superiority of aerial photography in fine weather (an advantage the Abwehr and Luftwaffe could only lament), the flood of information from the French Resistance, and the almost total control through Ultra decrypts of German military and naval messages, it was relatively easy to achieve knowledge of the Wehrmacht's defenses, location of units, special obstacles, numbers of tanks, and so on; they could also usually track the relocation of major army units. Being able to read the Japanese diplomatic and naval/military messages from Berlin to Tokyo also provided a general confirmation of German thinking. Yet, as the invading armies were to learn when hostilities commenced, the Germans could also create dummy bunkers, switch mobile coastal batteries from one cliff to another, and make rapid movements of tanks and trucks at night. Moreover, no simple counting of the enemy's strength could answer the really important question: how hard would the defenders fight, even against heavy odds?

Allied intelligence had one further massive advantage: its capacity to intercept and swiftly decode radio traffic between German commanders in the field and their headquarters. Just as the teams at Bletchley Park had gained much from reading messages between Doenitz and his U-boat captains when the tide turned in the Atlantic campaign, so did the code breakers gain valuable information from the surge in Wehrmacht traffic once the landings had taken place: who was reporting to whom, what orders were given out, what information about the strength and whereabouts of the Allied forces was communicated back to Army Group West, what indications there were that the deception techniques had worked. Decrypting even a quarter of this valuable data made a vast difference. It was important, therefore, that the German army relied such a lot upon emergency radio communications.[35]

Having French Resistance networks on the ground was an additional bonus for the Allied planners. Not only was the Resistance a source of important intelligence about local German forces, but also the widespread acts of sabotage compelled the diversion of hundreds of thousands of troops to secondary military activities such as guarding railway lines, searching houses, and the like. And because different Resistance cells were sent different special messages from London during the BBC's radio broadcasts, the Allies could add to the deception by markedly increasing the wireless traffic to the Pas-de-Calais area. Finally, of course, when the landings actually occurred, these small French units—equipped and trained by the Special Operations Executive—mounted damaging attacks upon bridges, roads, and, above all, telegraph poles and wires, forcing the Germans to make more use of radio signals.[36] There was, of course, no German equivalent in Britain to offer corresponding distractions.

All three of the above positive dimensions stand to the credit of the Western invaders. By contrast, Ramsay's planners could claim no credit for the first of the two "just-didn't-happen" aspects: the weather itself. The high tides, the storms coming out of the Atlantic, yet also the bouts of unusually placid weather that had so affected the outcomes of the cross-Channel invasions by Julius Caesar, William the Conqueror, Henry V, the Spanish Armada, and William of Orange had not changed their nature; transporting a large number of men and equipment across those unpredictable waters and landing successfully on the other side always contained risk.

So it was in early June 1944. There had been some fine days in late May, but by June 3–4 a large depression was sweeping in from the Atlantic. The meteorologists forecast low clouds, which would take away the air advantage, and rough seas, which would cause chaos on the beaches. On the fifth, miraculously, the storm abated a bit, the rains stopped, and a deeply torn Eisenhower gave the order to go (before retiring to write his letter). That night the enormous armada set out for France. Fortunately but understandably, this awful weather had caused the Germans to conclude that an invasion was not possible during the next few days; Rommel drove back to Germany for his wife's birthday (June 6), and von Rundstedt's headquarters in Paris dismissed warnings about coded BBC messages to the French Resistance that night as "foolish."[37] The six divisions of Army Group B, directly behind the landing area, received no alert signals at all.

Thus the ever-fickle weather gave Eisenhower's forces an absolutely crucial window of opportunity: by June 9 the artificial Mulberry harbors (to be discussed shortly) were in place and receiving enormous numbers of fresh troops, tanks, and trucks. Yet ten days later (the morning of the nineteenth, following a lovely calm day on the eighteenth), the whole Channel region was suddenly hit by one of the worst storms of the twentieth century, bringing all traffic to a halt, paralyzing aerial patrols and attacks, throwing eight hundred smaller craft onto the beaches, and eventually ripping apart the gigantic American Mulberry harbor off St. Laurent. The setback to the Allied timetable was colossal, and the boost to Rommel's chances of a counterattack considerable, until the gale moved on. Then the waters calmed, Allied aerial surveillance was restored, and the massive trans-Channel traffic resumed. But it had been a tense experience. Had that storm come on, say, June 10, the Royal Navy's Operation Neptune would have been a washed-out disaster, the Allied forces looking like beached crabs.

The weather was a natural force, not to be controlled by man. The other negative element in allowing Overlord to succeed was due purely to human decision—or in this case indecision by the defenders. The German failure to sweep the first Allied landing units off the beaches was due not just to the latter's clever deception techniques or heavy tactical airpower, though both were important. It was also due to an unusual indecisiveness among the Wehrmacht high command as to how best to respond to the inevitable Allied opening of the second front. Some of Germany's most experienced generals, including many who had faced Allied invasions before, were seriously divided, each camp offering genuine military reasons. There was an additional "spoil" factor here, namely, Hitler's own capacity to disrupt rational military-operational actions, either through his prejudice against tactical retreats or, as the war unfolded, through his increasingly drug-affected daily routine. General Guenther Blumentritt's bitter observation that Germany was going to lose the war "weil der Fuehrer schlaeft" (while the Fuehrer sleeps)—a reference to Hitler's not being disturbed on June 6 to authorize the release of the critical panzer reserve—was not the full story.[38] When and in what strength should the Wehrmacht unleash its riposte to the amphibious assault, wherever it took place along France's lengthy, difficult Atlantic shoreline?

In the whole of France and the Low Countries, the Wehrmacht possessed fifty-eight divisions by the summer of 1944, a significant rise from a year earlier. But the majority of those were static divisions, dug into places that they could (and would) defend very well, but lacking the trucks or even horses to be moved swiftly to another position. The key elements, therefore, were the dozen or so panzer or panzer-grenadier divisions, each with the battle experience to destroy any equivalent unit on the Allied side. Rommel, appointed by Hitler in January 1944 to defend this front, came to advocate a forward deployment. Throwing his immense energies into the further massive fortification of the Atlantic Wall, he wanted, as a corollary of that, his German armor to crush the invaders just as they were moving out of their beachheads. To him, the first three days would be vital: he felt it was too risky to rely upon a later, calibrated response from a distance, since Allied airpower might stop all such blows in their tracks. Army against army, he was sure, the Germans would win; army against army *plus* massive airpower, they would lose.

Much of the rest of the high command disagreed with Rommel's strong "outer crust" approach (including some, ironically, who had thought him too bold and reckless in his earlier campaigns, and now judged him too cautious). They—including not just von Rundstedt but also tank commanders such as Guderian and Geyr von Schweppenburg—were more confident in the Wehrmacht's tradition of launching massive and devastating counterattacks after an enemy's operation began to slow; one supposes that Anzio was an encouragement here. They also didn't like the idea of the panzer divisions being scattered along the coastline and so likely to be subjected to 15-inch warship shells and a very large number of 500-pound bombs from Marauders, Lancasters, and B-17s. Better to wait, they thought; let the invaders come in, and then drive them back to the wire-tangled beaches. Both arguments were about the threat from Allied firepower: would it be worse to take hits along the immediate shoreline from battleships or from an inland aerial offensive?

Hitler's compromise between his generals was the worst of all worlds: some mobile divisions to be allocated individually to the coastline, four to be held well back (and only released by direct order of the Oberkommando der Wehrmacht—essentially, himself), and four sent to the south of France. Weak everywhere, strong nowhere—just the opposite of what his hero Frederick the Great had always recommended. Had von Schweppenburg's four reserve armored divisions been positioned around St. Lô on June 6, right to the south of the Normandy beaches, they might well have pushed the American invaders and possibly also the British and Canadians farther to the east, back into the sea, under cloud-filled skies.

The Anglo-American-Canadian invaders were, clearly, very lucky. But they were also very smart. They were well orchestrated, they had command of the air and the sea, and they used deception and intelligence to a fine degree. In addition, they had both the weather and enemy indecisiveness on their side. But they also needed to be operationally competent, much more so than they had been in previous amphibious landings. And here, in the story of the assaults upon the Normandy shoreline, lay the second part of the challenge: getting up to 1 million men and 100,000 vehicles on *and off* the landing zones and then heading to Berlin.

Answering the million-men-landing question explains the Allied planners' choice of Normandy. Since Brittany was too far to the west and the Belgian/Dutch estuaries were too treacherous—and too close to a counterattack from Germany—the choice narrowed down to either the Pas-de-Calais region or Normandy, both of which could be given strong aerial coverage from English bases. While the German staffs were genuinely torn regarding each option, to the Allies a Normandy landing made more sense for several reasons: it was almost equidistant from all the large southern English and Welsh invasion ports, whereas a major landing near Calais would really pile up the successive waves of landing craft. Normandy gave space for Montgomery's insistence on landing at five beaches, not the original three; it gave the Allied navies more room for maneuver than in the Narrows; and it offered a promising chance to send an army westward from Normandy into the Cotentin Peninsula, take Cherbourg, and create a major link from France back to the millions of men and masses of munitions that could sail directly from America. All that was needed was to seize a part of western Europe and then move east into Germany. But that was much easier said than done.

The orchestration of the approach to the beaches, and the organization of the initial landings, still staggers the historical imagination. It would fill the rest of this book, and several more, to describe the landing plan in detail.[39] For example, the 171 squadrons of Allied fighters that would be aloft as the invasion unfolded all had their own patrol zones; they would be directed by RAF Fighter Command initially, and then for the first time ever air traffic control would switch to special units in the HQ ships offshore. The bombing of the German rear echelon was timed for around 2:00 to 4:00 a.m., just before the intensive bombing of the immediate coastal defenses, then a reversion by the heavy bombers to interdiction of the roads leading toward the Normandy beaches. As the vast armada came to its launching point, the troops also could not fail to notice the enormous bombardment of the German coastal emplacements by battleships, cruisers, and monitors. Ramsay wanted two full hours of heavy, shattering fire on the beaches before the destroyers and rocket-firing landing craft moved closer in, and they were to continue firing until shortly before the hordes of smaller landing craft, some carrying specialized tanks, most carrying platoons of nervous infantrymen and the specialized assault tools, approached the beaches. Even before them, there went the demolition teams, the bravest of the brave. How exactly does one demolish a spiky tetrahedron with wire and mines dangling off it? Along with those early units went another group of specialists, the naval "spotters," to scramble onshore to a nearby hill and control the fire from the warships.

As the troops and amphibious tanks came ashore (or drowned in the effort), their beach master was waiting. This was an obvious, necessary, extraordinary position, one confirmed by the experiences on the beaches of Sicily and Salerno: a Royal Navy captain became supreme director of the landing beach, moving the troops forward, ordering disabled vehicles to be pushed out of the way, sending the landing craft back to sea as soon as they had disembarked their contents. He functioned in a way like an old-fashioned traffic cop at a busy intersection, imposing order on impending chaos, heading off the potential gridlock. One of them, the redoubtable Captain William Tennant of the Royal Navy, had been in charge of getting 340,000 British, French, and Belgian troops off the beaches at Dunkirk in 1940; now, four years later, he was bringing even larger numbers back to the Normandy coasts.

A few hours before even the beach master and the obstacle-clearing crews had landed, the first units of the Allied airborne divisions had already been parachuted to their positions a few miles inland. Employing paratroops as a key component in a large-scale operation—as distinct from sending them on small special-ops raids—was an unusual and extraordinarily risky step: there was nothing (not artillery, not air support, not naval bombardment) that could be deployed to cover their exposed downward glide in the face of enemy resistance—as even the ultra-competent German paratroops had found when they landed in Crete in 1941. Yet despite the grave risks, Eisenhower's planners could see that the reward of a successful parachute strike would be enormous: the U.S. 82nd and 101st Airborne Divisions' capture of small towns a few miles inland from Utah Beach would have a great dislocating effect, and it was even more vital for the British 6th Airborne Division to seize the Orne River bridge and thus outflank Caen from the east.

Behind the landing force, stood out at sea and occupying different lanes from the amphibious craft and from each other, were the five Allied naval bombardment squadrons. They had, following the schedule mentioned above, begun firing on German coastal fortifications before the first craft reached the shore, then moved to fire at targets farther inland. From zone to zone, there was a marked difference of tactics. Hewitt went for a somewhat later, briefer bombardment (actually, only thirty to forty minutes), rather than Ramsay's solid two hours of preinvasion fire. Whichever system was preferred, it has to be said that the results were mixed. Solidly built and obliquely aligned concrete bunkers were not easy to destroy, even when fire control directors directed large salvoes onto their roofs. Testimony from captured Germans on the receiving end suggests that the greatest effect was caused by the earsplitting noise, choking dust, and confusion that the great shells brought with them; or, a little later, by the surprise of being caught by a battleship's salvoes while in a truck convoy 10 miles inland. It took a while for Wehrmacht field commanders to appreciate what those unarmed little spotter planes overhead were up to.

Above the beaches, as the full day came up, patrolled the Allied air squadrons. It was almost four years to the week that a small number of RAF Hurricane and Fairy Battle squadrons had tried to protect the retreating British and French military from Luftwaffe bombings and strafings as they streamed into the little boats off Dunkirk, just 120 miles northward upon the very same coast. Now the odds were totally reversed. Apart from the famous madcap run over the Allied beaches at

10:00 a.m. on June 6 by the Luftwaffe ace "Pips" Priller and his wing-man Heinz Wodarczyk in their Fw 190s, what German interference could come from the air? Farther afield, Allied heavy bombers contin-ued to plaster the enemy's communications networks running from the Rhineland into France and the Low Countries, while enormous num-bers of Coastal Command aircraft patrolled the Channel, the Western Approaches, and the Bay of Biscay. If the Allies had a problem in the air, it was most likely to be confusion and misidentification of other aircraft (at a mile away, a Fw 190 and a Mustang looked very similar) and thus the risk of friendly fire; hence the painting of three parallel white stripes on all Allied aircraft likely to operate over the D-Day beaches. No other air commanders ever enjoyed such a luxury of sur-plus power.

The most effective German defensive weapon was, curiously, their new "oyster" mine, which exploded when the waves created by an ap-proaching ship changed the water pressure. These were laid in large numbers by low-flying aircraft and small vessels. Despite the constant work of dozens and dozens of Allied minesweepers, those mines sank considerable numbers of warships and merchantmen and damaged many more, including Admiral Vian's flagship for the Eastern Task Force, the cruiser HMS *Scylla*. Many Allied vessels had to be towed slowly back to Portsmouth, crews waving at the reinforcing flotillas heading toward Normandy. By contrast, the German E-boats sent out to contest Operation Neptune were really the equivalent of kamikaze vessels, attacked at every opportunity by Beaufighters and Wellingtons, checked by massive destroyer screens (Canadian, Polish, and French as well as Royal Navy), and having their home bases torn apart by RAF Lancasters. Bomber Command might have had a mixed record over Berlin at night, but their 12,000-pound Tallboy bombs completely wrecked German-held harbors, bridges, and railway lines; when hit by 325 Lancasters on June 14, the concrete roofs of the naval station at Le Havre crashed down upon the fourteen hapless E-boats below. Low-flying Beaufighters terrorized every German-held port from the Chan-nel to Denmark.[40] At night, specially equipped Mosquitos took over.

Setting to sea from Belgian ports was also the choice of the remain-ing German destroyers. The brave assault by the four destroyers of the 8th German Destroyer Flotilla on the night of June 8–9—detected by decrypts, and smashed by an Anglo-Canadian-Polish destroyer force twice its size—was essentially the last formal sortie of the Kriegsmarine against the navies of the West. When the war had broken out almost five years earlier, a grim Admiral Raeder remarked that his underdeveloped navy at least knew how "to die gallantly." So they had done.

The U-boat flotillas ordered to attack the Allied landings were (as noted in chapter 1) also dispatched to a suicidal mission. By this stage Ultra was at last at full effectiveness, and because the submarines reached the Channel waters only after the landings had occurred, a colossal force of aircraft and surface escorts was waiting for them, all of them now equipped with advanced detection systems and horribly effective weaponry. The skies were full of Allied aircraft, the horizons full of frigates. Even the new schnorkel-equipped boats, although sinking a half dozen escorts in initial attacks, had no chance against such odds and, after heavy losses, all the U-boats were ordered out of the area.[41]

Because of such complete Allied preponderance in the air and near-total dominance of the sea, therefore, the only real question to be decided—as each side had recognized long before—was the battle on land: the fight on the beaches, and the further fight by the invaders to move deeper into France and toward Germany. A neutral observer might already have concluded that, given all the trump cards possessed by the Allies (airpower, sea power, logistics, Resistance, etc.), the odds were already heavily stacked in their favor. And that is surely correct. But it was precisely the challenge of landing and then moving on that Eisenhower and his senior officers worried about most. And it was upon the crushing of those Allied moves that the Germans pinned all their hopes, all their resources, and their sole strategy. As the events of June 6 unfolded, both sides were proven right, perhaps Rommel most of all, in seeing the first three days of fighting as being decisive.

There are in essence three different amphibious-landing parts to the Normandy landings on June 6, 1944. The best, at least from the Allied perspective, was that regarding the westernmost assault, on Utah Beach, by the 4th U.S. Infantry Division. Here was a case of victory being snatched from the jaws of potential disaster. Clouds were already obscuring the coastline before the explosions from the naval bombardment produced massive smoke; the guide vessels could no longer be seen. If there were beach directors (the Americans did not have named beach masters), they were in the mists. As the landing craft chugged toward the murky shore, they found themselves pushed by the tides to a mile south of their target zone and, helpfully, into much less strongly defended territory. Here the landings went extremely well, the rocket ships firing away, the amphibious tanks closing to the shore, the infantry platoons wading waist-high with rifles aloft in the face of very little opposition, and the B-26 bombers flying below the clouds to deliver their assaults. In rare and classic style (compare with Anzio), the 4th Division brushed aside the local opposition, pushed strongly inland, and, with much more wading through swampy land, gained 5 miles of ground by the end of the day. With about 21,000 troops and 1,800 vehicles ashore by that stage, Utah was not going to be dislodged easily, whoever came at it. The 8th and 27th Infantry Regiments lost, in total, twelve men. This was the smoothest amphibious operation the Allies ever made.

It was behind the Utah beaches that the 82nd and 101st U.S. Airborne Divisions made their early morning landings, the intention being to seize and hold inland towns such as Ste. Mère-Eglise until the main force arrived from the sea. Much has been made of the way in which the perverse winds carried the paratroopers in all directions, dropped them into the marshes, scattered and chiefly ruined their heavier equipment, and prevented them from supporting the glider-borne follow-up, resulting in high casualties for the latter. The 82nd Airborne's troops were indeed so widely strewn across the wetlands of the Merderet River that two-thirds of them were still missing three days later. But being "missing" was not the same as being ineffective. In fact, chaos became an unexpected advantage: the widespread and sporadic nature of the paratroops' landings caused immense confusion among the German forces behind Utah Beach and crimped any attempts to reinforce the beachhead defenders. In the course of all this localized fighting, a small group of paratroops ambushed a German staff car and killed (as it turned out) the commander of the German 91st Division, the chief reserve division for the Cotentin Peninsula. By the end of the day, elements of the 82nd Airborne were as far inland as the crossroads town of Pont l'Abbé, and this particular battle had been won.

All this was critically important because Ramsay's master plan had had to assume that the U.S. 4th Division would indeed get ashore and then clear of the beaches. The blunt fact was that the 90th Division was coming in close behind it, to be landed between June 6 and 9; then the 9th Division, to be landed between June 10 and 13; then the 79th Division, to pass through by June 30. (Similar buildups were planned for the other four beaches as well.) With four full army divisions put ashore, plus the two airborne divisions ahead, the American VII Corps (under General Joseph Collins) could conquer western Normandy. Had they been held on the beaches, the very size of their invasion force would have produced mass havoc and chaos; but it was not to be so.

The biggest part of Operation Overlord concerned the three landings by British and Canadian troops at the Gold, Juno, and Sword beaches. Perhaps the best words to describe these strikes from the sea would be adjectives such as *thorough, careful, fastidious,* and *well-orchestrated*—apart from the airborne operation, not very bold, showing again the British Army's deep respect for its old foe. Should the June 1944 Normandy campaign end in a disaster, the Americans could come again; there were four million more GIs waiting back home to cross the Atlantic if needed. The British could not afford another and bigger Dunkirk, Crete, Dakar, or Dieppe. Their manpower reserves were horribly overstretched as it was; this was their last big war. As a result, they were massively invested in deception, intelligence, command and control, signals, beach masters, mine clearance, specially designed tanks—whatever it took not to be pinned down on the shingle and suffer unsustainable losses.

Hence the attention the British gave to specialized armored units and unusual vehicles that would help them overcome Rommel's intricate and deadly beach defenses. The driving force here was an acerbic, determined visionary, Major General Percy Hobart, creator of the 7th Armoured Brigade (which would become known as the "Desert Rats") in the late 1930s, then demoted and retired, then rescued from obscurity by an angered Churchill,* and finally given the 79th (Experimen-

---

*The prime minister's October 1940 reproof to the War Office for not using Hobart's talents should be memorized by all commanders in chief and CEOs: "I am not at all impressed by the prejudice against him in certain quarters. Such prejudices attach frequently to persons of strong personality and original view. . . . We are now at war, fighting for our lives, and we cannot afford to confine Army appointments to officers who have excited no hostile comment in their career."

tal) Armoured Division and the necessary material resources to develop
what his own troops fondly referred to as "Hobart's Funnies" precisely
to deal with beach and field obstacles, a need that became even more
obvious after Dieppe.[42] The basic instrument for Hobart was the sturdy,
reliable Sherman tank or its later British counterpart, the Churchill.
The tanks were then converted in all manner of ways: amphibious
tanks with inflatable skirts that drove toward the shore, flail tanks
whose gigantic metal chains beat the sand and exploded enemy mines,
tanks with massive wire cutters or bulldozer blades, fascine tanks that
carried their own rolled-up metal or wooden bridges to allow the crossing of ditches and tank traps, flamethrower tanks just like those the
U.S. Marines were using in the Pacific, tanks that simply became ramps
for other tanks, and so on. Hobart was a genius, and the history of armored warfare had seen nothing like this. (In the same spirit, and frustrated by the slow movement of Bradley's forces along the narrow lanes
and high hedgerows of the ambush-prone Normandy bocage, an American sergeant named Curtis Culin produced another variant, the "Rhi-
noceros," with giant front teeth that could rip right through the hedge's
base and allow the regular tanks to race across open fields.)

Hobart's Funnies had their baptism by fire on the British-Canadian
beaches. It was by no means the case that they worked marvelously
on each and every occasion. How could they? No rehearsals off the
Scottish coast or Bristol Channel could match the murderous reality
of landing on an obstacle-strewn beach and being shot at from all
directions. All along the coast, those duplex-drive tanks (with the inflatable skirts) found the heavy tides slowing their progress, so they
were often overtaken by many of the landing craft carrying infantry
and other sorts of tanks. The British and the Free French at Sword
Beach probably had the best of it, since the 6th Airborne Division had
already seized the Merville battery while the enormous 155 mm guns at
Le Havre (which could have decimated any landing ship or close-in
destroyer) spent the morning in a rather foolish duel with HMS *Warspite* offshore. With the Royal Marine frogmen having dismantled the
beach obstacles, Hobart's obsessive search for problem-solving weaponry came into its own. As one British Army major at Sword recalled
in awe:

A German antitank gun took them under fire. The [bridge-carrying] Sherman drove right up to it and dropped its bridge directly onto the emplacement, putting the gun out of action. Flail tanks went to work clearing paths through the mines. "They drove off the beach flailing," Ferguson said. "They flailed straight up to the dunes, then turned right flailing and then flailed back to the high-water mark." Other tanks used cunning explosive devices (called bangalores or snakes or serpents) to blow gaps in the barbed wire and the dunes. Still others of Hobart's Funnies dropped their bridges over the seawall, followed by the bulldozers and then fascine-carrying tanks that dropped their bundles of logs into the antitank ditches. When that task was complete, the flail tanks could cross to the main lateral road, about 100 meters inland, and begin flailing right and left.[43]

These amazing machines, which would later assist the British-Canadian drive into the Low Countries and northern Germany, certainly contributed a great deal in getting the first amphibious units across the beaches and through the cramped and treacherous streets of the small villages that lined the Gold-Juno-Sword beaches. With a few exceptions, such as at Le Hamel, the opposition withered quickly; the "static" divisions consisted of some anti-Communist Russians, together with Lithuanians, Poles, First World War veterans, and sixteen-year-old boys. The safest thing for them to do, after token firing, was to surrender. Fighting from the concrete bunkers was no longer safe, for the advance commandos were dropping grenades through the apertures even as the naval support continued (and in one of the most spectacular naval shots of all time, the cruiser HMS *Ajax* got a 6-inch shell through the forward slit of a massive emplacement on Gold Beach and exploded the entire magazine). Here, at least, Rommel's "outer crust" crumpled swiftly. By day's end, the British had put 29,000 men ashore at Sword, suffered only 630 casualties, and taken thousands of prisoners. At Gold they put 25,000 men ashore at the cost of 400 casualties. These were well below all planning estimates of likely losses.

At Juno Beach the Canadians and British invaders had it much harder. The objectives were farther away, the defensive fire greater, the beach obstacles more intricate, and naval support much less. Still,

21,400 men were put ashore that day, with 1,200 casualties. This was the beach where the inflatable-skirted tanks came into their own, amazing Canadians and Germans alike as they arrived on the shingle, retracted their skirts, and then advanced and started firing. The seawall at Juno was considerably higher than at Omaha, but the Funnies went over it, cut through the barbed wire behind, and flailed through the minefields, with exhausted and overladen infantrymen trying to keep up. Then, perhaps as a natural reaction, most units slowed down, bivouacked, made tea, and fell fast asleep. Montgomery had planned for a very fast postinvasion move against Caen and right down the Orne River valley, but it didn't happen like that. The 6th British Parachute Division had made a stunningly successful capture of the Orne bridges, the three beachheads had held, and many more troops and tanks were pouring in. But within a week much tougher German opposition was being sent to the area, and there was to be no fast, expanding exploitation, as had happened in North Africa and Sicily. There was no second Dieppe, either, to the deep satisfaction of a now vastly grown Canadian Army.

The third D-Day tale was a far less happy one for the Allies: it was the dreadful, grinding slaughter of the American infantry units who landed (or sought to land) on Omaha Beach. The amphibious assault faltered horribly, with tanks, trucks, and overladen men tumbling to the ocean's bottom. The casualty list is dismal. Only five of the thirty-two amphibious tanks launched way out at sea came ashore, and thirty-two of the fifty howitzers sank with the flat barges that were transporting them. Most soldiers found themselves on the sand with only the weapons they carried, and pushed forward because it was impossible to return into the maelstrom at the water's edge. They then found the beach obstacles were virtually intact, and the crossfire from the gullies was obliterating. With most of their tanks sunk, the craft carrying their howitzers and cannon washed away, and many of the engineer teams drowned in the overlong watery crossing, the infantry and the Ranger platoons pressed on. The tanks that did land were hit within minutes, although a few bulldozer tanks hacked gaps in the defenses. American destroyers came in as close as 1,000 yards offshore—way too close to possible shore minefields—to fire at the German defenders.

By late morning the rising Atlantic tide cruelly narrowed the strip of beach still further, crushing new vehicles onto the damaged ones already ashore. The situation was so desperate that when Montgomery learned of it later in the day, he—like Clark at Anzio—briefly considered moving the American follow-on units from Omaha to the British-Canadian beaches; Bradley thought the same, and also wondered about moving the fresh waves of troops onto Utah Beach instead, though one suspects that either change would have led to unimaginable confusion offshore. In fact, the sheer press of numbers (Omaha had 34,000 men and 3,300 vehicles for the initial landing, with a similar total in reserve), the desperate daylong shellings by the U.S. Navy and bombings by the USAAF, and the remarkable work of junior officers and NCOs in reestablishing order eventually cracked the German coastal defenses. The troops got off the bloodied beaches and up to the bluffs, to hold a lodgement only a mile deep. Unless Rommel's panzers attacked (which they were not permitted to do), the 1st and 29th U.S. Infantry Divisions had made it, at bitter cost. They had not been thrown back into the sea. Still, when Eisenhower and Ramsay anxiously surveyed the beach from their warship later the next day, they were clearly much disturbed at how close it had been, and how confused and precarious things still seemed.

The U.S. Army suffered some 2,400 dead, wounded, and missing at Omaha. That grisly total is not large compared with losses in some Civil War battles, those on the first day of the Battle of the Somme, or the fatalities that occurred in the battles being fought at the same time on the Eastern Front, but it was the highest among the five landings that day, caused consternation and worry among the Allied commanders, and has provoked controversy ever since.

At one level, it is not difficult to explain why Omaha Beach was much the hardest to take and hold. The bluffs behind the beaches were considerably higher than elsewhere and could only be penetrated by going through the "draws" (small river valleys cutting through to the sea from the villages inland). The Germans had built numerous artillery and machine gun emplacements near the end of the ravines, but they were sited to fire out obliquely, covering the beaches to the right and left, rather than exposing these bunkers directly to the warships offshore. They were almost equally hard to spot from the air, whether by reconnaissance or in the actual bombing attacks. And the bluffs made any effort by tanks or other tracked/wheeled vehicles to get off the shoreline exceedingly difficult. Even with fine weather and calm waters, this would have been a difficult nut to crack, and no such favorable conditions prevailed. Prudence might have suggested that it would have been better to avoid the area altogether, but since the Allies had judged it vital to seize the larger Normandy beaches *and* to undertake a stroke into the Cotentin Peninsula to seize the harbor of Cherbourg, they could not allow a glaring gap to exist in the middle for the Germans to exploit; that was what had nearly ruined them at Anzio. The assault must take place.

Nevertheless, it is hard to accept the contention that the stark American losses at Omaha were just the result of rotten tides, hostile topography, and cunning German defensive systems. There is also evidence of poor battlefield management, combined with excessive self-confidence, which was never wise in a fight against the Wehrmacht. The naval bombardment was extremely short and thus very light (especially given the enfiladed position of the German bunkers). Admiral Hewitt's offshore control team had hoped for surprise, but that is hard to reconcile with the fact that at midnight more than a thousand heavy RAF bombers began massive attacks on German batteries along the Normandy coast and that by 1:30 a.m. the 82nd and 101st Airborne had already begun to arrive inland, or with the fact that the U.S. Navy's bombardment started twenty minutes *after* the British naval firing, which could be heard just up the coast, even though the British-Canadian landings were to take place an hour later. Bad luck twisted the knife: when the B-17s came in to crush the German defenses, they were frustrated by clouds. "Not a single bomb landed on the beach or bluff," observes Stephen Ambrose; cowsheds 3 miles (5 kilometers) inland took the hits.[44] The landing craft's rockets fell short, into the surf, and the swell seemed stiffer and higher at Omaha Beach than elsewhere.

With the wind increasing and the waves heightening, it was an act of folly for local commanders to order the duplex-drive tanks, the deep-wading Sherman tanks, and the craft carrying the howitzers to be launched more than 5,000 yards offshore. There were no LVTs (landing vehicles tracked); they were all in the Pacific. The Shermans that did get on the beaches fought as best they could, but here the Allies had no mine-clearing flail tanks, wire-cutting tanks, or fascine tanks. If they escaped drowning, the American troops were pounded on the beaches and at the entrance to the draws by the units from the German 352nd Infantry Division, which had been moved behind the Omaha beaches sometime earlier without, disturbingly, Bradley's staff being aware of it. The U.S. Army eventually got off the bloody shore simply by fighting to the top of the bluffs and forcing the Wehrmacht back. But not having other types of motorized equipment adds to the list of things that might have been done better, or even just tried.

Overall, the Longest Day gave the Allies an extraordinarily good result, a remarkable reward for what had been remarkable preparation, training, mobilization, and execution. At the end of that evening Ramsay, while noting his worries about Omaha and the western beaches, wrote in his diary: "Still on the whole we have very much to thank God for this day." Depending upon which sources are consulted, between 132,000 and 175,000 Americans, British, Canadians, French, and Poles had swarmed ashore that day, taking around 4,900 casualties—far fewer than the planning estimates. The front stretched over almost 50 miles; there were big gaps between some beaches, to be sure, but they were gaps that the Germans had no power to fill.[45] Their resistance had been much less than feared, in large part because their best troops were not at the beachhead defenses. Commando units and Canadian platoons found themselves fighting old men, Hitler Youth, and east European conscripts. However, this was not a good indicator of what was to come, especially not for the Canadians, who would soon find themselves attacking fanatically held German defenses around Caen, day after day, week after week, with microscopic progress. But that was a later story. Right now, in Churchill's words, "a foothold has been gained on the continent of Europe."

Yet striking swiftly some 5 miles inland, or grasping firmly onto hard-captured bluffs at the end of the first day, was not enough. A successful invasion required a massive follow-up effort; it demanded that more and more men, armaments, and supplies flow through the beachheads as the attacking armies spread out. And, as will be clear from the above sketch of the five landings, Normandy did not possess a major port between Cherbourg, in the far west, and Le Havre and Dieppe, much farther up the English Channel. Overlord therefore included as an essential component something that had never before been seen in the history of amphibious warfare: artificial harbors. At Churchill's instruction, Anglo-American construction teams were to create maritime havens, with the further requisition that they would also have to construct breakwaters that could float up and down with the tides.

The resultant Mulberry constructions were gigantic, multi-thousand-ton concrete caissons that nonetheless had enough ballast space within them that they could be towed on flat barges across to Normandy by the unsung heroes of the entire operation, Admiralty tugs, and then linked together like domino pieces into a giant breakwater (with, amazingly, roadways on top). At that stage, they were gently sunk into the shallow ocean bottom. There were also floating breakwaters and floating piers that acted as steel bridges from the sea, an absolute requirement since the tidal rise and fall was as much as twenty feet. Since they were not solid to the seabed, and thus were vulnerable to stormy weather, they were in turn protected by dozens and dozens of old merchantmen and warships (nicknamed "Gooseberries"), which were sunk in line to act as an outer barrier. Despite strong tides, Allied craftsmen and sailors began to anchor the two Mulberry harbors on June 9; on the day following, the Gooseberries were duly sunk. A staggering 1.5 million tons of steel and concrete were thus put in place.

Of course, there was much chaos at first, but actually far less than what would have attended thousands of light craft stranded on or grinding along an open beach. Outerwork harbors were a reassuring sight; watching U.S. trucks pouring along the artificial road on top of the caissons from ship to shore was even more reassuring. It told the average sailor and soldier that there really was a plan. Six days after D-Day itself, some 326,000 men and 54,000 vehicles had been brought across the Channel.

As noted above, on June 19 a massive Atlantic storm ripped much of this apart and dramatically slowed the daily infusion of troops, vehicles, and stores into Normandy. But it was not enough to alter the course of Overlord. Three days later, the flow of reinforcements resumed, and the clearing clouds allowed the USAAF and RAF and the heavier warships to resume their pounding of those German units at last directed to the invasion zone. By that stage, while the British-Canadian armies were placing maximum pressure upon Caen and drawing large Wehrmacht forces into mutual debilitation in and around the city, faster-moving American units under Bradley and, slightly later, Patton were pushing southward. By that time, too, the westernmost divisions under Collins had seized Cherbourg itself, a wonderful logistical resource once the German demolitions had been repaired, since American supplies and troop reinforcements could sail directly to the European theater, rather than going via Glasgow, Liverpool, and Southampton. Throughout June and July the Germans held on around Caen against repeated British and Canadian assaults, another testimony to their capacity for defensive warfare; but in so doing they drained their reinforcing divisions into a sort of Verdun-like grinding battle that, finally, they could not win. Meanwhile, the Americans were preparing Patton's U.S. 3rd Army to burst south, as it did in late July and early August. With Rommel badly hurt on July 17 (by a lone Spitfire shooting up his staff car) and his replacement, von Kluge, committing suicide after the July 20 assassination plot, plus Montgomery's 21st Army Group charging over the Seine and toward the Somme, the Battle of Normandy was over. On August 25 advance units of the Free French armies arrived in Paris, to immense jubilation. Two days later the Supreme Allied Commander himself, General Dwight D. Eisenhower, entered the French capital for the first time since 1932, when he had been working on the manuscript of General Pershing's aptly named *Guide to the American Battlefields in Europe*.[46] One wonders whether he still kept in his wallet a copy of his June 5 letter admitting defeat and full responsibility.

The Allied forces under Eisenhower had a long way to go to Berlin. Despite reinforcement of the original landing troops by another three million men, and despite an Allied aerial predominance that became greater and greater, relentlessly pounding the Wehrmacht in the field, devastating the supply lines from the Third Reich, and wrecking the industries and cities of the homeland itself, the German army fought on determinedly, sometimes aggressively (as in the surprise December 1944 counterattack in the Battle of the Bulge), and always with remarkable tactical efficiency. Nonetheless, this was no longer an amphibious war. It was an air-land campaign, an updated version of the Allied drive toward Germany between August and November 1918. The offshore battleships were no longer needed, the artificial harbors could be left to the tides, and the landing ships and their crews were sent off to the Pacific.

After the double-headed Normandy and Marianas campaigns of June 1944, the pace of amphibious landings in each hemisphere went to a different rhythm: tapering off in Europe, but building up in the Pacific. On August 15, 1944, the oft-postponed and diminished Allied invasion of the south of France (Operation Anvil, later termed Dragoon) took place, chiefly with French and American troops. By that stage the Wehrmacht was pulling out of France, so although this flanking strike, which was briefly resisted, was important in bringing Allied troops up to the southwestern borders of Germany and increasing the military pressures upon the Third Reich during the last eight months of the year, it had none of the significance of the great campaigns to the north and to the east. By September 1944, too, Red Army pressure across the plains of Poland had compelled the Wehrmacht high command to loosen its grip upon Greece and Crete. The next month British troops were reentering (without much opposition) battlegrounds from which they had been tumbled in that awful year of 1941. But this was no real amphibious operation, and the large Royal Navy force could savor an untroubled experience as they anchored off the Piraeus: another return from the sea, yet in waters where so many of its sister warships lay fathoms below.

There was much more enemy opposition when on November 1, on the other flank of the assault upon the Third Reich, a mixed bag of British Army units (English, Belgian, Canadian, Scots, Free French, Norwegian, and Polish) launched an attack on Walcheren, of all places. Seizure of the saucer-shaped island (which consisted of mere rimlands after RAF Bomber Command had blown up the outer walls) would free the Scheldt River and give access to the great harbor of Antwerp. Since the lack of a deepwater port in the southern North Sea was acutely affecting the supply situation for Eisenhower's armies, the taking of Walcheren this time around had a lot more strategic purpose to it than in 1809. By the end of the month that objective had been achieved, albeit after massive bombardments and with severe losses to the commando battalions.[47] Although the German garrison consisted chiefly of sick and recuperating troops, it held on weeks longer than Montgomery's planners had anticipated. Nowadays, the second Walcheren campaign is, like Dragoon and the return to Greece and Crete, a footnote to history. But the assault upon the Scheldt estuary positions is a reminder of how difficult such operations remained, even when employing overwhelming force and close to the crest of victory.

Amphibious Warfare and the Role of Planning

The Second World War witnessed the development and intensification of armed conflict in so many different dimensions—from armored warfare to strategic bombing to special operations—that it would be silly to claim that the amphibious campaigns themselves were the chief military expression of the 1939–45 conflict. They were, however, undoubtedly the most complex. The battle over the Atlantic convoys was a mixed struggle of sea power and airpower. The strategic air offensive against Germany was a contest between aerial forces. Blunting the Nazi blitzkrieg was essentially a bare-knuckle fight on land, with increasing airpower contributions. Only amphibious warfare, whether in Europe or in the Pacific, involved land, sea, and air operations in triangular harmony—or lack of it. Some historians describe the Normandy campaign as "triphibious," an awful-sounding word but not an inaccurate one.[48] A successful landing on a hostile shore was not only a massive tactical and operational problem in itself, but one that was contingent upon solving the command of the sea and control of the air problems first. It is a sharp reminder to authors of single-service campaigns in the 1939–45 conflict about how multidimensional the Second World War really was.

For these reasons it isn't possible to point either to a single breakthrough (like the long-range escort fighter in early 1944) or to a coincidental cluster of improvements in systems and weapons (as in detecting and destroying U-boats in the middle of 1943) that would organize the argument here on how the Allies eventually learned to land on an enemy-held shore. The circumstances varied from case to case: for example, the factor of surprise was important to MacArthur's leapfrogging campaign, and deception played a big role before D-Day, yet on Saipan and Guam, Iwo Jima and Okinawa, the entrenched Japanese garrisons just sat and waited for the assaults that were to come, with no space for maneuver. Bombarding the Normandy beaches for seven whole days, as was to occur at Okinawa, probably would have been disastrous—by the time the Allies landed, they might have faced twenty or more fully alerted German divisions on the other side of the hedgerows. Again, minute-by-minute orchestration of the beachhead was vital in many operations such as Normandy, but rather less so in Sicily, where the beachhead opposition was so small. Fluctuating clouds and fogs always made airpower more or less useful. An unusually low tide caused havoc at Tarawa; an unanticipated swell pushed the Utah Beach troops toward a safer part of the coast. There was no perfect recipe for a successful strike from the sea, however well one prepared for it.

Perhaps the best way to think about this, then, is to arrange the various aspects of amphibious warfare into first-order and second-order contributions. Thus, although offshore naval bombardments were impressive and on at least two occasions (Anzio and Omaha) really helped to blunt a counterattack upon a precariously held beach, only rarely did they destroy a determined enemy's capacity to fight back from secure bunkers.

By contrast, command of the air over the landing areas and the approaches to them was always critical, and here, by the beginning of 1944, the Allied advantage was becoming steadily more obvious, at least in the European theater. Moreover, the types of aircraft used for close-air support over the beaches (Marauders, Typhoons, Thunderbolts) were also well equipped to destroy the enemy's land forces during the breakout operations that followed.

Did the many improvements in amphibious weapons systems—ranging from Hobart's Funnies in Europe to the U.S. Marines' deployment of flamethrower and bulldozer tanks in the Pacific—constitute a first-order explanation for the increasing capacity to burst through an enemy-held shore? Yes, probably. Did the increasing employment of highly practiced advance units and of special forces such as the commandos and the Rangers help? Of course they did, and in some cases (taking the Orne River bridge, even the widespread scattering of the U.S. 82nd and 101st Airborne Divisions that so confused the German defenders) decisively so. There is also no doubt that the mass production of landing craft (whether the LVTs and infantry DUKWs, those carrying tanks or trucks, or those firing rockets) and of the equally important mother ships that brought them across the oceans to their offshore launch positions was absolutely critical. Indeed, take away those vessels and it is difficult to see how any of the larger landing operations described above could have occurred at all.[49]

There were other important aspects to amphibious warfare in play by 1944 that simply had not existed four years earlier. Having a command and control HQ offshore, separate both from the naval warships and from the landing forces, greatly reduced miscommunications and other errors. Having a single officer responsible for keeping order on the beach, and having much-improved wireless links back to the HQ ship for a forward directing unit to call in fire support, was another major advance. Having specialized underwater demolition teams and beach clearance groups was vital (just where did one dump all that barbed wire?). Having much smoother flow-through logistics greatly improved, though never fully eliminated, the predictable frictions involved in moving vast numbers of men, weapons, and supplies in the midst of battle.

But above them all, and orchestrating them all, sat an interservice, combined-operations organization (Ramsay's, essentially, and in the Pacific, Nimitz's) that had, after many setbacks and disappointments, figured out *how to do it*. If one had to point to a single feature of this story of the evolution of Allied amphibious warfare, it would probably be here, in the superior, even sophisticated ordering of the many moving parts. *Orchestration* is a term that might sum all this up. The multifaceted orchestra for a large and complex landing absolutely needed a conductor—and that could not be Churchill and Roosevelt, or the Combined Chiefs of Staff, or the powerful heads of the individual services. It had to be a different sort of person: an organizer, a planner, a problem solver.

The early American and British planners, operating under the shadow of Gallipoli and in times of acute fiscal severity, understood the special needs of amphibious operations from the beginning. The ideas that started with Pete Ellis and the other precocious Marine Corps planners in the 1920s, and a little later with the British Inter-Service Training and Development Centre in the 1930s, steadily metamorphosed via Mountbatten's Combined Operations Command and the USMC Marine Expeditionary Forces, and through the sobering experiences of Madagascar, Guadalcanal, Dieppe, and the Gilberts, into an end product that had its apotheosis in the twin Normandy-Marianas assaults of June 1944. Isley and Crowl put it nicely in their 1951 classic *The U.S. Marines and Amphibious War* when they observe that a close examination of those early, difficult operations "shows that the chief shortcomings were not in doctrine, but rather in the means to put existing doctrine into effect."[50]

It was simple, if one understood the basic operational principles. Amphibious warfare was a special kind of fighting that required many special ingredients: new interservice command structures, novel and often very weird-looking weapons systems, incredibly complex logistics, an extremely high level of battle training, very sophisticated low-altitude aerial support, and smart ways of directing masses of troops and vehicles on and off narrow beachheads. Getting all that right meant a very good chance of winning. But ignoring any of those critical requirements most likely would mean heavy punishment. And there was always the element of luck.

Had one brought back those early visionaries and planners to witness the first day or two of the June 1944 landings, they surely would have been staggered at the enormous size and intricacy of what was taking place before their eyes. But it is difficult to think of many features to which, conceptually, they had not given thought years before there was any prospect of realization. Like certain other strategic thinkers who had applied their great brains to the challenges posed by modern war, they were prophets before their time.

CHAPTER FIVE

HOW TO DEFEAT THE
"TYRANNY OF DISTANCE"

The Japanese offensive plan, and action, had profited greatly by the strategic advantage of Japan's geographical position. . . . [T]he outcome of their rapid conquests was that they had covered Japan with concentric rings of defence that provided formidable obstacles to any countermove toward Japan that the Western Allies attempted.

—B. H. LIDDELL HART, *History of the Second World War*

[Admiral] Koga's intention [was] to give battle anywhere along the new defence line. Without the initiative and any control over the timing of operations, [his plan] was as good as any that could be devised, but its most obvious weakness . . . was that the American carriers could descend on any single part of the line and overwhelm it before it could be effectively supported.

—H. P. WILLMOTT, *The Great Crusade*

Like other massive campaigns of military conquest in modern times, most notably those directed by Hitler or by Napoleon, the extraordinary Japanese offensives of 1941–42 were within a relatively short time to become the victim of what one historian has termed "the tyranny of distance": that is, they failed to recognize the natural limits upon human endeavor imposed by geography.[1] The Japanese high command overreached itself. It would have been difficult to convince the frightened inhabitants of Darwin and Brisbane (or their equally nervous equivalents in San Francisco and Bombay) of that geopolitical fact in the spring of 1942, when what had been an incremental Japanese challenge to the Anglo-American hegemony in East Asia and the Pacific during the 1930s finally burst into a violent, all-consuming fury of aggression coming ever closer to their shores.

Yet it was true. Japan, which had never been beaten in war before, went too far. As the present chapter will show, however, this was not simply an innate matter, like the waxing and waning of a tide. In contrast, Japan's vast empire was brought to its end by external means, chiefly through the impressive deployment of American resources, personnel, and weapons systems. The two combatants did not spar in a fixed space, like a boxing ring, but across a vast geographic arena in which the exploitation of distance, time, and opportunity by each combatant's leaders and planners was just as important as the morale of their fighting forces and the quality of the weapons. There was an invaluable bonus to be secured by the side that best understood, and best prepared for, the peculiar geopolitics of the gigantic Pacific sphere. In this situation, the maritime-oriented Americans had a clear advantage over a Japanese military leadership fixated upon land campaigns across China.

It was scarcely surprising, however, that imperial Japan should seek to expand abroad once its centuries of self-imposed isolationism were replaced after the 1868 Meiji Restoration by a drive to modernize, though not liberalize; in this regard, it was simply imitating earlier successful colonial powers, especially those of the West.[2] An impressive turn toward industrialization and export-led growth massively increased its productive capacities and provided the basis whereby it alone among the states of Asia and Africa possessed, by the 1920s, a modern army, navy, and air force—in fact, among the best in the world. All this created a critical and growing dependence upon foreign foodstuffs and raw materials and spurred Japan's nationalists to argue for acquiring secure sources of supply, like those enjoyed by the British Empire, the United States, and the Soviet Union.

The real measure of comparison for Japan—in fact, the only comparison—was Great Britain, the other politically coherent island state that broke beyond its coastal limits to become something much larger, a nation that had organized its navy and army to ensure control of distant sources for supply of the homeland. Since getting that tin, rubber, iron ore, copper, timber, and (especially) oil was essential, the only question remaining for Tokyo was whether those sources could be obtained with the acquiescence of the established powers or had to be gotten against their wishes, which obviously raised the risk of war. The debate among the Japanese political and military/naval elites was, therefore, not whether their country should become another world actor on the lines of Britain and Germany but how and when to achieve that natural goal. A long-standing member of the Japanese house of notables such as Yamagata Arimoto, looking back over the previous five decades from 1920, might occasionally have wondered at the transformation of Japan's international position since his youth—wondered not just at the transformation of medieval cities such as Yokohama to look ever more like Southampton and Baltimore, but also at the steady array of overseas acquisitions by a nation that had virtually cut itself off from the world for over three centuries. Yet neither he nor anyone else in power questioned Japan's right to expand.[3]

The transformation was remarkable. The 1894–95 war with China had not only crushed the disorganized armed forces of that decrepit empire but also brought to Japan the strategically vital island of Taiwan. In 1904–5 Japan shocked the world by defeating czarist Russia, at sea (Tsushima) and on land (Port Arthur, Mukden), thus gaining both southern Manchuria and the Korean peninsula. The most remarkable battle of that war was the overwhelming victory of the Japanese navy over the Russians at Tsushima (May 1905); Admiral Tojo's achievement inspired the service for decades to come. Nine years later, invoking in a very liberal sense the terms of its alliance with Britain, Tokyo took advantage of the outbreak of the First World War in Europe in 1914 to seize Germany's eastern empire, that is, the treaty concessions in northern China, Shantung and Tsingtao, plus the Central Pacific island groups of the Carolines, Marshalls, and Marianas—obscure enough at the time, but absolutely vital in the Pacific War thirty years later.[4] At this stage, Japan's geopolitical position was very strong. With the European Great Powers embroiled in war and Woodrow Wilson's America trying very hard to stay out of things, Tokyo had a free hand across the entire region.

Yet Arimoto and his fellow nobles were, in temperament, not unlike their British aristocratic equivalents or the Prussians after 1871:

they knew when to stop, when to display moderation, and, in Japan's case, when to compromise with great powers who in military-industrial terms were much bigger than they were. In 1915 the Japanese Foreign Office had made its infamous "21 Demands" upon China (which would have led to a virtual Japanese suzerainty) but backed away quite swiftly following vigorous American diplomatic protests. It was surely more important to get international recognition of its takeover of the German colonies, which it duly did at Versailles. There was also great prestige for any Asian nation that occupied a prominent place among the "big five" in the Paris negotiations, and to take equally prominent membership in the League of Nations.

The older Japanese elites were, therefore, as willing as those in other governments to settle for compromise solutions in the Washington treaties of 1921–22: all participants accepted recognition of the territorial status quo across the entire East Asian and Western Pacific region, promises about the nonfortification of bases, and, in particular, very strict limitations on the number, size, and tonnage of the world's leading navies. But that was probably the limit of Japanese willingness to proceed by negotiation. The refusal of the West to agree to a racial equality clause regarding all peoples at Versailles, the British abandonment of the Anglo-Japanese Alliance in 1922, the passing away of the old *genro* (the traditional aristocrats, somewhat like the English Whigs), the rise of new nationalist ideas about the country's special culture and special place in the world, the radicalization of the junior army officer corps, and the inherent weaknesses at the very top of the political system pushed Japan away from being a status quo power to being a revisionist power, just a few years ahead of Hitler's Germany and Mussolini's Italy.*

The revisionist events followed helter-skelter upon each other: the Kwantung Army's coup in Manchuria from 1931 onward, the seizure of those massive lands, the decision to defy world opinion and leave the League of Nations in the same year (1933–34) as Hitler's Germany did, the notification in 1935 that Japan would no longer consider itself as

---

*The Japanese emperor was descended from God and was not to interfere in politics. But he was also the supreme warlord, and the army and navy (theoretically) reported only to him. The services were not to be controlled by civil authorities—so the power of the cabinet and the Diet (parliament) in matters of war and peace was limited.

being bound by the Washington and London naval treaty restrictions, the adhesion to the Anti-Comintern Pact and the large-scale move into China proper in 1937, and the various efforts (including "accidental" bombings of their warships) to push the West out of East Asia. By 1940–41 there occurred another major coup, the takeover of French Indochina from a very reluctant but helpless Vichy regime, a move that in turn not only threatened Chinese Nationalists along a new flank but gave Japanese naval airpower control over the South China Sea and beyond.

Up to this point, things had gone extremely well for Japan.[5] Then came the decision by Roosevelt in July 1941, supported by the British and the Dutch, to freeze all Japanese commercial assets, essentially cutting off Japan's consumption of oil (88 percent of which was imported). The empire of the Rising Sun could either buckle under or strike out to gain its needed energy supplies and other vital war materials. If it did not strike, its economy would grind to a halt, and its China campaign would crumble. The air force's planes would sit quietly on the runways, and the navy's warships would remain at their moorings. Unless it resisted this foreign pressure, the argument went, Japan would be tumbled back to its pre-Meiji, medieval condition.

Accepting second-class status was simply inconceivable to this generation of Japanese military and naval leaders. Virtually all of them— Admiral Nagumo (who was to head the Pearl Harbor operation), General Tojo (who pushed so hard for the war), the brilliant Admiral Yamamoto (who headed the navy in 1941)—had been young officers in the First World War; they had witnessed Japan's advances then, and had seen it recognized as a great power in the Versailles territorial carve-ups. They could only go forward, to fulfill the national destiny and end the West's dominance in East Asia. The military logic followed from this political and economic rationale. Thus the compelling commercial need to gain the oil and rubber and tin of Sumatra and Malaya led to the operational decisions to strike at Hong Kong, the Philippines, Borneo, Java, and Singapore, eliminating American, British, and Dutch armed forces and seizing their bases. And the strategic need to defend

THE JAPANESE EMPIRE'S EXPANSION AT ITS PEAK, 1942
Like the German conquests in Russia, another case of overstretch.

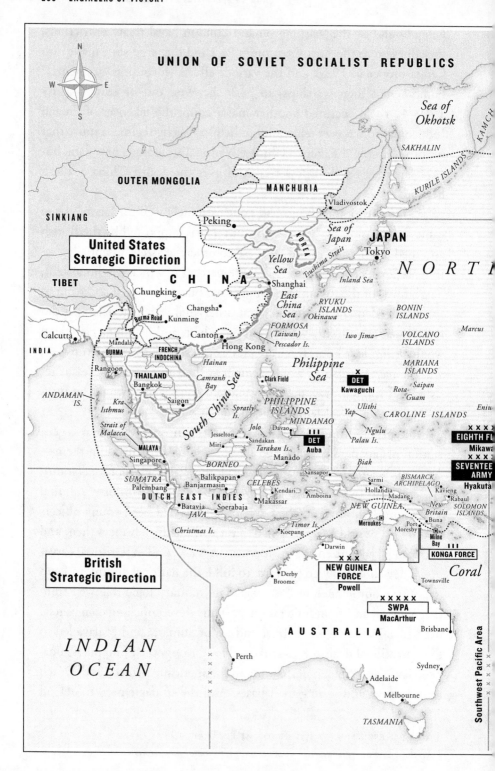

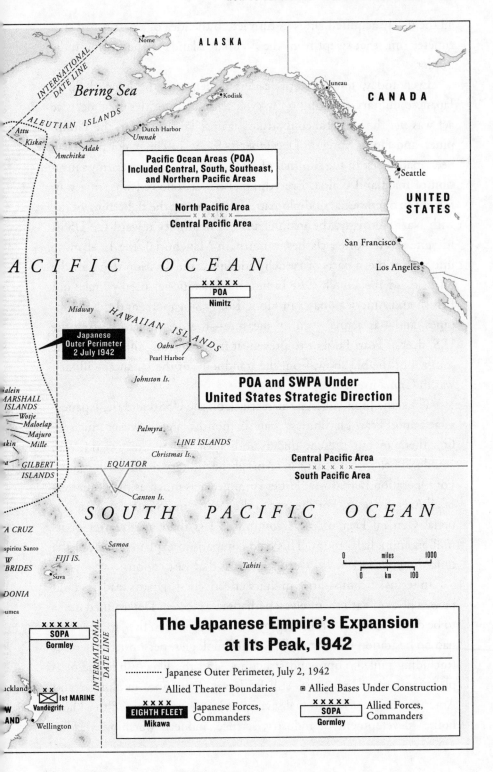

ALASKA

Nome

Bering Sea

Juneau

CANADA

Kodiak

ALEUTIAN ISLANDS

INTERNATIONAL DATE LINE

Attu
Kiska
Adak
Amchitka

Dutch Harbor
Umnak

Seattle

Pacific Ocean Areas (POA)
Included Central, South, Southeast,
and Northern Pacific Areas

UNITED
STATES

North Pacific Area
x x x x x
Central Pacific Area

San Francisco

P A C I F I C   O C E A N

Los Angeles

x x x x x
POA
Nimitz

Midway

HAWAIIAN ISLANDS

Japanese
Outer Perimeter
2 July 1942

Oahu
Pearl Harbor

Johnston Is.

POA and SWPA Under
United States Strategic Direction

Palmyra

LINE ISLANDS

alein
MARSHALL
ISLANDS
Wotje
Maloelap
Majuro
akin
Mille

Christmas Is.

EQUATOR

Central Pacific Area
x x x x x
South Pacific Area

a
GILBERT
ISLANDS

Canton Is.

S O U T H   P A C I F I C   O C E A N

A CRUZ

spiritu Santo
W
BRIDES

Samoa

FIJI IS.
Suva

Tahiti

0   miles   1000

0   km   100

DONIA

umea

x x x x x
SOPA
Gormley

INTERNATIONAL DATE LINE

The Japanese Empire's Expansion
at Its Peak, 1942

uckland

x x
1st MARINE
Vandegrift

Wellington

Japanese Outer Perimeter, July 2, 1942

Allied Theater Boundaries   Allied Bases Under Construction

x x x x
EIGHTH FLEET
Mikawa

Japanese Forces,
Commanders

x x x x x
SOPA
Gormley

Allied Forces,
Commanders

W
AND

all the newly acquired lands in turn led to the idea of a garrisoned "perimeter rim" that swept from the Aleutian Islands to the Burma-India border.

Even today, few Western readers understand the logic chain of Japanese military thinking or Tokyo's sense of priorities. The decisive act was *not* the simultaneous strikes against Pearl Harbor, the Philippines, and Hong Kong on December 7–8, 1941. The major move had been made back in the summer of 1937, with the Japanese army's invasion of mainland China. Everything else that followed was, in a way, merely an operational or diplomatic consequence: the tightening of ties with Nazi Germany, the maintenance of neutrality toward the USSR (despite severe border clashes in northern Manchuria), the 1941 move into the southern parts of French Indochina, the decision to go for the oil fields of the Dutch East Indies, the operational need to take out British and American bases in Hong Kong, Singapore, and the Philippines, and—as a final security measure—the decision to destroy the U.S. fleet at Pearl Harbor to prevent it from blocking this drive to the south. But the big show, from the standpoint of the Japanese military, was in China itself.

Thus, the June 1941 German attack on the USSR and the Japanese attack upon Pearl Harbor less than six months later, however much we link them in our present understanding of the Second World War, could not be more different in terms of the military resources allocated. For Operation Barbarossa, Hitler committed as much as three-quarters of all Axis divisions. For the war in the Pacific and Southeast Asia, Imperial General Headquarters committed less than one-quarter of its million and a half troops. This bald comparison explains much of the unfolding of the 1941–45 struggle across that vast region.

In terms of comparative military effectiveness, Japan's leaders could be well pleased with themselves, perhaps too pleased. Determined never to be dominated or intimidated by Western practices, the Japanese had had no hesitation in copying foreign technologies, personnel structures (including ranks), command structures, and the like. While the people's standard of living was still far lower than those in Europe and America, the massive national savings of Japan were directed to a "hothouse" development of industry, science, modern warships, steadily improved aircraft, and extraordinarily high levels of training in their officer corps.

Like all interwar militaries, the Japanese forces possessed varied strengths and weaknesses. The army had been the dominant service since the successful Russo-Japanese War and was oriented toward the Asian continent, with a particular concern about a coming war with the USSR. Thus the Americans and British were of much less interest to the generals, until, of course, their economic blockades crippled the Japanese war machine. The army was large and disciplined, and both its choice of weapons systems and its training pointed to its vision of future conflicts. There was little effort to follow Liddell Hart, Fuller, Guderian, Tukhachevsky, and other Western advocates of fast armored warfare, for where could one deploy main battle tanks in Asia when no bridges were strong enough and few metalled roads existed? On the other hand, there was an increasing interest in preparing for and practicing landings from the sea, river crossings, and jungle and mountain warfare in Southeast Asia. Taking distant Pacific islands was not on their mind. Nor was there any point in creating an expensive long-range strategic bombing force, since what modern industrial targets were there to bomb—the wooden shanties of Shanghai? The fishing wharfs of Vladivostok?

The army's relations with the Japanese navy were fractured, and this wasn't just because of those typical interservice quarrels of the 1920s and 1930s about budgets. It was a far deeper quarrel about the nation's strategic purpose, since the navy thought a large entanglement on the Asian continent was folly and, having thoroughly imbibed the doctrines of Alfred Thayer Mahan, called instead for a focus upon the threat from the Western sea powers. The army, by contrast, was imbued with the Prussian military tradition and particularly impressed by the Elder Moltke's three swift land victories (1864, 1866, 1870–71) that led to the unification of Germany. Since neither side would concede to the other's viewpoint, the two services tended to go their own ways. While this threatened to make any larger military policy incoherent, it helped the navy in other respects. Above all, it meant that it could develop its own naval air arm, with a carrier fleet and suitable planes. Because the navy was much more invested than the army in twentieth-century technologies, and therefore more knowledgeable than their army equivalents about America, Britain, and Germany, some naval leaders (above all Yamamoto, who had been a naval attaché in Washington) were concerned about Japan's overall economic disadvantage. Still, when the decision for war was made in late 1941, the services felt they were ready.[6]

The amazingly wide Japanese tide of conquest lasted a mere six months, from December 1941 until June 1942. By that time, the Japanese Fifteenth Army had reached the border between Burma and Assam, in British India; it could go no farther. The carriers of the Imperial Japanese Navy had struck hard against Ceylon and the inadequate British naval forces in the Indian Ocean during early April 1942 and had the capability to move farther westward, against Aden and Suez; but they were pulled back from that intriguing possibility (which the army never liked) for operations elsewhere.[7] The navy's slightly later foray toward Australasia was blunted by the Battle of the Coral Sea (off southeast New Guinea) on May 7–8—where neither side won decisively, but which prompted the Japanese to withdraw northward for a while. The Japanese army's move from the north shores of New Guinea toward Port Moresby in the south was held in the mountainous jungles by hastily assembled American and Australian divisions under General Douglas MacArthur. Most important of all, the prospect of a Japanese drive across the Central Pacific, seizing Hawaii and thus threatening the American West Coast, was crushed at the vital carriers-only battle near Midway Island on June 4, 1942. Pearl Harbor was in large part revenged in the waters hundreds of miles west of Hawaii.

Such setbacks, however, were not regarded by Imperial General Headquarters in Tokyo as disastrous, perhaps not even as serious. In fact, while the naval leadership might be yearning to finish off the Allied (chiefly American) fleets once and for all, the dominant army faction could view the situation in Southeast Asia and the Pacific with some equanimity. Japan's armed services had done what was wanted, which was to tumble the hated Americans, British, and Dutch out of their own future Greater East Asia Co-Prosperity Sphere. They had seized the absolutely critical oilfields of Java, Sumatra, and North Borneo, which meant that Tokyo's most important strategic mission could be pursued more energetically than ever—that is, the subjugation of

China and achievement of unchallenged primacy over mainland East Asia. Amazingly, the Japanese army had deployed a mere eleven of its fifty-one divisions to achieve this vast array of conquests across the Western Pacific and Southeast Asia. All that was necessary now was to strengthen the outer perimeter rim with a series of island strongpoints and beat off any impertinent Western counterattacks. Once American and British noses had been bloodied, the effete democracies would recognize this strategic fait accompli and negotiate a peace in a year or two's time.

There is one further spatial and geopolitical point to be made about Japan's enormous territorial expansion during 1941 and 1942. It offered a fine example of what Liddell Hart had termed "an expanding torrent," that is, an attack whose arc steadily widened the farther that advances were made. But Liddell Hart had thought of such an operational expansion as being carried out by a relatively small group of fast panzer units in western Europe, breaching an enemy line and then spreading out for a farther 100 to 200 miles.[8] In the Pacific, as in Russia, the distances were to be measured in many thousands of miles. When the Japanese expeditionary forces moved toward Alaska, Midway, Hong Kong, the Philippines, the Gilberts, the Solomon Islands, New Guinea, Thailand, Malaya and Singapore, North Borneo, the Dutch East Indies, and beyond—to Burma, perhaps northern Australia—those few divisions were dispersing themselves across vast distances, while most of the army's troops were pushing into central and southern China.

Logically, then, the farther that these Japanese units advanced across the Pacific and through Southeast Asia, the thinner became the *density* of occupying troops in captured territory. As we have seen above, this was not unlike the Wehrmacht's contemporaneous dilemma. Hitler had insisted that the Nazi "torrent" expand to the north (Leningrad), center (Moscow), and south (Stalingrad) of the vast Russian plains, in addition to holding the Balkans, controlling all of western Europe, and maintaining a foothold in North Africa. Yet the distances involved across western Russia were nowhere as great as those between the Aleutians and Burma, and Hitler possessed far larger military and industrial resources when the counterthrusts came.

Thus, by the summer of 1942 Imperial General Headquarters in

Tokyo had overstretched itself and put most of its troops in the wrong place, but it didn't recognize that. It had achieved enormous territorial gains, its homelands were intact, and the booty from its conquests was pouring back home. While the Allies had held the Japanese advances on various fronts, there were no breakthrough counterattacks; not yet.[9] So to Imperial General Headquarters, things were not serious. The setbacks at the Coral Sea and in the jungles of New Guinea against MacArthur's forces were unsettling, and the losses of four carriers at Midway were regrettable. But there were ample resources to make some farther if less dramatic advances, perhaps probing the Assam borders, or advancing down the Solomon Islands chain, to make the perimeter ring more solid. MacArthur's slow progress in Papua and the tangled battles taking place around Guadalcanal seemed obscure, distant, and not of great import. In sum, by the closing months of 1942, alarm bells were not ringing; even a year later, one suspects that their sound was a very distant toll to Japan's continentalist generals, schooled in the Prussian tradition that control of the mainland was the essence of a proper grand strategy.[10]

Allied Strategic Options in the War Against Japan

Thus, because Japan had more or less achieved what it wanted, and because the military leadership at Imperial General Headquarters did not particularly need to go very much further, the onus was upon the Allies to alter things; it was they who had to take the offensive and then compel a Japanese defeat. This, in essence, was the strategic logic of the entire war in the Pacific and East Asia from the summer of 1942 onward, and it was also the basic assumption of the leaders and planners at Casablanca six months later. But how and where and with what means did one crush the Japanese Empire?

Moving from a defensive strategic posture to an offensive one is always a complex challenge, even to the most efficient and imaginative organizations, and in this case geography also made the Allied task one of extreme difficulty. The blunt cartographical fact was that the home islands of Japan were a very long way from *any* enemy takeoff point, unless it was from Siberia/Manchuria (but Stalin, fighting for his life, had no intention of opening up a second front while the Wehrmacht was still a thousand miles or more inside the Soviet Union). Thus, the turn of the tide in the Pacific was fundamentally different from that in Europe. Challenging though it was in military-operational terms, the task of crossing the English Channel to destroy Nazi domination of Europe was understandable and realizable, and was an operation that had been well rehearsed by the landings in North Africa and Sicily/ Italy. By contrast, the defeat of Japan could not be planned for until the Combined Chiefs of Staff had decided upon the takeoff point—or points.

Ruling out the Siberia/Mongolia option, then, the Allies could choose from four attack routes from the perimeter to the Japanese core, since Tokyo's expansionist drives from 1937 to 1942 fanned outward in so many different directions.*

The first alternative was to base the counteroffensive chiefly upon mainland China, the theater closest to Japan and most engaged in the fighting. The second would involve the recovery of Southeast Asia, that is, Burma, Thailand, Malaya/Singapore, French Indochina, Borneo, and the Dutch East Indies. The third would be to build upon the American-Australian command structure in the Southwest Pacific under MacArthur and push northward from Australia to New Guinea and the Solomon and Bismarck archipelagos to the Celebes and the Philippines themselves, which would then become a springboard to Formosa. From here, Allied armies might possibly link with Chiang Kai-Shek's mainland forces or turn to assault Japan directly, from the south. In this scenario, Luzon and Formosa would resemble Britain's own position as the launchpad for the Allied conquest of Nazi Europe. The fourth and final option would be to drive across the wide expanses of the Central Pacific, recovering the island groups that Japan had seized in 1942 (the Gilberts) and taking the empire's important man-date islands (the Carolines, Marshalls, and Marianas) as stepping-stones

---

*On the map, an advance upon northern Japan via Alaska and the Aleutian Islands seems a fifth option (it is, after all, the great-circle route for airliners flying from New York to Tokyo today). But the appalling and continuous storms and mists of the North Pacific would heavily curb America's airpower advantage and make large-scale am-phibious operations virtually impossible.

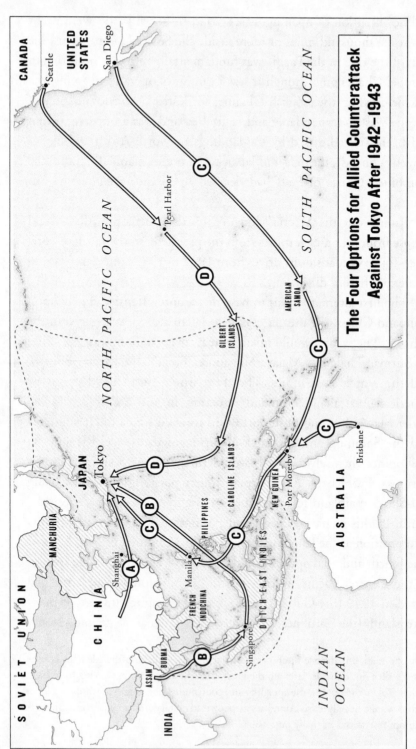

The Four Options for Allied Counterattack
Against Tokyo After 1942–43

THE FOUR OPTIONS FOR ALLIED COUNTERATTACK AGAINST TOKYO AFTER 1942–43
Given the ferocious weather of the Aleutian Islands of the North Pacific, there were four possible routes for an
Allied advance upon Japan. The Central Pacific one turned out to be by far the easiest.

to Iwo Jima, Okinawa, and the invasion of Japan itself. It would be much more of a "blue water" strategy, and would be run chiefly by the navy and the air force, not the army, at least until the actual invasion of Japan itself.

Of course, it was not a zero-sum game; the options were not mutually exclusive. All four theater options had legitimate claims, which will be discussed below. More sensibly, it would have been militarily stupid for the Allies to commit to only a single line of advance in the Asia-Pacific theater, for that, in turn, would concentrate the Japanese defenses. Applying pressure in all four fields of conflict would not only disperse the enemy's resources but also allow a switching of the more mobile Allied forces if a weak spot was detected.

Ultimately, one of these return routes proved to be decisive, but it is worth taking some time to examine the other three options, not just to understand better why they were less vital but also to see what light those operations shed upon the overall challenge of defeating Japan. All three pinned down enormous numbers of Japanese troops and aircraft (and, in the case of the Southwest Pacific campaign, Japanese warships) that otherwise would have been free to contest the chief American line of attack.

Advancing upon Japan via the China theater made a lot of sense at first sight. Although this was where the bulk of the Japanese army was fighting, it was also a place where the United States probably could build up a fair-sized air force, and it was difficult for the deadly Japanese carrier-borne air squadrons to reach. It was important for America to give support to the Chinese Nationalist armies, because, practically speaking, keeping China in the war consumed so many Japanese divisions. There were also significant sentimental aspects at play here as well. President Roosevelt was a strong supporter of the U.S.-China relationship. The American general "Vinegar Joe" Stilwell was Chiang Kai-Shek's chief military advisor. Claude Chennault's group of glamorous volunteer airmen, the "Flying Tigers," was there. So were a large number of American missionaries, teachers, and traders. From the viewpoint of the Army Air Corps planners, aware that Doolittle's squadron had actually landed in China after making its daring raid on Tokyo in April 1942, it was obvious that certain Chinese air bases were close enough to allow the strategic bombing of Japan's cities and industries. Moreover, American bombers based there would also be able to interdict Japanese maritime routes down the China Sea as they headed for Southeast Asia.

So why was the China option not taken, or taken only for a while, and with inadequate resources? The answer again is chiefly explained by geography. Inland China was simply too far away from the U.S. productive base and therefore too hard to supply in large numbers. Since all the waters of the Western Pacific and the Indonesian archipelago were dominated by the Japanese navy, the only Western material assistance could come from British India, over the "hump," that is, the enormous mountain peaks of the eastern Himalayas, and at a disproportionate logistical cost—even just carrying the gas supplies for U.S. bombers in China caused massive wear and tear and was highly uneconomical. Nor would this work for the deployment of U.S. ground forces. Since the consumption levels of an American army division, let alone of an army corps, were fantastically greater than those of any other armed force in the world, there was no way of putting such large units into southern China and supplying them by air.

Furthermore, given the Japanese high command's obsession with winning whatever the cost, it was bound to pour fresh divisions against U.S. and Chinese Nationalist forces and in particular against identified B-17 or, later, B-29 air bases in the south and west of the country—and, for once, the Japanese lines of supply would be shorter and easier. Looking back, and with the privileged knowledge of the massive American punch that was being assembled in the Central Pacific, it is hard to understand the Japanese army's continental thrusts in 1943 and 1944, driving the Chinese Nationalist forces farther southwest, and in turn sucking in its own brigades. This handed an advantage to the United States, for while it could not put heavy forces onto the Asian continent, it certainly could do enough to distract the bulk of the Japanese army. By giving a degree of support to the Chinese Nationalist government, in the form of vital munitions and medical supplies, plus Stilwell and other military advisors and the later B-29 squadrons, the United States kept the Chinese resistance going and helped to pin down millions of Japanese troops on the mainland. And as 1944 unfolded, those enormous forces found themselves being steadily cut off from their supplies by American submarines and aircraft.[11]

The Southeast Asia route was another alternative that had its advocates, especially in London and Delhi. It *was* an important theater, and
not just for emotional and symbolic reasons, such as the desire (especially Churchill's) to regain Singapore. It was a significant region of the
overall war, and attacking there did mean that Britain was showing a
commitment to the struggle against Japan and not leaving everything
to the Americans and Chinese (with Australian contributions). It would
pull a lot of Japanese soldiers and aircraft from the Pacific realm, and
the recovery of Malaya's tin and rubber and of the oil from Sumatra,
Java, and North Borneo would have brought Japan's war machine to a
halt. Moreover, and for the logistical reasons described above, it would
be impossible to do anything significant in the China theater unless the
Allied airfields in Assam and northern Burma were developed and
properly protected.

But the campaigning by the forces of Southeast Asia Command
under Mountbatten and his predecessors suffered from two great disadvantages. The first was that starting off the counteroffensive from the
Assam/Burma border and then advancing through such inaccessible
places as Imphal and Kohima left the armed forces a very long way
from Tokyo Bay, and the topographic, logistical, and health obstacles
would have remained immense even if far greater resources had been
allocated to this theater than they had been at Casablanca and the other
critical Allied conferences. The nature of this vast jungle-clad terrain
simply excluded certain forms of warfare that were commonplace elsewhere. There were no real industrial targets for strategic bombing (unless one wanted to blast Rangoon and Singapore), and even the
increasingly efficient tactical air squadrons, which did have targets,
were always last in line to receive the newer types of planes and equipment. There were no great tracts of open ground, as in North Africa
and the Ukraine, where the allocation of, say, 250 Sherman tanks would
have made a difference; there were few metallic roads through the rain
forests, and the wooden bridges were simply too weak for tanks—often,
getting even a medium-sized artillery piece across a river required the
use of ropes and pulleys. The airborne divisions were allocated to Europe. Special Forces in the jungles of Burma, like Wingate's legendary
Chindits and Merrill's Marauders (their American equivalent) played a
distracting role but could not sway the outcome. Even when Mount-

batten's regular units were finally strengthened in late 1944, the jungle continually blunted his matériel advantage. The seasonal monsoons were a nightmare that bogged down everybody's armies. Not even William Slim, arguably the best general the British Army produced in the war, could find an easy answer to flooded river valleys, impenetrable jungles, and the curse of tropical diseases; fighting the Japanese was easy by comparison.[12]

After early inconclusive operations, his forces did smash the Japanese garrisons in Burma and push all the way down the Irrawaddy in 1945. But the advance would have been far swifter had the British been able to make a series of amphibious hops along the coastline, akin to the increasingly accelerated ones MacArthur's units made from the Solomons to the Philippines in 1944. Yet since the landings at Anzio, Normandy, the Marianas, Leyte Gulf, and Iwo Jima always got greater priority for landing craft, such plans for Southeast Asia were repeatedly postponed, if not canceled altogether. So jungle battles were the only way to go, and as American and Australian forces had found out in New Guinea, they were painfully slow. It was only in April and May 1944 that the Fifteenth Army won its major victory at Imphal-Kohima, but that was still on the Indian side of the border. The British would not take Mandalay until March 1945, and Rangoon only by early May. By the time they could seriously plan for ambitious forward moves upon Malaya and Singapore, the Japanese had shifted their remaining aerial and naval forces to the Pacific, Okinawa had fallen, and the Americans were preparing for the invasion of the home islands. The great novelist John Masters could, in his classic war memoir *The Road Past Mandalay*, rhapsodize about "the old Indian Army" bursting down the (at last) tarmac roads that led into Rangoon's northern suburbs in the late spring of 1945, but to the Japanese—and the Americans, too—they were headed in the wrong direction. The Pacific War was won elsewhere.[13]

These critical factors of distance, logistics, and topography all meant that the Southwest Pacific theater clearly offered a relatively better route for the defeat of Japan than did the China and Burma options. Here the recovery campaign could begin just a short distance north of Australia's ports and air bases, and the American flow of troops and supplies could come fairly directly via Hawaii, Samoa, Fiji, the northern Australian harbors, and Port Moresby in southern New Guinea. Moreover, in the first year of the Pacific War this region simply had to be given a lot of attention by the Allies, because it was here—Papua, the Coral Sea, the central Solomon Islands—that the Japanese seemed intent upon cutting the American-Australian line of communications. As each side committed more and more troops, aircraft, and warships into the battles for Guadalcanal and New Guinea as 1942 unfolded into 1943, the urgency for a victory became an important factor, calling for even more resources. In addition, there was the powerful and intimidating figure of Douglas MacArthur himself, who (with enormous media and congressional attention) insisted that the fulcrum of the war against Japan was situated in his command.[14]

The New Guinea campaign on one hand and the Mediterranean campaign on the other possessed the same basic strategic profile: neither of them was headed directly at the enemy's heartland, but both built up such operational and political momentum that they could not be stopped. And, fortunately for the Grand Alliance, there were enough resources coming to the fronts by late 1943 and 1944 that the European and the Pacific campaigns could afford both the direct thrusts toward France and Honshu *and* the hugely expensive indirect campaigns as well.

There were some further strategic and operational advantages to be derived from the U.S. commitment to the Southwest Pacific theater. Until the Central Pacific island groups were captured and turned into advance bases, this more southerly region played to America's productive strengths. It could send an uninterrupted flow of supply and support diagonally across the ocean from its West Coast factories. Even if the Solomon Islands were a long way from San Diego, they were also a very long way from Yokohama. Second, because the battles were small-scale as compared to the Somme or Kursk, the campaign did not attract the attention it deserved in Imperial General Headquarters, which kept feeding limited reinforcements to the Southwest Pacific, though never enough to knock back the Allied advances. Finally, as this theater gradually sucked in more and more midsized Japanese forces, it also helped to blood the newly recruited Allied units, steadily building up their experience both of jungle warfare and of landings from the sea. Thus, for example, the First U.S. Marine Division, which came ashore in

1942 at Guadalcanal, was three years later still in the thick of the fighting at Okinawa, and had fought many more battles in between. Army general Bob Eichelberger, whom MacArthur sent to take Buna (New Guinea) with orders to seize it or not come back alive, led his units all the way from the villages of Papua to the northern Philippines and was preparing them for the great amphibious invasion of Japan when the atomic bombs went off. Unsurprisingly, his later memoir was called *Our Jungle Road to Tokyo*.[15]

But, given the astonishing and accelerated pace of Admiral Nimitz's planned thrust out of Pearl Harbor and through the Central Pacific by late 1943, and the comparative shortness and ease of that route,* it is difficult to argue that MacArthur's 1943–44 campaigns along the north shore of New Guinea had primary strategic relevance, even when the American-Australian divisions began to quicken the pace of things by leapfrogging over Japanese military strongpoints and thereby isolating them. By the time MacArthur's troops had finally taken Biak Island (August 1944) off the northwest shores of New Guinea, the Marianas had already been seized and the preparations for softening up Japan's inner defenses via the seizure of Iwo Jima and Okinawa, and by strategic bombing, had begun. It might fairly be argued that most of the Philippines could have been left neglected, and that Southwest Pacific Command could have joined much earlier in the northward turn toward Japan. It was not to be. As Liddell Hart noted, "Political considerations, and MacArthur's natural desire for a triumphant return to the Philippines, prevailed against such arguments for by-passing the great islands."[16]

Perhaps the stronger operational reason for the Chiefs of Staff agreeing to support the Southwest Pacific drive so fully was not so much that they accepted MacArthur's personal imperative but that they appreciated that these twin advances kept Tokyo perpetually off-balance—the best example being that of the Japanese main fleet rushing backward and forward over hundreds of miles between Biak and

---

* By "ease," I mean not that the actual operations were easy but that the targets themselves were small specks in the Pacific Ocean and, once the garrisons were isolated by superior American air- and sea power, they could either surrender or die where they stood. This was not like the much more open-ended assaults upon, say, Anzio and Normandy.

the Marianas in June 1944 as news (some of it erroneous) of the enemy's moves kept suggesting that the main threat was coming from the other direction.[17] With the Japanese navy beginning to run short of fuel and its aircraft being destroyed both around Rabaul and across the Philippine Sea by the forces of Southwest Pacific Command, it could hardly be said that MacArthur's forces played an insignificant role, but their greatest function overall was to distract the enemy leadership, wear down enormous numbers of Japan's armed forces, and give Nimitz and Central Pacific Command a vastly improved chance of going directly for the homeland.

So the three alternative theaters all contributed much to the defeat of Japan. The totals of the respective Japanese personnel losses (169,000 in India and Burma and a massive 772,000 in the South Pacific, compared to 296,000 in the Central Pacific) tell us that.[18] And by the end of the war large Japanese garrisons were strewn across Southeast Asia, the Dutch East Indies, isolated Pacific islands, the southern Philippines, Manchuria, and, above all, China: millions of men, protecting the wrong approach routes and weakening Japan's central defenses.

Fulfilling "War Plan Orange"

Since it was most unlikely that Japan would have been brought to its knees through Allied campaigns from mainland China and Southeast Asia, and since the route from Guadalcanal and Port Moresby via Manila was so laborious, there might seem a powerful inevitability about the remaining geopolitical option: an American drive across the Central Pacific in a series of thrusts that went from east to west, and virtually all north of the equator. Many authors are insistent: "In the end, only one campaign counted: Nimitz's thrust of the U.S. Navy across the Pacific to bomb Japan. Every other campaign, however bitter, however desperate, was either ancillary to this or a sideshow."[19]

Further geopolitical considerations confirm that assumption. By 1941–42 it was obvious that only the United States could end Japanese military domination of this vast region; neither Britain nor the USSR, both embroiled in the titanic struggle against the Third Reich, could possibly divert the resources necessary to deal with Japan. What was more, the munitions base of the powerful American war machine was increasingly to be found along the West Coast, roughly between thirty and fifty degrees north, which was exactly Japan's latitude. Since there was no intervening land directly between San Francisco and Tokyo harbors, and since an Aleutian route possessed the weather-related difficulties mentioned earlier, it made sense to advance along an axis that was somewhat south of the Tropic of Cancer, from the Hawaiian bases to the Gilbert and Marshall Islands, and then to the Carolines and Marianas, before turning northward to the Japanese mainland.

There was also the great benefit of America's interrupted possession of Hawaii, which, with its air bases, dockyards, and repair works, had an inestimable strategic capacity to control the Central Pacific, both as a bulwark against further Japanese expansion eastward *and* as a gigantic staging point for any return. Had Japan ever succeeded in seizing the Hawaiian islands, even if it planned no further moves toward the West Coast of the United States, its possession of bases on Oahu could have set back an American counterstroke for years and made the operational connection to Australia much more difficult. Retained in American hands, however, Hawaii was the sortie point for the occupation of the Gilbert and Marshall Islands to the southwest, and for everything that followed after that.

In a practical sense, too, the losses and gains of the Pearl Harbor and Midway battles helped to push the American armed forces further toward a Micronesian strategy. The devastation of many of the slow First World War battleships at Pearl had compelled the U.S. Navy, to a large extent unintentionally, into a carrier-centric force, a shift of weapons system that the Coral Sea and Midway contests so quickly confirmed as wise. But the lesson that those forceful aircraft carrier admirals such as Marc Mitscher and William "Bull" Halsey drew from the Pearl Harbor catastrophe was that it was always safer for the fleet to be at sea rather than constricted in port; in an age of sudden, swift airpower, a protected harbor actually offered very little protection, but was an invitation to attack. Yet if American carriers and their escorting cruisers and destroyers were to remain for lengthy periods on the high seas, they needed to create a sophisticated and mobile "fleet train" to follow and nourish the warships wherever they went. In the nineteenth century the Royal Navy had solved its supply problems by use of a vast network of naval bases and coaling stations, from Portsmouth and Gibraltar all the way to Sydney and Hong Kong. The twentieth-century U.S. Navy was to pursue a different logistical plan, which made large-scale operations across the Central Pacific possible for the first time ever. Still, the existence of the fleet train only gave the navy's flotillas longer "legs" while at sea; it could not itself form bases for long-range strategic bombing or for massive invasion forces. That required strategically located islands, to provide airfields and harbors. Fortunately, there were a small number of them, to the west and west-southwest of Hawaii, available for the taking.

The final reason a Hawaii-based operational drive through Micronesia was preferred by the navy was because the area commander would be an admiral (Nimitz) and not the difficult and imperious MacArthur. This was of vital concern to the chief of naval operations, Fleet Admiral King, a tough, unrelenting character determined to see the U.S. Navy remain an independent service, neither reduced to a fetch-and-carry force for MacArthur's drive from New Guinea to the Philippines nor playing second fiddle to the British, the American air force, or the Marshall-and-Eisenhower wing of the U.S. Army in support of their Atlantic, Mediterranean, and Normandy/Germany campaigns. Unsurprisingly, then, Central Pacific Command received most of King's attention and, when the new fast carriers and battleships were launched, the bulk of the navy's warship allotments. It also received many of the desperately required landing craft. By late 1943, the tools were at hand to carry out that move across the Central Pacific, which was designated War Plan Orange.[20]

What were those tools or fighting systems? The first, chronologically, was the U.S. Marine Corps, with its plan to seize bases in Micronesia— that is, the Marshall, Caroline, and Mariana island groups, which Japan had taken over from imperial Germany during the First World War. Any competent general staff, confronted with the geostrategic nightmare of getting 6,000 miles across the Pacific, would have recommended a stage-by-stage method, like moving across a sort of gigantic chessboard. The planners at Newport and Quantico had thought this through well before Pearl Harbor, and an impressive number of mid-level officers had worked on weapons systems and tactics that could turn an amphibious-warfare theory of seizing difficult, prickly coral reefs into a viable operation, thus converting those islands into strategic assets.

The second factor was the evolution of fast carrier groups. The U.S. Navy had been swift to understand the strategic and operational implications of the early British experiments of launching aircraft from a reconstructed "flattop" warship, and thus had pushed ahead in the 1930s with the building of a few capacious carriers that had no equal in fighting punch—except for their equivalents in the Japanese fleet. The carrier enthusiasts were constantly blocked, or at least slowed down, by the much larger pro-battleship lobby in the service, but, there being no independent American air force at the time, they did not suffer the fate of the Fleet Air Arm in the bruising Royal Navy–Royal Air Force procurement battles of the interwar years. When war came, therefore, the U.S. Navy had dependable carriers with decent planes, which gave them a good fighting chance against the Japanese—who had excellent carriers with excellent aircraft. And the spectacular attack on Pearl Harbor opened everyone's eyes to the amazingly destructive reach of carrier-borne airpower. The type of ships that had allowed the Japanese to hit Hawaii could, in their American version, obviously also be used to lead the counterattack westward.

The third factor was the introduction of the long-range B-29 Superfortress, a bomber so advanced that not only could it carry an immense bomb load thousands of miles but its maximum altitude (30,000 feet) made it inaccessible to enemy fighters and antiaircraft shells. No doubt the American reconquest of the Western Pacific could have crept forward, in an indirect way, from New Guinea to the Celebes to the northern Philippines, perhaps then to the Chinese coastlands, then to Okinawa, allowing the construction of further airfields from which the B-17 Flying Fortresses and B-24 Liberators could increase the aerial assault upon Japan. But a new bomber with unprecedented range and destructive power, that is, the Superfortress, together with the astonishingly suitable position of the Marianas, combined to offer an enormous shortcut to a war that many planners originally thought might go on until 1946 or 1947 and would have to include a giant Allied invasion of mainland Japan.

The dropping of the atomic bombs cut short all those invasion scenarios, but absent the B-29s and the capture of the Marianas, when would those bombs have been dropped, and where from? The B-29 was an excellent example of what Hitler termed a "wonder weapon," yet by 1943–45 only the United States could afford to build such an aerial monster. The electrical wiring and the aluminum for a single plane were probably equivalent to that needed to construct a squadron of Messerschmitts.

The fourth factor was the units who built the bases, the installations, the assembly points, and the roads that carried the fight forward—in this particular case, the U.S. Navy's Construction Battalions, the "Seabees." Although created by the exigencies of this particular war, the job this new force did would have been recognized by military men of all ages. It is difficult to imagine a military victory without engineers, but all too often historians of grand campaigns take their work for granted and assume that troops, fleets, and air squadrons can be moved long distances by the stroke of a pen on a large-scale map.[21] Yet to troops on the ground, however well equipped, mountains, rivers, swamps, deserts, and jungles geographically determine the nature of the battles that are to be fought. Nowhere was this more true than in the vast expanses of the Pacific Ocean. In sum, it needed Marine Corps amphibious doctrine and practice, the fast carrier groups, the B-29s, and the Seabees to cross the Pacific together. Just how those parts grew up and then came together forms the rest of this chapter, together with the somewhat separate story of a fifth element, the U.S. submarine service.

Hitting the Beach

The evolution of the U.S. Marine Corps's form of amphibious warfare can be compared in many ways with the experiences of Allied amphibious warfare in Europe (see chapter 4), although there were also significant differences due to force organization and, above all, the geography of the Pacific. The Corps's story is one of almost unremitting combat, from their landings at Guadalcanal in mid-August 1942 to the end of the resistance by the Japanese garrison on Okinawa in late June 1945. It was a story of learning the hard way, of fighting those same natural obstacles of jungle, weather, and disease that the British were encountering in Burma and MacArthur was dealing with in Papua New Guinea, and of fighting an enemy that would never surrender. As late as July 1945 American theater commanders in the Pacific were chafing at the slow progress being made in taking Okinawa, but how exactly did one make *fast* progress against such a dug-in enemy? Eventually MacArthur's Southwest Pacific Command, after exceptionally laborious fighting during the first two years, figured out that they would win the war much more swiftly if they leapfrogged into positions where the enemy wasn't. Yet there were always some places—the Solomons, the Gilberts, the Marianas, Iwo Jima, Okinawa—that they had to take, however strongly held. This was the marines' saga.[22]

Why was it that the United States Marine Corps (USMC) came to occupy this special, legendary place in the annals of war? The simple answer was that the marines were the only people in post-1919 America who took a keen, active, and progressive interest in amphibious operations. Just as the RAF and U.S. Army Air Corps had to assert the claims of the value of strategic bombing to justify themselves as an independent service following the Great War, so also had the Marine Corps to explain why it was needed when American military budgets were being trimmed so hard after 1919. The result of their efforts was to be seen in a combination of convincing warfare doctrine, improved technological and logistical assets, and well-trained specialist units that forever identified the Corps with massive and effective assault from the sea. It also studied the recent past for helpful lessons; not for nothing do the two historians of the first comprehensive book on the U.S. Marines and amphibious warfare begin with an illuminating comparative essay titled "Success at Okinawa—Failure at Gallipoli."[23]

Viewed retrospectively, it is rather remarkable that the amphibious operations concept survived at all. The USMC's mother service, the U.S. Navy, remained obsessed about future wars on the high seas between battle fleets and had no desire for a secondary role in supporting landing forces. When the Washington Naval Treaties of 1921–22 forced further fleet cutbacks, the navy could only regard the marines as a rival charge upon its limited budgets, even as, curiously, its war plans against Japan required expeditionary forces to seize bases in the Pacific. The U.S. Army, for its part, emerged from the First World War with a deep resentment of the great publicity given to the Corps's fighting on the

Western Front and an angry feeling that its own monopoly on land warfare had been seriously challenged. What was more, there were senior marine officers who wanted to preserve that larger battlefront posture and feared being pushed back into a pre-1917 limbo, carrying out ancillary missions in distant waters. Finally, the drive for economies shriveled the Marine Corps to less than 15,000 officers and men, hardly enough for the menial duties of providing ships' guards, protecting embassies, and holding American bases in the Caribbean. By 1921, marines were even deployed to guard U.S. mails against robbers, and must have felt a long way from the Halls of Montezuma.

But the amphibious warfare doctrine persisted throughout the interwar years and slowly became an operational possibility for two reasons, one strategic and the other more personal and fortuitous. The strategic reason was, quite simply, the prospect of a future war with Japan. Planning for a conflict to resist Japanese expansionism in East Asia and the Western Pacific had been carried out by American strategists since the first decade of the century. The elimination of the German navy in 1919, plus the parallel recognition that an Anglo-American war was highly unlikely, meant that the only feasible great-power enemy could be Japan. Its extensive gains of Central Pacific islands—the Marshalls, Carolines, and Marianas—during the war, its stubborn negotiation stance over battleship numbers and fleet bases at the Washington Conference, and its continued *pénétration pacifique* of China all enhanced American suspicions. What was more, the modern Japanese navy was the only rival large enough (assuming British nonhostility) to justify the size and expense of the U.S. Navy. Unsurprisingly, then, the chief of naval operations warned the commandant of the Marine Corps as early as January 1920 that War Plan Orange would henceforth determine the navy's plans and programs, and the Corps should prepare accordingly for possible amphibious operations across the Pacific.[24] Although shrinking numbers of personnel made this utterly impossible in practice, the strategic statement was there and would not go away as long as suspicions of Japan's intentions remained. The marines' claim to be a special fighting force, not just a second-level gendarmerie, relied upon the Japanese threat.

That clear identification of the future foe in turn pushed a small number of individual planners and midlevel officers into problem solving: how, practically, would they take the fight to a Japanese foe 5,000 miles across the Pacific? Although he was certainly not the only American pondering on that problem, Major Earl H. "Pete" Ellis of the USMC began to figure it out in a serious of reworked memoranda between 1919 and 1922. It would have to be a specialized service—the larger, heavier army couldn't do it, and probably wouldn't be interested. It had to be the marines, which was fortuitous because all naval/amphibious operations are under naval command, and the USMC was also the advanced base force of the navy. And, as Ellis put it in his famous July 1921 memorandum on advanced base forces in Micronesia, it had to involve specific tasks: "the reduction and occupation of the [Japanese-held] islands" and "the establishment of the necessary bases therein."[25]

By the same logic, the service charged with carrying out this task—as the USMC was formally defined in an army-navy agreement of 1927—had to prepare to implement it at ground level, in the precarious approaches to an enemy-held shore, and in the establishment of a secure beachhead upon it. And the early USMC planners understood that specially trained amphibious units required special platforms and special weaponry (this was now about 340 years after the Spanish marines' operations to capture the Azores). Ellis's own early memoranda called for landing craft with bow guns, for specially equipped signals troops, for demolition experts to neutralize beach obstacles and minefields, for marine aviation to strafe the beaches. Some other ideas were going to be less useful—motorboats towing barges, for example—but the man was remarkably far-seeing, his mysterious death in the Caroline Islands in 1923 making him even more romantic and intriguing.[26]

Even at the time of Ellis's death, other innovative Marine Corps majors and colonels were also at work trying to figure out how to get things done. This was not easy. Annual exercises along America's own southern shores, in Panama (Culebra), and in the Caribbean, all of which were important in the longer term but which initially threw up enormous shortcomings, confirmed the gap between theory and reality when trying to land on a shallow beach or into mangrove swamps. Millett's wonderful history of the USMC has a tart comment upon the "fiasco" of the winter 1923–24 exercises off Culebra: "The Navy coxswains did not reach the right beach at the proper time; the unloading of supplies was chaotic; the naval bombardment was inadequate; and the Navy's landing boats were clearly unsuitable for both troops and equipment. The exercise, however, identified enough errors to keep the Corps busy for fifteen years."[27]

On the other hand, the glaring need to have better equipment itself stimulated a constant search for new ideas and techniques. As early as the mid-1920s, observers strolling along the banks of the Potomac River or the Hudson might have seen an amphibious tank crossing those waters, yet another product of the eccentric tank designer J. Walter Christie, whose greatest contribution to the Second World War (see chapter 3) would be in designing the suspension and chassis of a vehicle that one day would become the Soviet T-34. Christie's novel, river-crossing tank foundered in less protected (offshore) waters, but the idea of building an engine-powered, forward-shooting, amphibious vehicle would reappear more than a decade later, to the Allies' advantage, often as the LVT. In much the same way, great dissatisfaction at the inadequacies of early prototypes of motored landing craft simply caused more imaginative Corps officers to search for newer ideas, one of which was the inventor Andrew Higgins's flat-bottomed boat, originally designed for use in the bayous of the Florida Everglades. Its later version, the landing craft vehicle, personnel (LCVP), would carry hundreds of thousands of soldiers and marines to the beaches from 1942 onward.

Two rather contradictory points emerge. The first is that this was never a dream progression for the U.S. Marines, but a constantly impeded and all too frequently suspended pursuit of a doctrine ahead of its time. It was not just that there was little money for developing this form of warfare or that there were massive obstructions from certain circles in the army, the navy, and even the Corps itself; it was that penury, plus other calls upon the service, plus the politics of isolation frequently put the whole idea into suspense—operations in China and Nicaragua essentially depleted the Corps of its active units for long periods at a time. For almost all of the 1930s, the basic doctrinal manual was most appropriately termed the *Tentative Manual for Landing Operations*. Even as late as 1939 the USMC commandant agreed that they should stress the marines' base-defense role when asking Congress for additional funds, since expeditionary forces would be regarded as "interventionist."[28] The more positive aspect, however, was that, having established a doctrine of amphibious warfare and persuaded the Joint Chiefs to recognize the Corps's special role in implementing it, survival was ensured even when the tools of war were missing. All that was needed was further Axis aggression and a consequent heightening of public alarm, both of which were to be forthcoming. All military organizations benefit from having real or perceived enemies.

But there was of course a second amphibious army in the Pacific during this war, namely, the U.S. Army itself. This is easily explained by the fact that the army was already substantially in that theater, that is, in the Philippines and Hawaii, when the Japanese attacked in December 1941. MacArthur's presence there, the fact that he insisted that he lead the Allied riposte against Japan, and the equally important fact that the American Joint Chiefs didn't want him anywhere else meant that considerable army divisions would be sent to the Pacific. Then there was the practical matter that the Marine Corps, even if expanding rapidly, was far too small to do all the land fighting in the Pacific. Since neither the navy nor the army would admit to allowing the other service to have the supreme command across this massive theater, there came about a compromise: MacArthur would run the new Southwest Pacific Command, a heavily army-run enterprise (though with considerable air and naval elements), and Nimitz would run Central Pacific Command as a U.S. Navy fiefdom, although with substantial army participation. It was clumsy—wouldn't they spend more time competing for resources than combining to beat Japan? asked Churchill at Casablanca—but on the whole it worked.

It was not easy for the army. A whole generation of its officers and NCOs whose focus had been upon European-style fighting had to start thinking about using landing craft rather than battalions of heavy tanks. Some units had begun to experiment with amphibious warfare techniques in the late 1930s (chiefly crossing rivers), but it was nonetheless hard for an army officer to think that he had to learn from the marines, or borrow equipment, or on some occasions be under a U.S. Marine Corps general. And the army's expansion in numbers after December 1941 was so much larger than that of any other service in absolute terms that it was bound to have lots of raw divisions fighting alongside, say, the marines' more experienced 1st or 2nd Divisions (as in the Marianas campaign). But the army learned fast at this challenge of fighting across coral reefs. The apotheosis probably came at the great battle for Okinawa, where on April 1, 1945, two marine and two army divisions (covered, it might be noted, by more than forty aircraft carriers of different types) came ashore together near the small village of Hagushi. To the Japanese defenders, it made no difference which of the oncoming battalions, debouching from similar landing ships into similar landing craft, belonged to which enemy service. The invaders were all part of a vast amphibious force coming from the sea to the land.[29]

Controlling the Oceans and the Skies Above

All sorts of aircraft contributed to the increasing American control of the skies over the Western Pacific from 1943 onward, and it would be wrong not to acknowledge the importance of the squadrons of P-38 Lightnings, P-47 Thunderbolts, B-25 Mitchell medium bombers, B-24 Liberators, and so on. Yet it is fair to argue that the two greatest contributions to gaining aerial supremacy over Japan in those vast spaces west of Hawaii and north of New Guinea—a distance much larger than that from Ireland to Ukraine—were the *Essex*-class aircraft carriers and the F6F Hellcat fighters designed for work from those warships. They tipped the balance partly because they came into full service in that critical year of mid-1943 to early 1944, just when the American counterattack was building up speed, but also because they complemented each other, the Hellcat being a great defender of the carrier system that had launched it, as well as a terror to the enemy.

The story of the evolution of the U.S. fast carriers goes back to the early 1920s. American observers had been deeply impressed by the Royal Navy's original action of converting certain older vessels, such as halfway-built battle cruisers, from gunships into flattops—vessels stripped of their entire superstructure to become horizontal takeoff and landing ships that would release fighters, high-altitude bombers, torpedo bombers, and dive-bombers, which could fly several hundred miles to hurt the enemy's ships and then return to the mother craft. The renowned Admiral William Sims, generally regarded as the father of the U.S. carrier service, told a congressional committee as early as 1925: "A Fleet whose carriers give it command of the air over the enemy fleet can defeat the latter. . . . [T]he fast carrier is the capital ship of the future."[30]

While the Royal Navy's carrier service fell behind during the inter-war years, in Japan and America—neither of which had an independent air service as a "third force"—the fleet air arm advanced, despite the usual prejudices of pro-battleship senior admirals. The endless miles of ocean beyond their coastal cities fueled this need to have carriers and their aircraft on the high seas for long-range protection of their respective homelands. In the early to mid-1930s, worried about a surprise attack out of the blue, some elements in the Japanese navy pitched the case for the universal abolition of carriers, fearing that this revolutionary weapons system could inflict more damage upon their country than it could inflict upon others—as indeed turned out to be the case. But the notion was rejected in the international naval disarmament negotiations of the 1920s, and so Japan felt it had little else to do but build more carriers and better aircraft.

The Japanese naval air arm and their designers were extremely good, and their capacity to take a Western design and improve on it was remarkable. By 1930 their shipyards had produced the 60-plane *Akagi* and the 72-plane *Kaga,* both extremely fast and capacious; there was nothing like them in the West, though the United States was catching up. At the end of that decade those early but still effective prototypes were being reinforced by the *Soryu, Hiryu, Shokaku,* and *Zuikaku;* and more carriers were to come. The ultra-efficient head of the navy's Bureau of Aeronautics, Admiral Isoruku Yamamoto, and the dynamic air wing commander Minoru Genda (who would later lead the Pearl Harbor attack) pushed for the carriers to be liberated from their original, static, homeland-defense role into a much more aggressive, freewheeling one. Yamamoto also pushed for increased production of newer Japanese aircraft, especially the formidable Mitsubishi Zero. By early December 1941, when its six large carriers struck at Pearl Harbor, it was the best naval fast carrier force the world had ever seen.[31]

But the war had just begun and the Americans would be coming back with carrier forces in such numbers as defied the imagination. The U.S. Navy, too, like the British and the Japanese, had been grappling with what turned out to be the largest technical-logistical-material problem of all, namely, if the power, speed, size, and shape of *all* aircraft in the 1920s and 1930s were exploding upward, as they were, far larger decks for the carriers' tops were needed, far larger storage space to put the aircraft below, far more fuel holdings for eighty or more aircraft, far bigger antiaircraft protection systems, far more room for a vastly expanded crew, and so on. What was needed was a ship that was at least 850 feet long and had a propulsion system powerful enough to move the giant vessel at high speeds (over 30 knots) for days, even weeks. Designers would have to think through the challenges of creating a hull and a landing/takeoff deck that would be stable in all weather conditions and when fully laden with well over 5,000 tons of aircraft and fuel. Finally, the enormous losses suffered by the Royal Navy to Luftwaffe attacks in the Norway and Crete campaigns were a wake-up call to admiralty designers everywhere to double or even treble the number of antiaircraft guns (and crews) on board ship. All this made for very complex and highly expensive weapons systems.

The American navy was therefore asking the president and Congress to give it an enormous injection of funds for a vast expansion of military-naval-aerial muscle. And by the end of the 1930s the worried political machine responded, slowly but surely, as the clouds of war drew closer. It wasn't too difficult for the service to suggest that fast carriers, independent of the slower battle fleet, could keep hostile forces a long way away from the homeland; even American isolationists bought that argument. With all the new money being poured into aircraft and ships, the productive capacity of an entire continent started to come out of the Depression, and did so, fortunately, with very large reserves of raw materials, capital, and skilled labor. Thus, an entire new class of larger aircraft-carrying warships would be built, as late entrants into the war—but not too late.

Usually two of the navy's seven fleet carriers were stationed in the Atlantic before and even after Pearl Harbor; the *Wasp,* on a brief loan in 1942, helped the British by steaming eastward from Gibraltar and flying Spitfires to reinforce Malta. The other five were in the Pacific and, when not in West Coast ports, normally operated out of Pearl— fortunately, they were absent on December 7, 1941. With five battleships sunk at their moorings, and others damaged, the remains of the big-gun navy were pulled back to California. The early clashes in the Pacific War were therefore borne on the American side by those prewar carriers, whose names are now saluted in history: *Lexington, Saratoga, Yorktown, Enterprise,* and *Hornet.*

These ships certainly blunted the Japanese advances, but at a terrible cost. The *Lexington* sank after sustaining heavy damage in the Battle of the Coral Sea in May 1942, and the *Yorktown* was destroyed in the aftermath of the great carrier clash at Midway a month later. Shortly afterward, the *Wasp* was torpedoed while on escort duty to Guadalcanal, the *Hornet* was torn apart by Japanese bombers during the Battle of Santa Cruz, and the *Enterprise* was worn out by all the steaming and fighting and sent home for a refit. For some time in early 1943, therefore, the U.S. Navy had only one carrier operating in the Pacific, the USS *Saratoga*, until it was reinforced, remarkably, by the new British fast carrier HMS *Victorious*—a nice compensation for the American loan of the *Wasp* to Mediterranean convoy duty. The two boats worked well together, with their aircraft interchangeably using either deck.[32]

But change was coming. There were and are many claimed "turning points" in the Pacific War, ranging from Midway and Guadalcanal to Leyte Gulf, but a strong if less well-known contender is May 30, 1943, the day that USS *Essex* steamed through the entrance to Pearl Harbor. She was the first of the brand-new, tough, powerful, and sophisticated carriers that were to put an enormous stamp on the nature of this wide-ocean war. The new *Yorktown* (following the U.S. Navy's tradition of transferring to a newly launched craft the name of a ship lost in battle) arrived in late July, along with the first few of the new light carriers. But it was the *Essex*-class carriers that were to command the scene.[33] Given that this was the first boat of a near-revolutionary design (radar-controlled gunnery and detection systems, armored hangars, side elevators to save space, enormous turbines to power a speed of over 30 knots, and 90 to 100 aircraft), the designers and shipbuilders had worked miracles. Her keel was laid down in April 1941, she was launched in July 1942, and commissioning took place in December 1942. Thirty-one others in this class, the later ones with improved facilities, were on order, with twenty-four gigantic naval shipyards mobilized to produce America's new navy.* This reinforcement came just at

---

*Actually, only twenty-four carriers were completed; the other orders were canceled as the war's end came into sight. As with the B-29s (below), and unlike the cavity magnetron, there were vast teams of engineers involved, not a single group.

the right time. The new *Yorktown* had arrived at Pearl Harbor less than four months before the Gilberts operation.

The new carriers were going to Nimitz's Central Pacific Command, not to the Southwest Pacific, to the dismay of Admiral Halsey, who possessed a fleet quite solid in battleships and cruisers but still weak with carrier airpower as he strove to render support to the army/marine drive through the Bismarcks and New Guinea. Overall, though, the relative hiatus in fighting in the Central Pacific gave the American admirals and planners at Hawaii time to experiment with a different tactical carrier doctrine. Eschewing the rule that U.S. carriers (each, of course, with escorts) should operate singly—to avoid a cluster of them all being sunk in one fell swoop, as had happened to the Japanese at Midway—the more ambitious and aviation-minded officers argued that it was precisely in clustering those warships that the navy's punch would be greatest. Hundreds of aircraft from many carriers could deal a much bigger blow to an enemy stronghold or battle fleet than could a single carrier, and they would reinforce each other in beating off any Japanese aerial assault.

This change of doctrine was to be tried out first in hit-and-run raids on smaller targets rather than going at a major enemy force or a big base such as Rabaul. This was smart, since the carriers and most of their crews were new and untested, as were the Hellcats and most of their pilots. Just as with the Duke of Wellington's cautious early movement in the campaign to reconquer Spain in 1808–14, so Nimitz also wanted to move ahead cautiously across the Pacific, with plenty of space to fall back. On August 31, 1943, Task Force 15 (TF 15), built around the new *Essex* and *Yorktown* and the light carrier *Independence,* and protected by a fast battleship, two cruisers, and ten destroyers, dealt heavy blows to Japanese airfields and installations on Marcus, a small island base much closer to Japan than to Hawaii. The attacks were not pressed through after the first morning's raids and the force commander pulled his ships back to Hawaii as swiftly as possible, but that was the point: hit, then run. On September 1 another battle group built around two light carriers attacked Baker Island, east of the Gilberts, with the air strikes being followed by an occupation and the construction of an American air base. Those two light carriers (*Princeton* and *Belleau*

Wood) then joined the new *Lexington* in a re-formed TF 15 and struck at Tarawa in the Gilberts on September 18.

The latter attack was not so successful: enemy camouflage on Tarawa frustrated most of the bombing attacks and caused Imperial General Headquarters to realize that they had another problem to deal with apart from warding off MacArthur's advance on the great base at Rabaul. The Gilberts garrison was ordered to be defended to the death, which was ominous for the invading marines, especially since aerial photography had spotted neither Tarawa's reefs nor the many hidden defensive emplacements. Overall, these small-scale raids were useful trial operations for American troops, officers, and staff planners, and also confirmed the larger argument that advancing toward Japan along two axes was, at least for the next while, the best way to go. The final experiment, a six-carrier aerial assault on Wake Island on October 5, 1943, designed to force the resident Zeros to rise to give battle and then be shot down by the new Hellcats, also allowed TF 14's commander to experiment with working all six carriers together or splitting them into subgroups.[34]

The Gilbert Islands battle for Tarawa and the smaller Makin in November 1943 is rightly remembered as a Marine Corps endeavor, but it might have been far worse for the leathernecks without the actions of the carrier groups in covering these amphibious landings. Land-based airpower, which worked so well for MacArthur and Halsey's thrusts along the northern shores of New Guinea, couldn't be that effective in Micronesia, for there simply weren't enough islands that could be converted into workable bomber bases, and the distances between them were so immense. The fast carrier groups therefore had to assume the main responsibility for aerial protection, and not just for the Gilberts operation but for the later attacks upon the Marshalls, Carolines, and Marianas as well; the carriers, the great historian of the Pacific War writes, were "the cornerstone" of operations.[35] By this stage, a half dozen *Essex*-class carriers and another half dozen *Independence*-class light carriers had arrived in the Pacific. It is amazing: in January the U.S. Navy had had only a single carrier effective in these seas, yet by November it had an armada. The tide was now turning very rapidly.

The greatest battles for command of the skies were actually not over the Gilberts themselves but way to the southwest, where just a short while earlier Halsey's TF 38, with the lumbering *Saratoga* and the juvenile *Princeton,* followed a few days later by a three-carrier reinforcement (TF 50.3) from Nimitz's command, devastated a large number of Japanese cruisers and destroyers at Rabaul and damaged around half of the Japanese carrier-borne planes that had been flown in to reinforce the land aircraft there. When 120 land-based Japanese attackers descended on Task Force 50.3, they were pushed off in less than an hour. Squadrons of air combat patrol Hellcats, plus the deadly proximity fuses of the new 5-inch guns on the carriers, were able to hold their own against massive aerial assault even without battleship and cruiser protective fire.[36]

Thus by the time the marines battled across the coral reefs of the Gilbert Islands in November 1943, an enormous carrier-borne aerial umbrella was being put in place. The figures are worth recounting, partly because this was the very first of Central Pacific's coordinated interservice operations, partly because it was the template for operations to come. The landing force for Tarawa was accompanied by three battleships, three heavy cruisers, five escort carriers (for close-in land attacks), and twenty-one destroyers; that for Makin, in the north, by four battleships, four heavy cruisers, four escort carriers, and thirteen destroyers. Most important of all, a group of four fast carrier task forces—thirteen carriers in all—gave protection by prowling around the wider seas, raiding other Japanese bases, and occasionally returning to launch air strikes in support of the landings. The fighting on land was awful, but American naval control was never in doubt.

Command of the air over the Gilberts was ensured, not just because of the American forces massed there, but also because of what had occurred in the Southwest Pacific. MacArthur's drive past Rabaul and the enormous and painful aerial battles over those islands not only distracted the Japanese admirals and planners but badly cut into the available numbers of their aircraft and (most important) their skilled pilots. The Japanese battle fleet remained gigantic, mainly untouched, and raring to fight, but without airpower, those great latter-day dreadnoughts could not strike decisively. How could they, when Japanese land-based airpower was weakening fast, and when most of the remaining carriers now had inadequate, weakly trained aircrews?

The battle partner to the *Essex*-class carrier was the Grumman F6F

Hellcat, a single-engine fighter of great toughness, speed, and consistency. A fully worked-up fleet carrier would usually go into action with thirty-six fighters (the Hellcats) to support thirty-six dive-bombers (SB2C Helldivers) and eighteen torpedo planes (TBF Avengers), as well as provide aerial defense for the warships if Japanese planes counterattacked.

The dive-bombers and the torpedo bombers have their own important place in the Pacific struggle between 1943 and 1945, but the F6F played a particularly critical role. It was designed by Grumman engineers to replace the earlier F4F Wildcat, itself a tough and reliable fighter that could absorb a good deal of damage and still get back to base. Since every air force by 1941 was ratcheting up the power, speed, and firing capacity of its aircraft—the newer Spitfires, Mosquitos, the 190s, Thunderbolts, Typhoons—it was prudent of Grumman to do the same. Aerial scraps between Wildcats and Zeros in the first year of the war were a real learning experience for the Americans, because the Japanese plane had better maneuverability and a faster rate of climb. But Grumman and the U.S. Navy had moved on even before hostilities: the F6F contract was signed in June 1941, the original Wright engine was replaced by the more powerful Pratt and Whitney Double Wasp engine shortly afterward, and the first aircraft of that type had its maiden flight on July 30, 1942.[37] The parallels with the almost contemporaneous story of the Rolls-Royce Merlin engine replacing the less-powerful Alison in the P-51 Mustang (chapter 2) are remarkable.

The Hellcat borrowed and improved upon the defensive armor of the Wildcat: the cockpit, the windscreen, the engine, and the fuel tanks were all heavily protected, so it was hard to shoot down. (One aircraft, after heavy dogfighting around Rabaul, flew back to its carrier with two hundred bullet holes in it.) But what was different now was that the newer plane could outclimb, outdive, and outturn any Japanese plane, even the Zero, which was starting to fall behind in the race for airspeed and fighting power. The first Hellcats joined the *Essex* in February 1943 and were able to take advantage of the lull in the Central Pacific to engage in intensive practice and experiment with further improved features. On November 23–24, just as the Gilberts operation was under way, carrier-borne Hellcats claimed thirty Zeros destroyed for the loss of one of their own. This was no fluke; the majority of the Japanese aircraft later destroyed during the Battle of the Philippine Sea were victims of the Hellcat's superiority in the air, as were those shot down over Rabaul, the Philippines, Iwo Jima, and Okinawa. By this time the Japanese air forces, like the Luftwaffe, had very few experienced fighter pilots left, but the difference in casualty rates remains astonishing. In contrast, American planners in late 1944 were beginning to wonder whether they would actually be receiving *too many* highly trained aircrews.

F6F Hellcats flew more than 66,000 combat sorties during the war, almost all from the decks of carriers, and destroyed 5,163 enemy aircraft at a cost of a mere 270 of their own. There was nothing like it elsewhere in the war. According to one statistician, Hellcats accounted for 75 percent of all aerial victories recorded by the U.S. Navy in the Pacific. This tough, adaptable fighter—of which 12,275 were built in total—thus proved to be a perfect fit in the package of weapons systems put together to defeat the Japanese Empire.[38]

So was set the scene for the next, intermediate operations of 1944—those through and around the Marshall and Caroline island groups—before the climactic assault upon the Marianas themselves.[39] It is interesting to note that the more the American advances took place via these remote Micronesian atolls (as distinct from the campaigns in the more enclosed waters of the Southwest Pacific), the less interrupted they were by Japanese airpower. None of Japan's land-based planes could go very far into the vastnesses of the Central Pacific, and attempts by their carrier-based aircraft to disrupt American amphibious operations faced the obvious hazards—depletion by lurking U.S. submarines, and heavy counterattacks from flocks of aircraft operating from a dozen or more enemy carriers.

The apotheosis of the U.S. fast carriers' decisive role in the Pacific War was the June 19, 1944, clash between the Japanese and American fleets way to the west of Saipan, on whose beaches the amphibious forces were unloading and fighting their way inland. It is known officially as the Battle of the Philippine Sea, but more colloquially to the carrier pilots as the "Great Marianas Turkey Shoot."[40] This battle of long-distance naval airpower was a reaffirmation of Midway, and of the U.S. Navy's strategic decision to push across the Central Pacific with carriers and amphibious forces. Both at the time and afterward, the

American commander Admiral Raymond Spruance was harshly criticized for adopting far too cautious a tactical plan—like Jellicoe at Jutland, as one critic has put it—by ordering his fleets westward of the Marianas during the day, then pulling them back and closer to the islands overnight.[41] This seems hardly fair, since protecting the American amphibious forces on Saipan (where Spruance hastily landed his reserve division when intelligence informed him that Japanese naval forces had left their bases) was his first priority, *not* sinking an enemy battleship or three; that could wait.

In the enormous aerial attacks and counterattacks that followed, the Japanese navy's surface vessels were again mauled by Mitscher's four fast carrier groups. Overall, Japan lost around 480 planes, or about three-quarters of those deployed. The imbalance was by now amazing: the Japanese suffered air losses ten times those of Task Force 58. The loss of trained pilots, in addition to those already killed around Rabaul, was beyond repair, and Japan's great battle fleet looked like more and more of an anomaly. Yamamoto's dream of punishing America and forcing it into accepting a Japanese East Asian domain was no more. That extraordinary cadre of naval airmen, who had ravaged Pearl Harbor and sent the *Prince of Wales* and *Repulse* to the bottom of the ocean in a few hours, was now itself shattered.

It could indeed be argued that this June battle by the carriers was more significant strategically than Leyte Gulf itself (October 1944), although an even greater number of warships and aircraft would be involved on both sides there, but for a less significant position. Although this book's analysis falls away after the capture of the Marianas, it is worth noting that the Japanese entrapment plan for repulsing U.S. forces from any invasion of the Philippines—the Sho-I plan—rested upon Admiral Ozawa's carrier force decoying the far larger American carrier task forces to the north, and thus away from the great battleships picking their way through the various straits to surprise and eliminate the U.S. amphibious landings at Leyte. And Ozawa's force was a true decoy, for by November 1944 it possessed barely a hundred planes and scarcely any trained aircrews. Probably the Japanese navy's battleship admirals did not mind making a tethered goat out of Ozawa's vessels, for if the decoy plan worked well, they would at last have a decisive victory. As it turned out, although the Americans (especially Halsey) were initially fooled by the gambit, they had enough resources to hurt the Imperial Japanese Navy on all fronts; Ozawa's four carriers went to the bottom, as did three battleships and nine cruisers. It was now October 1944, and the Japanese position was crumbling.

Supporting the carriers in all these campaigns were the ships and crews of the support squadrons. Just as Caesar needed his cook in the conquest of Gaul, the fast and wide-ranging flotillas that surged out of Pearl Harbor needed their modern equivalent, the superbly equipped fleet train of supply ships and especially the speedy and specifically designed oil tankers without which Spruance, Halsey, Mitscher, and the others would have had a much more restricted radius of action. Mini-convoys of tankers, heavily screened by destroyer escorts, were always at sea, a relatively short distance behind the major battle forces. They rarely get much attention from military historians, but these tankers, as much as any other jigsaw piece, contributed to defeat the tyranny of distance. To the embarrassment of the Royal Navy, it found that when it finally dispatched its modern battleships and carriers to the Far East in 1944–45, it had a very "short-legged" navy by comparison. Their raids on Japanese installations in Burma and the Dutch East Indies went fine, but their lack of refueling capacity (and hence their need for U.S. help) for wider operations in the Pacific allowed Admiral King to veto their operations in those waters until very late in the war. The Admiralty had, after all, designed warships to fight off the coasts of Norway and in the Mediterranean, while the Americans had been planning for wide, oceanic war—and had got that right.[42]

Reducing the Japanese Homeland

The seizure of the Marianas would not have meant so much in the Pacific War had it not been for that separate but linked American endeavor, the development of the greatest of the Allied long-range strategic bombers, the B-29 Superfortress. This extraordinary plane was the apotheosis of the Mitchell-Trenchard belief that "the bomber will always get through," now not because it would fight its way against hostile fighters to chosen targets, as in Europe, but because it would fly higher and faster than any Japanese defending aircraft could reach. The B-29 weighed about twice as much as the B-17 or the Lancaster and, with a pressurized cabin, its crew could fly at very high altitudes at a speed of 350 mph—just catching up with it would exhaust a defense fighter's fuel tanks—and then unload a terrible retribution of heavy, splinter, or incendiary bombs upon the hapless population below. It was not an even fight, as in the Battle of Britain; it was, rather, brutal aerial devastation. The B-29s' most historic acts involved dropping the atomic bombs on Hiroshima and Nagasaki that finally induced the hard-necked Japanese military leadership to surrender, but their most destructive act was the low-flying firebombing of Tokyo in March 1945 that killed 130,000 people, shortly after the Anglo-American bombers had inflicted a similar number of casualties at Dresden. Between 1940 and 1942, often standing beside a row of burned-out houses in East London, Winston Churchill had warned the Axis powers that their turn would come. But even he, with his immense imagination, had no idea of the size of the future aerial devastations.

Where did the B-29 come from?[43] The specifications for this enormous aircraft had already been issued in 1938–39—just like so many of the weapons systems we associate with the middle years of the war: Lancaster bombers, Spitfires, Sherman tanks, landing craft, and T-34s, all of which needed time for testing and development before full deployment. By the Second World War, heavy bombers were steadily taking over the role of battleships and cruisers in placing economic pressure upon the enemy, but they were also, awkwardly, taking almost as long as a major warship to be designed and constructed. The story of the Superfortress is an example of this disturbing but natural law: the more sophisticated the instrument being built, the greater the number of teething troubles.

The B-29's enormous wingspan of 141 feet and huge takeoff weight of 120,000 pounds had implications for air base design and runway breadth and length. Most airfields, including all those in Europe, were simply too small to take it, so it could only be deployed on the new, artificial grounds that were being erected on Pacific atolls. But those engineering challenges for building the tarmacs did not compare with the task of constructing the beast itself. To get such a weight off the ground and then up to 35,000 feet required extraordinary technical devices—almost all untested—and stretched the ingenuity of even Boeing's superb design and development teams. (And they were teams, dozens and dozens of them; no single super-clever designer here, like Barnes Wallis for the Wellington.) For example, how exactly did one construct a pressurized cabin for so lengthy an aircraft when it had crew in both the nose section and in the rear, but an extended central bomb bay section that should not be pressurized? It is not difficult to understand why it took a long time to figure out the compromise solution, a long pressurized "crawl tunnel" over the bomb bays. How did one perfect the revolutionary central fire control system (CFCS), directed by an analog computer that corrected for wind, gravity, airspeed, and so on?[44]

The biggest problem the Boeing engineers and subcontractors faced was with the Wright R-3350 engines themselves. All the early ones were inadequate and unreliable. The contrast with the Merlin in the Mustang and the Pratt and Whitney in the Hellcat is intriguing. In those cases, a superb replacement engine solved the problem of powering a potentially wonderful aircraft. In this case, the B-29, a bold and terrific tool of war, had to overcome not only technical challenges such as cabin pressurization but the fact that it was underpowered and underperforming. The Wright engine cowling for the B-29 was wrong, the flaps were wrong, and the engines would overheat and catch fire. The second prototype had a fire during testing and crashed into a nearby meat factory, with mass casualties, including all the crew. The greater the pressure to get the plane perfected, the more numerous and serious the setbacks. Engineers at the giant Wichita plant, ordered to get four entire groups upgraded and completed, ironically referred to their work as the "Battle of Kansas." They were not wrong. There was serious discussion of canceling the program before Hap Arnold—as we have seen, desperate about the inadequacies of Allied strategic bombing in Europe and grasping for any straw—ordered work to go on. December 1943 and January 1944 really were the hinge period for Allied strategic bombing.

Work did go on, and the planes came through, due to hundreds of minor adjustments by the Boeing engineers. The pilots, often desperately scared, described the first twenty seconds of flight after takeoff as being "an urgent struggle for airspeed" as the crews willed their massive craft to gain altitude.[45] Sometimes their prayers were not answered; several of them crashed as they left Saipan because an engine failed just as the heavily laden plane made its urgent efforts to gain altitude. In fact, it was not until after 1945 that a much better Pratt and Whitney engine solved this problem. Even the famous *Enola Gay*, lumbering off Tinian with its critical atomic bomb for Hiroshima in the hold on August 5, 1945, came dangerously close to failing at the end of the vast runway before wobbling slowly into the night.

When it had gained its desired height, however, the B-29 was normally untouchable and virtually indestructible. What could get at it? Its problems were its own hypertechnological demands, and many more of this aircraft were lost to operational causes (engine failure, air pressure loss) than were shot down by the enemy. Yet once three or four air groups of Superfortresses were safely launched, inaccessible to enemy fighters, the punishment of Japan commenced.

The biggest question Air Force planners had was where to site the giant aircraft. Roosevelt's original instruction was to fly the planes out of bases in southern China, themselves supplied by larger bases in India. That general idea simply wouldn't work in the case of the energy-consuming B-29s, which would have to fly from Seattle to Hawaii to Australia to Assam, over the hump to a base in Chengdu Province, and then on to bomb Japan. But then someone would also have to bring to Chengdu the bombs and fuel and machine gun shells, as well as the construction battalions, the cement, the Quonset huts, the wiring. It would be like reinforcing the Allies' Italian campaign by flying planes from Florida to Bahia to Freetown to Capetown to Zanzibar to Aden to Cairo to Calabria. There had to be a shorter way.

Yet in June 1944 American planners felt they had to try the China option. Although production problems halved the number of B-29 groups that could be based out of India, the scheme went ahead. On June 5, 1944, ninety-eight B-29s flew from India to attack Japanese railroad repair shops in Bangkok; it must have been an extraordinary surprise to local Thais and the Japanese garrison alike. And on June 15, 1944, forty-seven Superfortresses, which had indeed flown across the eastern Himalayas and refueled in Chengdu, bombed the Imperial Iron and Steel Works at Yawata—the first attack on the Japanese homeland since the Doolittle raid more than two years earlier.

But June 1944 was also when the marines, army, and Seabees took the Marianas, thus diminishing the point of keeping up the hugely ex-

pensive, logistically tortuous campaign of bombing Japan from southern China. These B-29 attacks upon Japan out of Chengdu Province
did not cease immediately; they could continue to hit the foe's homeland for another half year before the Mariana bases were ready, and in
doing so provided important operational experiences.

So they continued to attack, though only with about two raids each
month, from July until December 1944. On November 21, for example, 61 B-29s took off from Chengdu to bomb Japanese targets; but
three days later 111 B-29s attacked Tokyo from the more convenient
Mariana bases. The last B-29 bombing assault upon Japan from China
took place on January 6, 1945, after which the squadrons were relocated to the Pacific.[46] The crews had had one of the toughest aerial as-
signments of the war, and their very presence drove Imperial General
Headquarters to order its troops farther and farther into central China,
thereby pulling them away from the altogether more strategically significant American drive through the Pacific. The original American intention to station bombers in these distant places was to render support
to Chiang Kai-Shek's shaky regime and to find a smart way of eliminating Japan's industrial base and cities. But perhaps the greatest benefit
from Operation Matterhorn was to further increase the Japanese army's
"continentalist" tendency, leaving it, by 1944–45, with well over a million troops in the wrong place at the wrong time.

The later story of the B-29 attacks upon Japan takes us out of our
time frame, though even a brief synopsis of the bombing campaign
from the Marianas and onward would confirm its main point. The first
mission from the islands against Tokyo, on November 24, 1944, took
place while Americans back home were celebrating Thanksgiving. The
Japanese authorities, by contrast, were giving out instructions about
conserving food and water, and forming neighborhood air raid watches.
From the turn of the year onward, the aerial attacks intensified. After a
while, the B-29s' ferocious field commander, Lieutenant General Curtis LeMay, decided that their high-level bombing was inflicting insufficient pain and that it was probably unnecessary to fly at such altitudes
because Japanese antiaircraft defenses were much weaker than those he
had experienced in Europe. Without consulting Washington, he had
the aircraft stripped of much of their heavyweight armament and of
their remote-controlled sighting equipment to make more capacity for

fuel and a newer type of bomb—a jelly-like incendiary deliberately designed to burn Japan's vulnerable wooden cities.

On March 9–10, 1945, some 333 bombers set out from the Marianas, flew over the fighting on Iwo Jima, and proceeded to devastate Tokyo in the greatest firestorm of the entire war. During the next days Nagoya, Osaka, and Kobe suffered the same fate. LeMay was certainly tearing apart Japanese industry and much else besides; in these few weeks, Toland calculates, "forty-five square miles of crucial industrial areas had been incinerated." Two million buildings were razed to the ground overall, and thirteen million Japanese civilians lost their homes. Strategic bombing worked.[47] The great moral problem, as with the Allied bombing of German cities at the same time, was that this destruction of enemy war industries was also taking the lives of hundreds of thousands of civilians, chiefly women, children, and older people. By this stage of the conflict, though, not many among the victors were raising St. Augustine's questions about proportionality in war. To a large degree, the A-bomb devastations of Hiroshima and Nagasaki were epitaphs to the earlier, larger airborne devastations of Berlin, Dresden, and Tokyo.

Building Bases Across the Pacific: The Seabees

The story of the Seabees is about a man who created a team, which in turn created a gigantic organization that brought American military-industrial power—in the form of cement, tarmac, steel girders, electrical wire, rubber, glass, bulldozers, and lighting equipment—across the 7,000 miles of the Pacific Ocean to the outlying territories of Japan. It is a tale that is equivalent to that of Hobart and his Funnies, and that of Harker and Freeman in rescuing the P-51 Mustang from death; it ranks with Barnes Wallis and his creation of the Wellington, the "Dambuster" skipping bombs, the Tallboy and Blockbuster bombs. Ben Moreell was one of those neglected middlemen who made Allied grand strategy *work*.

Moreell was a civil engineer turned naval officer who became the only noncombat serviceman in the United States Navy to achieve a full admiral's rank.[48] After majoring in engineering at Washington University, he immediately joined the navy during the First World War and showed his extraordinary talents for construction and development of military bases, catching the attention of the young FDR, then assistant secretary of the navy. In the early 1930s Moreell was sent to study at the Ecole Nationale des Ponts et Chaussées in Paris, the premier place in Europe for military bridge and road construction. In December 1937 Roosevelt appointed him to be chief of the Bureau of Yards and Docks, and also the chief civil engineer of the navy, a brilliant double appointment. One of the first things Moreell did was to urge and organize the construction of two giant dry docks at Pearl Harbor. The American battleships and other craft that were damaged on that fateful morning of December 7, 1941, were thus able to be partially repaired at the home base; they did not need to be towed, powerless, to San Diego. Ships badly damaged in the Solomons, Gilberts, and Marianas could also limp into Pearl Harbor. This was a man who was thinking ahead.

But if Moreell planned protective or defensive construction, he also thought of what would be needed to carry the fight back to the enemy. In late December 1941, he recommended to Roosevelt the establishment of naval construction battalions that would be recruited from the building trades, and whose officers could exercise authority over all lower-ranked officers and men assigned to these units. The idea of putting Civilian Engineer Corps officers over regular navy and marine units caused an immense fuss in Washington in early 1942, but Moreell got his way. Thus were the Construction Battalions (CBs)—the Seabees—born on March 5, 1942. Moreell already had their motto ready: "Construimus, Batuimus" ("We Build, We Fight")—for he had conveniently got a ruling that these were *fighting* men and therefore, if they fell into enemy hands, could not be executed as armed civilian guerrillas.

Getting skilled construction workers into the Seabees in early 1942, when every service was screaming for personnel, was a major stroke. According to the U.S. Naval Historical Center's account, "the first recruits were men who had helped to build the Boulder Dam, the national highways, and New York's skyscrapers; who had worked in the mines and quarries and dug the subway tunnels; who had worked in shipyards and built docks and wharfs and even ocean liners and aircraft carriers. . . . They knew more than 60 skilled trades." Their average age was thirty-seven, though some of them sneaked in even though they were over sixty.[49] Later in the war Moreell was not allowed to skim off America's skilled workforce to the same degree, but by that stage he and his team had created an impressively sophisticated system of training centers and advanced base depots through which all new recruits passed before being sent off to the front lines.

By the time peace arrived, 325,000 men had enlisted in the Seabees, and in total they had constructed more than $10 billion worth of infrastructure, from Trinidad to Londonderry, from Halifax to Anzio.* They created and manned the steel pontoons that allowed the Allied armies and their supplies to come onshore in Sicily, Salerno, and southern France. Ten thousand Seabees of Naval Construction Regiment 25 came ashore on the Normandy beaches, along with their U.S. Army Engineering equivalents, to demolish Rommel's steel and concrete obstacles—German engineers had built the fabled Atlantic Wall, and American engineers took it down. Seabees manned many of the landing craft for the first waves of troops and tanks, then towed in and anchored thousands of pontoons. At Milford Haven, they assembled the extraordinary Mulberry harbors that were to shelter the Allied beachheads. Their biggest logistical challenge in the European war was to rebuild the ports of Cherbourg and Le Havre, which had been thoroughly devastated by German demolition teams—yet the first American cargoes were being landed in Cherbourg only eleven days after the capture of that city. Building the pontoons to cross the Rhine was relatively easy for them. While all this was going on, Moreell was brought in to negotiate a settlement to the national strike of oil refinery workers in 1943; a year later he was asked to be the administrator to the bituminous coal industry after it came under federal control. His organizational capacities were immense.

The greatest achievements of the Seabees were in the Asia-Pacific theaters, where 80 percent of the entire Naval Construction Force was located. The raw statistics are staggering: in the Pacific alone, these artificers of victory built 111 major airstrips and 441 piers, tanks for the

---

*The fabulous steel-drum musical culture of the Caribbean is a direct result of the arrival of the fuel drums, found at the outskirts of all the new airfields that the Seabees built across the islands. The gigantic Carlsen airfield on Trinidad had no fewer than eighty paved runways; one can imagine how many empty steel drums were around by the end of the war when the American planes flew away.

storage of 100 million gallons of gasoline, housing for 1.5 million men, and hospitals for 70,000 patients. The war stories are even more eye-opening. The first battalion ever sent to the Southwest Pacific—to the disease-ridden island of Bora Bora—had just enough time to erect the fuel tanks that were to be used by U.S. ships and aircraft during the Battle of the Coral Sea. Seabees went ashore with the marines at Guadalcanal and spent day and night repairing the bomb craters of Henderson Field as well as bulldozing Japanese emplacements. They hopped all the way with MacArthur's Southwest Pacific Command from Papua New Guinea and the Solomons, via New Britain, the Admiralty Islands, Hollandia, the Celebes, and then the central Philippines.

When MacArthur at last fulfilled his promise "I shall return" and marched ashore through the shallow waters for photographic effect in October 1944, he gripped the news editors' attention. But that photo op was possible only because the Seabees had skillfully operated the pontoon barges and causeway units that brought the army—and the photographers—ashore in the first place. Very soon, 37,000 men of the Naval Construction Force were spread across the Philippines, building the main fleet bases, submarine bases, airfields, repair facilities, roads, housing, and hospitals for the gigantic military force that would then leap northward to Japan itself.[50]

The invasion leap did not happen, for Hiroshima and Nagasaki brought the war to a close much more swiftly than most planners expected. The instrument of victory was a B-29 carrying the first atomic bomb, which took off on August 5 on its fateful overnight flight from one of the Tinian airstrips, so recently laid out by the Seabees, so recently captured by the marines, and in turn protected by the fast carriers. The symbiosis is remarkable.

The Seabees didn't just build things; they fought and took casualties again and again. During the marines' troubled assault on Tarawa, these engineers had to figure out, under heavy fire, how to get landing craft, and then tanks, across low-lying coral reefs. The Seabees took casualties in so many battles simply because they had to be in the second wave that stormed the beaches, at Guadalcanal, Sicily, Anzio, Saipan, Normandy, and elsewhere. But this of course was Moreell's original point: they were fighters as well as construction workers, because before one could dismantle an enemy's coastal blocks and barbed wire, one might have to kill the soldiers lurking behind them. In the Pacific theater alone, the Seabees earned more than 2,000 Purple Hearts; they lost around 200 men in combat, and significantly more in dangerous construction jobs. Their stamina was remarkable. Fifteen hours after the marines had wiped out the remaining Japanese garrison at Tarawa, the bomb-cratered airfield was working again.

When the Marshalls were taken in February 1944, the coconut-fringed Majuro atoll was converted into a major fleet base with all facilities (including a U.S. Navy officers' club), while nearby Kwajalein was converted into a gigantic airfield and repair center, with far fewer creature comforts (for the army officers). The Carolines were the logical next step, because they had a superb position for supply lines to the intended invasion of the Philippines. By September 1944, the Seabees were constructing facilities there, including Pelelieu, which had been fought over so hard, as well as on the island of Morotai, which was MacArthur's key stepping-stone between northwest New Guinea and the southern Philippines.

The crown jewels of the Western Pacific, that is, the Marianas, had been taken a couple of months earlier. The Seabees' role at Guam, Saipan, and Taipan in June and July 1944 was possibly their greatest during the war. Guam was important symbolically—the first American territory recovered from the enemy—and Saipan was to be turned into another huge naval base for the U.S. Pacific Fleet as well as an air base. Tinian, however, was the main prize. Though it was small, it was also relatively flat and low-lying, more like a massive wheat field in East Anglia than the spiny highlands of New Guinea or even the nearby jagged cliffs of Saipan. It had airfields already built on it, and space for more.

Tinian's chief problem—at least as viewed by the invasion forces—was the difficulty of access. One could come ashore at the best beaches, at Tinian Town itself, but that was where the Japanese had the fiercest defenses. To the north were the improperly named White 1 and White 2 beaches, which were tiny in width and guarded by low coral cliffs. There was, for example, only room for a maximum of eight LVTs at a time, compared with the ninety-six in each wave at Saipan, so the chances of being either stranded on the coral (as at Tarawa) or blown apart in a narrow funnel of a beach were significant. Yet the invasion, on July 24, 1944, turned out to be surprisingly easy, with far fewer casualties than at Saipan and Guam. The Americans put 15,000 men on White 1 and White 2 that day, about the same time as the Red Army reached the Vistula, Bradley broke out in Normandy, and Mark Clark's armies started pushing north of Rome. The Axis defenses were all crumbling at the same time; imperial overstretch, a sort of collective geopolitical "metal fatigue" (as the engineer Moreell might have thought of it), had set in.

Tinian was another superb example of how to do amphibious warfare. American forces staged a major but phony landing effort at the more predictable target of Tinian Town, together with a big naval bombardment, and as with the Normandy operation, deception paid a big reward. Japanese attention, like Hitler's and Jodl's, went to the wrong place. Then the marines came ashore, one company at a time, through the narrows of the White beaches; a whole regiment was on land within three hours, and with complete surprise. But all this only happened because the Seabees had devised and then operated special movable ramps that could get the fighting troops and their equipment over the outlying reefs.[51]

That was only the beginning. The major work was the construction of the airfields that would allow the systematic destruction of Japan's war capacities. Within a few weeks of the capture of the Marianas, the air bases—that is, tarmac runways built on foundations of ground-up coral, plus the wiring, control towers, Quonset huts, perimeter fences, engineering sheds, and radar stations—were being built. In all, the Seabees built five major air bases on the Marianas: one on Guam, one on Saipan, and no fewer than three on Tinian. Each of those five bases had a four-group wing capacity for the incoming B-29s. Tinian therefore could contain twelve groups. The Superfortresses began to arrive in mid- to late October 1944, while the Seabees were still expanding the base facilities. As noted above, the first raid against Tokyo, by 111 B-29s, was flown on November 24, 1944, out of the Marianas. From this time onward, endless streams of high-flying B-29s, glinting brilliantly as the Pacific sunshine reflected off their silver bodies, headed north to punish the enemy.

The Silent Service

The fifth piece in the Pacific tool kit was the U.S. submarine service, whose history is much less well known than that of the marines and the B-29s, partly because by definition the submarines were far less visible. American submarines in the Pacific War were chiefly independent actors. By their very nature, submarines are silent and subversive, hiding in the depths and surfacing for air under vast, lonely skies. In this respect, Germany's massed wolf packs in the mid-Atlantic struggle were very much the exception. In the later stages of the war, it is true, the increasing capacity of American cryptographers to read Japanese naval codes meant that the commander of Pacific submarines at Pearl Harbor could instruct three or four boats to converge on an enemy battle group. Generally, though, submarines were lone rangers, didn't indulge in frequent reports to headquarters, and liked it that way.[52] Thus it is no surprise that the U.S. submarine service wasn't intimately connected with that four-part operational and technical story outlined above, the quartet of Marine Corps amphibious doctrine, fast aircraft carriers, B-29s, and Seabees, which combined to force their way across the Central Pacific.

But American subs did play an important role in the collapse of Japan, and very often mixed their independent ghost-rider roles with coordinating actions that helped the broad advance across these wide waters.[53] To begin with, since aircraft reconnaissance was limited by range, they were often deployed as the "eyes" or forward scouts of the U.S. Navy, spotting Japanese warships coming out of the San Bernardino Strait, or others heading to reinforce Rabaul. This shadowing role was not very attractive to aggressive submarine captains unless they were also allowed to attack, but it was invaluable to Nimitz in Hawaii and Halsey with the Southwest Pacific Fleet. In late May 1944, the USS *Harder* was sent to patrol and report from the harbor of Tawi Tawi, which she did with great care; but she also managed to sink no fewer than five Japanese destroyers in four days, provoking the Japanese fleet to sail early and head toward what became the Battle of the Philippine Sea.[54] Other U.S. submarines, during the multiple battles at Leyte Gulf later in the year, spotted the enemy's main battleship and cruiser squadron steaming through the Palawan Passage and alerted Halsey to prepare his response (they also sank two heavy cruisers and damaged another before retiring). And at the close of the war it was another pair of submarines that first detected the giant battleship *Yamato* on its one-way suicide run toward Okinawa.

As the American war effort in the Pacific reached its crescendo, the submarines were to play additional ancillary, but again extremely valuable, roles. For example, they could act as navigational guides or "course correctors" for the streams of USAF bombers heading toward Japanese cities in obscure weather conditions. They ferried supplies and agents to assist the resistance movements in the Philippines and returned with new intelligence about the enemy's forces. They also, when needed, acted as a sort of lighthouse for amphibious forces approaching enemy-held beaches in the hours before dawn, a technique that had been perfected earlier by a specially trained squadron of Royal Navy submarines during the North Africa and Sicily landings.

Finally, this jack-of-all-trades craft was used to rescue American airmen who were forced to ditch their damaged planes as they returned from raids upon enemy targets. This useful service was already in operation by the time of the Gilbert Islands operation in November 1943, and by the end of the following year U.S. submarines had picked up no fewer than 224 downed aviators. By the time Curtis LeMay's B-29s were flying off the Mariana air bases, three subs were on rescue duty for every aerial mission; another 380 aviators were picked up during the war months of 1945, saving lives, returning skilled airmen to the force, and immensely boosting fliers' morale.[55]

But the greatest feat of American submarines in the Pacific War was the decimation of the warships of the Imperial Japanese Navy and, even more, of the Japanese merchant marine.*[56] Because the U.S. submarine service knew that it, like all the other branches engaged in the Pacific War, would have to grapple with the tyranny of distance, it had constructed large fleet submarines, usually displacing 1,500 tons and a few

---

*British and Dutch submarines also took their toll, chiefly in the Indian Ocean and the waters surrounding the Dutch East Indies, but the American share of the sinkings was overwhelming. During the war U.S. forces sank 2,117 Japanese merchantmen, totaling a massive 8 million tons of shipping, and 60 percent of that (5.25 million tons) was atttributed to American submarines. Anglo-Dutch subs sank another 73 merchantmen, of 211,000 tons.

going up to 2,700 tons, with ample food and fuel supplies, even some refrigeration, giving a boat a range of 10,000 or more miles. They were nothing like the small European submarines that prowled around the shallow waters of the North Sea and the Mediterranean, and even Doenitz's later U-boats were considerably smaller than the American subs in size.

On the other hand, at war's outbreak the U.S. submarine service probably had the worst torpedoes of any major navy: the Mark 14 torpedo, which seemed an impressive instrument of destruction, had not been rigorously tested, since the cost ($10,000 each) was a greater concern to the Torpedo Bureau at Newport, Rhode Island, than its reliability. Even when war broke out, commanders were cautioned to use them sparingly. There were forward U.S. submarine bases in the Philippines (soon shifted to northern Australia) and at Pearl Harbor, and American naval strategy counted upon them doing significant damage, but for a long while no one appreciated that they were using faulty weapons.

These early American torpedoes had not one but three faults. The first was that the metal contact pin at the head of the torpedo, which was supposed to detonate the mass of TNT upon impact, was simply too weak to do the job. The frustration of submarine captains, who had perhaps spent many an anxious hour maneuvering to get in a good position and fired a spread of torpedoes, only to hear a dull bang against a Japanese hull, was understandable—but the cause was not known for quite a while. The second weakness was even harder to understand. The U.S. submarine service had come into the war with torpedoes possessing highly sophisticated magnetic mechanisms; that is, a torpedo fired at an enemy ship would detonate nearby, and the destructive power of a nearby explosion in water would be greater than a direct hit. Yet the Bureau of Naval Ordnance and the jealously monopolistic weapons development branch at Newport had never tested this ingenious system with a live warhead, nor had they considered that the magnetic field of a target cruising off the Marshalls might be different from that of one off the shores of Rhode Island, where the tests were done. Third, many of the torpedoes that were fired in the early period of the war passed well under their targets because the Mark 14 torpedoes had a tendency to run far deeper than their setting dictated. Often, then, either the torpedo missed entirely or the firing pin malfunctioned upon impact. In either case there was no kill.*

But how was a submarine commander in distant and dangerous waters to know any of this, when the so-called experts back home were convinced all was well? In fact, only in the summer and fall of 1943 did the Navy scientists at Newport work things out. For almost two years, commanders who complained that their torpedoes were "duds" or were going wide of the target were told to get closer before launching. Only after tough and experienced submarine captains such as Dudley W. Morton and Roy S. Benson had told Rear Admiral Charles A. Lockwood, the Central Pacific commander of submarines—often in purple language—that things were badly wrong did the situation begin to change. The simplest way to test the defects was to fire torpedoes at an isolated stretch of cliffs near Hawaii and watch the dismal results. In June Nimitz ordered the return to contact torpedoes, not magnetic-field ones; the depth control mechanism was improved in September 1943, and the contact pin was replaced by a much more reliable one. It had taken the U.S. Navy a good twenty-one months to produce a reliable torpedo, and most of the initiative came from regional commanders and captains, not the home authorities. This improvement did not happen in the Southwest Pacific because the officer in charge there, Captain Christie, had earlier played a role in developing the magnetic-field torpedo and was loath to see it go—until ordered to do so, right at the end of 1943.[57] Allied forces had their own large share of unimaginative, obstructionist bureaucrats, as well as of visionaries.

A detailed account of the American submarine campaign in the Pacific after June 1944 is not needed here, but its impact is not in doubt. Roskill reports, "Although every year of the war had seen an increase in the rate of loss inflicted upon the Japanese merchant navy, it was not until November 1943 that it rose steeply."[58] Despite the intense battles of 1942, the merchant marine's net loss was a mere 239,000 tons; by 1943, it was a dangerous 942,000 tons; and by 1944 it was a colossal 2,150,000 tons. Entering 1945, the Japanese merchant marine

---

*This was not solely an American blunder. HMS *Ark Royal*'s Swordfish torpedo planes were first equipped to attack the *Bismarck* (May 1941) with acoustic torpedoes; when they failed, the Royal Navy reverted to contact torpedoes, with a stunning result.

was about one-quarter the size it had been when Pearl Harbor was attacked. Mass starvation loomed, as did industrial collapse.

The obvious thing to note about the above statistics is that the surge in sinkings took place from the second half of 1943 onward, through 1944, and then petered out in 1945, as by then there were hardly any targets to sink. In consequence, virtually all of the top ten American submarine aces in the Pacific War made their kills after June 1943, almost the exact opposite to the story of their German U-boat counterparts in the Atlantic campaign.[59] There were many other factors to explain the enormous damage that the U.S. submarine service wreaked upon Japan in the last two years of the war (for example, improved radar, more powerful warheads, better intelligence, better air-sea cooperation), but the single most important one was the creation of torpedoes that were reliable, efficient, and deadly. Something that worked came about because certain problem solvers in the middle had pushed things forward.

After the fall of the central Filipino island of Luzon in January 1945, and though Manila itself was not fully secured by MacArthur's forces until March 4, Admiral Halsey reported that "the outer defenses of the Japanese Empire no longer include Burma and the Netherlands East Indies; those countries are now isolated outposts, and their products are no longer available to the Japanese war machine."[60] This was true, but in fact those Japanese sea-lanes were being systematically cut (and eventually would be totally interdicted) by the U.S. submarine service, at far less cost than fighting on land. When the war ended, some enthusiasts claimed that the subs alone could have brought an import-dependent Japan to its knees without the A-bombs or the massive landings planned for November 1945.* That of course is a hypothetical, but it shows the long way that the U.S. submarine service had come since the unsuccessful days of 1942 and much of 1943.

Any account of submarine warfare in the Pacific also has to note the

---

*LeMay and his USAF bomber commanders claimed that the B-29s also could have forced the Japanese to surrender by prolonging the mass firebombing attacks. Perhaps yes, perhaps no. The only thing that seems certain is that the starving of the Japanese nation by submarine blockade or the blasting of Japanese populations by low-altitude bombing would have caused far greater losses of life than did the A-bombs dropped at Hiroshima and Nagasaki.

amazing and catastrophic failure of the Japanese navy to use its own craft and to protect its own merchant marine. This is all the more puzzling, since Japan's navy had from 1868 almost uniformly followed the Royal Navy's best practices and ought surely to have drawn tactical lessons about underwater warfare from both the First World War and the current Battle of the Atlantic. The problem was not in the boats themselves—a considerable number of the Japanese submarines were, like their American equivalents, very large, with great cruising range, and possessing the famous "long lance" torpedoes—but in their deployment. There was blatant neglect by the naval high command of the submarine's natural role as an independent and aggressive commerce destroyer; later, there came the further distortion of that role by Imperial General Headquarters. The battleship-dominated naval leaders were obsessed about using submarines to detect and destroy Allied warships, not their merchant marines. In this, they had some successes early in the war, especially in enclosed waters such as those in the Solomon Islands; the *Wasp* was sunk under precisely such circumstances, and a Japanese sub finished off the *Yorktown* after it was badly damaged at Midway.

But when U.S. antisubmarine warfare tactics improved, when miniaturized radar reached the aircraft and warships of the Pacific Fleet, when the ocean was flooded with dozens and dozens of new destroyers, and when the new fast battleships and fast carriers became the core of the American navy, the prospects of causing large U.S. warship casualties fell away. Japan's huge subs (some of them designed to carry and deploy spotter aircraft, and several even to be submerged mini aircraft carriers) were easy to detect on the surface by radar and underwater by sonar; they were noisy and difficult to maneuver, and because their hulls were weak, they were easy to hit mortally. No "schnorkel" fast boats were developed along German lines, and a primitive form of radar was placed on the first boats only in mid- to late 1944. Air-sea cooperation was feeble. Before and during the war, Japanese yards constructed 174 oceangoing submarines, of which 128 were lost. Morale in the service was low, since its officers knew that they had no support in high places, and the service steadily shrank until it was a second-class instrument of war, in distinct contrast to the sub forces of the German, British, and American navies.[61]

It is therefore unsurprising that Japanese submarines were an indecisive factor in the Pacific fighting. Of the sixteen deployed during the Leyte Gulf battles, for example, only one managed a kill, of a destroyer escort. By that time, however, the Imperial Japanese Army was truly mangling the effectiveness of the submarine fleet by insisting that they act as cargo vessels, bringing food and munitions supplies to bypassed island garrisons. The more garrisons that were isolated by the American leapfrog tactics, the harder the submarine fleet had to work in this capacity, seriously diverting its skills and assets. Occasionally they were ordered to carry out symbolic acts, such as the offshore shelling of Vancouver Island in 1942. What if those same boats, with their powerful torpedoes, had sat off the harbors of Portland and Long Beach and inflicted the devastation in those waters that Doenitz's U-boats were doing up and down the eastern seaboard in those same months? As it was, the Japanese submarine fleet sank a mere 184 ships during the entire war. The official American naval historian of the war, S. E. Morison, usually gentle in his comments, felt forced to describe Japan's U-boat policy as "verging on the idiotic."[62] It is difficult to disagree.

While the Japanese submarine forces were frittered away, the surface navy showed an unbelievable myopia toward that most vital of maritime tasks, the protection of maritime commerce. In this regard, nothing could be more different than the attitude of the British and Japanese admiralties. Apart from confronting the few German and Italian battleships, and pushing through relief convoys to Malta and Egypt, the greater part of the Royal Navy was dedicated to getting Allied merchantmen safely across the high seas to the home island. By comparison, the Japanese simply assumed that the early conquests in the Pacific and Southeast Asia gave them control of the waters in between. Any insolent Allied intruder would be detected and sunk (and quite a few were, though the number of kills reported back to headquarters turned out to have been vastly inflated). They had nothing like Max Horton's Western Approaches Command and no equivalent of RAF or RCAF Coastal Commands. They had no statistical or operational research department for analysis of the unfolding campaigns. There was very little weapons development: the depth charges they were using at the end of the war were roughly the same as those employed at the start of the conflict. Miniaturized radar, Hedgehogs, homing torpedoes, and hunter-killer groups were all missing.[63]

By 1943–44, the Japanese naval high command was realizing that they had to cluster their merchant ships into convoys and provide a screen of escorts; but their tactics remained primitive. A sudden sinking of one of the Japanese merchant ships caused the escorts to rush around in all directions, haphazardly dropping depth charges, so the best thing the U.S. sub commander could do after firing his torpedoes was to lie on the ocean bottom (or at least go very deep) and wait for a few hours before resuming the stalking. Furthermore, following the great Philippine Sea and Marianas battles of June 1944, the American service commanders—especially Rear Admiral Lockwood at Pearl Harbor—had enough boats to group their submarines into small wolf packs. By September 1944 three such groups, each of three boats, were operating all around Formosan waters, including in the Straits, inflicting havoc on merchantmen and their destroyers alike. Apart from dropping dozens and sometimes even hundreds of depth charges—only a few of which hit these underwater predators—the Japanese navy had no other response, no new form of countermeasure. The statistics of Japanese merchant ship losses above show the extent of this abject failure.

The American Surge

Because the newly engineered tools of war, from torpedoes to landing craft to B-29 bombers, took so relatively long to produce in large numbers and reliable standard form, the American counteroffensive in the Pacific evolved much later than did the Allied comebacks in Europe: the Battle of the Atlantic was won, North Africa occupied, Sicily occupied, and southern Italy occupied before even the Gilberts operation (November 1943) began. But once the Americans had assembled and tested their newer systems, they struck at remarkable speed. It was a tribute to the Americans' astonishing lift capacity that, at roughly the same time as the Allied counteroffensives began in North Africa and Europe, it could also be initiating a number of amphibious comebacks in theaters of war 8,000 miles away from Morocco. To the United States, the "Germany first" principle had never meant that they should hold back from aggressive actions in the Pacific once Japan's own expansion had run out of steam, as had occurred as early as the Battle of Midway in June 1942. The main question, given the sheer distances involved between almost any two strategic points in the Pacific, and given the parallel need to train many more active divisions, receive many more aircraft and warships, and produce the all-critical landing craft in far larger numbers, was where the first recovery operations should be attempted. Until the highest authorities had made that decision, the most sensible thing to do, in the first instance, was to strengthen Allied positions in the Pacific and to acquire what Ronald Spector nicely described as "the seizure of Japan's strategic points."[64]

The least significant of America's counterblows, strategically, was the recovery of a couple of the Aleutian Islands (Kiska and Attu) from the Japanese garrisons that had been dropped off there as a sideshow to Midway. Even that reconquest did not take place until May–July 1943, amid constant fogs, and after a short while the Japanese abandoned those uninteresting atolls. Its navy offered resistance for half a day (the desultory Battle of the Komandorski Islands), then pulled back. Its army was not waiting at the beaches, so all American landings here were unopposed. The merit of this exercise was that it gave raw U.S. troops (100,000 of them) an opportunity for landing on distant shores, and the massive naval support—including bombardments by three battleships—an early chance to see how hard it was to do much damage to enemy troops holed up in distant hills. Its demerit was that it gave next to no chance—though there was some final resistance by Japanese troops on Kiska—to experience some fighting against an enemy determined to drive the landing forces back onto the reefs.[65] Once having taken these storm-beaten, mist-clad islands, the American Joint Chiefs were content and turned elsewhere.

Altogether more significant strategically were the earlier American-Australian counteroffensives in the Southwest Pacific theater, that is, in New Guinea and the Solomon Islands (especially Guadalcanal). General MacArthur's dogged series of amphibious landings and moppings-up of nearby Japanese garrisons was different from the Allied assaults against the shores of North Africa, Sicily, Italy, and France, just as they were, operationally, often a different tale from Central Pacific Command's efforts to capture the Gilberts, Marianas, Iwo Jima, and Okinawa. On the whole, the landings in the Southwest Pacific theater occurred at points along the enemy's coastline that were *not* protected by troops, gun emplacements, and minefields. Of course the Japanese responded promptly and viciously to such Allied intrusions, but the local garrison was often a long way from where MacArthur landed, so the sheer physical problem of getting Allied troops from ship to shore in the face of massive resistance was rarely experienced during these campaigns. Even in the bloody struggle for Guadalcanal (August 1942–February 1943), the greatest gain came from being able to put 11,000 out of the 19,000 U.S. marines onshore on a single day without immediate opposition, thus enabling them to consolidate lines inland, grab the airfield, and then, encamped under the trees, to cause the Japanese to consistently underestimate the size of the American garrison for the rest of that grim battle.[66]

Landings without direct opposition, at Guadalcanal and elsewhere, had their own important lessons for future assaults. While the marines were probably better prepared than any other forces for amphibious operations, that preparation had all been done in theory and as summer exercises. They, along with MacArthur's army divisions in New Guinea and New Britain, were to benefit enormously from experimenting with preinvasion intelligence gathering, beachhead control, logistical follow-up, fire support (if necessary), aerial patrols, and, in general, effective command and control when they landed unopposed. They also were beginning to learn how to deal with coral reefs and mangrove swamps, with rain forests 8,000 feet in altitude, with constant mists, and with very high rates of sickness, especially dysentery. It really was useful to begin a counteroffensive in places where the enemy wasn't. It was better to push a mile or five inland before the enemy rudely showed up.

There is one other illuminating point of comparison with the European amphibious operations of this time. In September 1943 Admiral Nimitz created a new operational command to be responsible for the seizure of the Gilbert Islands, the first step forward in the Central Pacific thrust. The story of the U.S. Marines' attack upon Tarawa in late November is recorded earlier in this chapter, but as a landing roughly contemporaneous to the Sicily/Italy operations, it deserves a cross-reference here. Eisenhower took no chances in committing 478,000 troops to Sicily. In the Gilberts, another massive invasion force would be deployed. While the smaller island of Makin would be assaulted by 7,000 U.S. Army troops, a full 18,000 men of the 2nd Marine Division were to take Tarawa—and at long last Pete Ellis's vision would unfold, but in far greater dimensions than he imagined. Supporting the amphibious forces for the Gilbert operations was an armada of brand-new and older battleships, fleet and escort carriers, squadrons of heavy cruisers, dozens of destroyers, 850 carrier aircraft, and 150 USAAF medium bombers. This naval force was already the largest fleet in the world. And all this for an island group (defended by a mere 3,000 naval troops on Tarawa and 800 on Makin) that was not even part of the critical external perimeter as defined by Imperial General Headquarters.

Hence the postmortems over the shocking American combat losses in the Gilberts operation are at least as numerous as those over Dieppe, often to the point of forgetting that the islands *were* captured. Still, it was a sobering experience. Neither the distant naval bombardment nor the aerial attacks did much to harm the low-lying Japanese bunkers or to knock out many of the garrison's guns. Command and control was ragged. Rear Admiral John R. Hill, in charge of the southern (Tarawa) operation, was located in the battleship USS *Maryland,* whose salvoes kept knocking out its own radio communications. The bombardment ceased too early, just as at the Somme and Gallipoli, so the defenders were able to resume their posts and pour fire on the attacking troops. Worst of all, the water level across the coral-filled lagoon was unusually low, which meant that the landing craft were stuck on the outer reef, and consequently the marines had to wade 700 yards to the shore—and were slaughtered. The very green army units on Makin made it to the beaches, but then were pinned down in the coconut groves. Finally, with reserve battalions brought in for both battles, the Americans prevailed, assisted greatly by the Japanese habit of suicidal counterattacks near the end. But the cost of the three-day battle for Tarawa was high: more than 1,000 American dead and 2,000 wounded, all to capture an island of less than 3 square miles. The press photos of dead marines floating in the water or lying across coral reefs was the most startling evidence yet to the American public that the Pacific War was going to be long and hard.

What a difference from the seizure of the Mariana Islands, a mere eight months later, but by then the Marine Corps had figured out its amphibious-warfare techniques, the fast carriers were ranging around, the Hellcats were aloft, the Seabees were constructing gigantic runways made of bulldozed coral, and the B-29s were soon to be flying in. The marines at Tarawa had paid a price, but it would never need to be repeated.

Thus, a mere nine days after the Normandy landings, an enormous force of American troops (127,000 of them, chiefly marines) began to land on the Mariana Islands, thirteen time zones eastward from the English Channel. This was the single most important amphibious operation in the Pacific War, and far more threatening to Japan than MacArthur's hops along the north coast of New Guinea on his way to the Philippines. By taking Saipan and also Guam—the first American possession to be reoccupied—Central Pacific Command was able to acquire, at last, air bases for the long-range strategic bombing of the Japanese homeland. Once the islands were seized and the surrounding waters made safe, there was very little Tokyo could do about it. This operation, coming as it did between the strikes against the Caroline Islands and the Battle of the Philippine Sea, was critical.[67]

The early operations by the U.S. carrier squadrons around the Marianas had given Nimitz's forces command of the air. The two-day naval bombardment was of a much heavier variety than that carried out off Omaha Beach, and more in the measured British style; the distant firing by the nine brand-new American battleships didn't do much, but when the slower battleships and heavy cruisers came closer inland, some of them to within 1,500 yards, the shellings began to count. Still, even close-in firing could not crush well-built positions, and the Japanese were able to inflict lots of damage upon the experienced but mislanded 2nd Marine Division, while many amphibious craft (i.e., the LVTs) could not surmount the beach obstacles. But the Americans battered their way in. They had 8,000 marines onshore in the first twenty minutes, and 20,000 men on Saipan at the end of the first day; in that regard, it was somewhat like Utah Beach in Normandy. Supporting aircraft were overhead, and the battleships were firing from close-in positions. The beaches were clearly marked out by color codes. Eventually there was a separate channel for return traffic: empty landing craft, damaged vessels, and hospital ships with their dead and wounded. There was no external interference, due to the smashing of Japanese naval airpower in the battle of the Philippine Sea.

Operation Forager was not a perfect job. Amazingly, there still was not an interservice HQ ship like the *Largs* or *Bolulu;* overall command for ship-to-shore movement rested with a very competent Commodore Thiess in a large patrol craft, but that was not the same and certainly would not have worked off Normandy.[68] The LVTs were feebly powered and, as noted, often could not reach their destinations over craggy ranges of coral and rocks. Naval shelling was much less useful than the aerial strafing of the beaches, but even the latter was ragged and often distracted to ancillary missions. What was needed (and would not arrive for another year) were escort carriers specifically tasked for beachhead support and not called away for possible high-seas battles. The Japanese garrisons fought as vigorously as ever and would never surrender; it took three days of tough combat to secure the northern beachheads of Saipan, during which one particular U.S. Army division (the 27th New York National Guard) faltered and was sharply criticized by the overall commander (a marine general), causing an interservice row. Yet the results were not in doubt, and the island fell on July 9. A little later all opposition on Guam was eradicated. Within another several weeks, the Seabees had started construction of those ultralong B-29 airstrips, their bulldozers tolling the end for many Japanese cities and for the Japanese Empire itself.

Perhaps we can forgive Samuel Eliot Morison, that patriotic Harvard professor turned into the official U.S. naval historian by his former student FDR and sent off to the Pacific War, for his proud statement summing up this operation just a few years after the war: "Added together, 'Overlord' in Europe and 'Forager' in the Pacific made the greatest military effort ever put forth by the United States or any other nation at one time. It should be a matter of pride and congratulation to the American and British people that their united efforts made June of 1944 the greatest month yet in military and naval history; that simultaneously they were able to launch these two mighty overseas expeditions against their powerful enemies in the East and the West."[69]

Rokossovski and Konev, battling their way across the Dnieper against two dozen Wehrmacht divisions, may have ranked this achievement differently, but in the history of amphibious warfare operations Morison was surely correct. Perhaps the shadows of the Spanish Royal Marines were there at the appropriately named Caroline and Mariana island groups, once the possessions of Philip of Spain. The Allies had at last figured out how to land on an enemy-held shore.

Amphibious warfare was at the center of everything, and by 1944 the landing operations were becoming ever larger and more ambitious. Realizing that his Southwest Pacific Command was in danger of being relegated to secondary status and determined to lead an American return to the Philippines, MacArthur had begun to accelerate his pace of advance from November 1943 (Bougainville) onward, ignoring and isolating the great Japanese base at Rabaul, and skipping along the north shore of New Guinea. But the advance forces of his command did not reach the far end of that massive island until July 1944, and by that time so much else had happened, mainly in the Central Pacific theater.

The rest of the Pacific War thus consisted of an unfolding of American amphibious power against tenacious and suicidal Japanese resistance: in the Philippines from December 1944 to March 1945, at Iwo Jima from February to March 1945, and at Okinawa from April to June 1945. On April 7 the first P-51 long-range Mustangs flew from Iwo Jima bases as escorts to the B-29s attacking Japanese cities. By this stage in the war, however, there was little for these high-flying, long-range fighters to do; the bombers were razing those cities to the ground with impunity.

With the surrender of the Third Reich on May 7, 1945, the Second World War became, essentially, amphibious warfare against Japan. Of course, British Empire troops under Slim were racing into Rangoon and preparing to leap farther south, to Malaya and Singapore, and there was still mopping up to be done elsewhere in Southeast Asia (Borneo, Mindanao). Fighting continued unabated in much of mainland China. American submarines and long-range bombers were ripping the Japanese economy to shreds. But the most important operation strategically—following the occupation of Saipan and Guam—was the taking of Okinawa, then to be converted into an enormous forward base for the final assault upon Japan. In a narrower, operational sense, too, the Okinawa campaign was significant and symbolic, justifying

Isley and Crowl's play upon the many contrasts between this assault from the sea and that at the Dardanelles (see page 222). So much about the nature of amphibious warfare had changed since those Gallipoli landings thirty summers before—the carrier raids, the B-29 bombings, the special equipment of the landing forces. Yet the ancient difficulties remained, above all that of getting onto beaches that were manned by an enemy, be it Turks or Japanese, determined to give no ground. By the end of the formal fighting in late June, American casualties on Okinawa totaled 49,000 men (12,500 killed), by far their heaviest campaign loss in the war in the Pacific, and a grim hint to Nimitz and MacArthur of what was to come.

Well before the shooting stopped on Okinawa, American planners had been preparing for the greatest amphibious operation of all time: that on the home islands of Japan, to take place possibly as early as November 1945. The figures involved in this assault were going to be enormous, surpassing Overlord by an impressive degree: some 500,000 men, arranged in four gigantic army corps, all American, were being organized to overwhelm the southern, vital island of Kyushu, to be followed by an even larger amphibious force against the Tokyo region in the early spring of 1946. To that end, most of the U.S. military units, warships, and air squadrons that had fought in the European theater were now directed toward the Pacific. In the event, these strikes from the sea were not necessary. Atomic bombs dropped on Hiroshima (August 6) and Nagasaki (August 9) forced the end of the Pacific War. To the millions of American servicemen, and to the smaller number of Australian, British, Canadian, Dutch, French, Indian, and New Zealand units already in or being sent to the Far East, the relief felt at the end of the war was palpable; almost every memoir mentions that feeling, the slow ebbing of strain. The fighting men of the western lands had had enough of being thrown onto a hostile shore. It was time to go home.

Larger Thoughts

The taking of the Mariana Islands in the middle of 1944 was not just another step forward in America's war in the Pacific, or merely a nice coincidence with the invasion of Normandy. In strategic terms, it was the key to the Central Pacific campaign, and a far more significant move toward the defeat of Japan than any other action across this vast theater. For it was those Mariana outcrops—Saipan, Guam, and especially that nondescript flat island of Tinian—that gave the U.S. Air Force a massive and indestructible aircraft carrier for the annihilation of Japan's industries and cities. Little else in the Pacific counted in that same positive way. Holding on to Hawaii counted, for the obvious reason that virtually everything went through there. Midway and Guadalcanal counted in 1942, for it was there that the expansionism of the Japanese navy and army, respectively, was blunted. Tarawa counted in late 1943, negatively, because it taught the United States the real costs and possibilities of amphibious operations. But the capture of the Marianas was a true turning point, comparable more to Normandy or Stalingrad-Kursk than to the battles for Egypt, Sicily, and Rome.

Taking those islands in June 1944 did not, of course, mean the war against Japan was over. Many events were to occur in the thirteen months that followed—the assault upon the Philippines, the enormous, sprawling, and intensive battles of Leyte Gulf, Iwo Jima, and Okinawa, and the relentless firebombing of Japanese cities and industry, culminating in the dropping of the two atomic bombs in August 1945. Yet much of that would have been significantly harder, perhaps altogether impractical, without the Marianas bases. When Mussolini learned of the Anglo-American landings in Italy, he famously declared, "History has seized us by the throat." The same might be said of the American capture of the Marianas. After learning of the loss of those islands and the irrecoverable damage inflicted on their naval and air forces around Rabaul and in the Battle of the Philippine Sea, the Japanese cabinet resigned. It was time to go; time, even, for some of the Japanese leaders to secretly communicate with Moscow about terms for surrender.

Recently the scholar James B. Wood posed the question of whether defeat was inevitable. His thoughtful book complements much of the analysis presented in the pages above, albeit from a Japanese perspective. By "defeat," Wood does not mean the total crushing of August 1945; rather, he asks whether the Japanese leadership might not, by different policies, have left Tokyo with an inner "rim" of possessions, a reasonably self-sufficient economy, and a continuing status as a great power, thus achieving a solution reasonably close to the one Admiral Yamamoto desired.[70]

Meticulously, then, Wood goes through the checklist: What if the Japanese navy had convoyed its merchant marine from the beginning? What if its own formidable submarines had been deployed to cut Allied mercantile communications? What if it had paid even more attention to naval air forces and less to battleships? What if Imperial General Headquarters had appreciated that it could not conquer about a quarter of Asia and the Pacific with a mere dozen divisions? What if, especially, Japan had been content with building up a really formidable inner defensive rim (the Marianas to the Philippines to Borneo) instead of pushing arrogantly toward Midway, the Aleutians, the Solomons, Australia, and northeast India? After all, if it had not been punished so badly in the first two or three years of the war by trying to defend its outer ramparts, could it not have retained the capacity to blunt American assaults to the point that some "non-total-war" compromise was possible, as was common in all of the great-power conflicts of the eighteenth and nineteenth centuries? Why not pursue that option, with smarter operational and tactical policies?

Wood poses a set of important questions, and his work has received proper respect. But the answer to them has to be no. Once the United States had been treacherously mauled at Pearl Harbor, there was no going back for the Americans, no compromise, nothing but the unconditional surrender of the aggressor nation. Japan, counting only its frontline aircraft, warships, and army divisions and seeing them all so strong, had chosen to attack a country that had ten times its economic power (in terms of GNP) but which had not yet mobilized for total war. When that country was fully energized, the defeat of Japan would come, because America had no reason to compromise. Curiously, both Yamamoto and Churchill knew that. The former talked of America as a "sleeping giant, yet to be aroused"; the prime minister fondly referred to a "gigantic industrial boiler" that was not fully stoked up.

But gigantic productive power means little in wartime unless it is harnessed and its resources are directed to the right places. Total steel output means nothing at all until it is directed toward well-designed

Essex-class carriers. Aluminum and rubber and copper mean nothing until they are given to the B-29 construction program. Skilled workers mean nothing until Ben Moreell organizes them. Flat-bottomed Everglades boats mean nothing until the marines convert them to landing craft. Sophisticated torpedoes mean nothing until someone figures out why they are not working and fixes the problem. Long-range carrier operations mean nothing until there are long-range oil tankers. Somebody—some organization, some team given a free hand to experiment—has to come up with solutions and then put them into practice.

PROBLEM SOLVING IN HISTORY

The German prisoners of war marching through the streets of Moscow in July 1944.

On July 17, 1944, an enormous force of approximately 57,000 German soldiers—a full twenty abreast—marched along the main streets of Moscow.

The photograph of this dramatic scene must remind viewers of similar Wehrmacht units marching through the Arc de Triomphe four summers earlier. But look more closely. This was not Hitler's dream fulfilled. It was Stalin's calculated revenge.

354 ENGINEERS OF VICTORY

When Roman emperors defeated the Germanic tribes, they brought their captives back to the imperial capital to give the applauding populace the physical proof of the great victory. So was it also in Moscow in July 1944. The 57,000 marching troops, carefully guarded on each side by Russian riflemen, were the survivors of the German Fourth Army that had been beaten, surrounded, and "*Kesselschlacht*-ed" during Operation Bagration. The German barbarians were again being displayed, in their full humiliation, before being dispatched to execution or captivity.

Masses of Japanese were never displayed in such a way because very few of them surrendered, and the American and British Commonwealth armies preferred to organize their own victory parades through Rome, Paris, Brussels, and the other captured capitals. Different expressions of triumph with a single meaning. At the June 1944 collapse of Axis perimeters in four different theaters of the war, there was no doubt about the result of this titanic conflict, and the victors could already prepare for the ceremonials of May 1945 and August 1945.

One way of beginning to understand how the Western Allies, at least, turned defeat into victory is through a detailed perusal of the diaries of Field Marshal Lord Alanbrooke, Churchill's chief military aide during the Second World War. From his handling of his own army divisions' retreat through Dunkirk after the disastrous Battle of France in June 1940 to his appointment as Chief of the Imperial General Staff, from his place in all the important Anglo-American conferences on their coalition strategy to his monitoring of how his beloved British armed forces received so many further batterings (Greece/Crete, Tobruk, Singapore) before the long hard road to recovery, Alanbrooke has provided later generations with a candid account of how the war slowly turned from Axis triumph to Allied victory. When in 1957 and 1959 the patriotic historian Sir Arthur Bryant produced his two-volume edited version of those diaries—titled *The Turn of the Tide 1939–1943* and *Triumph in the West 1943–1945*—the reader could already appreciate Alanbrooke's intellectual powers and capacity for wide-ranging though acerbic judgments.[1] But it was not until the more extensive and even more candid (because unexpurgated) edition of *War Diaries 1939–1945* appeared in 2001 that one gained a fuller grasp, naturally from Alan-brooke's own perspective, of the many wearying demands upon his time: handling the army, handling his fellow British Chiefs, handling the Americans, and above all handling his own mercurial, difficult, and brilliant War Lord, Churchill.[2] Masters and commanders, indeed.[3]

Yet what is also striking about the diaries is the extent to which Alanbrooke so frequently delves below the higher politics of the war to wonder about whether the machinery and personnel for an eventual victory over the Axis were good enough. Perhaps it was the experiences of Dunkirk and Crete that haunted his mind, for he clearly worried each day about whether the middle levels of the British Empire's sprawl-ing machine could withstand the terrifying demands placed upon it, and whether the Grand Alliance could withstand further setbacks. His concerns (though never defeatist) turned out to be absolutely invalu-able. They restrained Churchill from bizarre ideas about a large-scale invasion of northern Norway, they tempered the unrealistic American desire to invade France in 1943, and, above all, they established a ruth-lessly pragmatic litmus test for all proposed new operations: "Where will it be? Can it be done? Who will do it? Are there enough forces, equipment, training?" Alanbrooke, though he would have hated the comparison, had a very Leninist approach to things—his "who, whom" practicality was of the essence.

As it turned out, all that had to be done was done, in due fulfill-ment of the Casablanca operational directives, albeit at heavy cost and with many a setback in every theater of the war. In the middle of fight-ing in so many regions of the globe, on land, at sea, and in the air, it must have been hard to recognize when the tide was turning, when the setbacks were fewer and the victories more plain. It is noteworthy that not until July 25, 1943—as Mussolini surrendered, the Red Army blasted its way around Orel, and Patton took Palermo—could Alan-brooke allow in his private nighttime diary that there might have come about "at least a change-over from 'the end of the beginning' to 'the beginning of the end.'"[4] Yet for another full year he would feel deep unease about the setbacks in Italy, the defeats in Burma, the ineffective-ness of the bombing campaigns, and the planned invasion of France at what he believed might be a premature moment. Like another cautious realist, Eisenhower (see page 249), he worried to the last minute that the Normandy operation could turn out to be a horrible, massive disaster. Thus in June 1944 it was an emotional moment for Alanbrooke to stand again on the French soil from which he and his divisions in the BEF had been expelled four years earlier.[5] By then, even this cautious, acerbic, tough-minded man could see that the end of the struggle was in sight.

This awful war would claim many more military and civilian victims between then and August 1945. Still, it is time now to turn to some general reflections of how and when and why the tide changed in the five great campaigns analyzed above. Each chapter has tried to recapture how it looked to the problem solvers without assuming that the tide would turn. Now we can look back at the whole.

No single or monocausal argument can be drawn from the analysis. If an overall judgment had to be made, it would be to caution against our instinctive human desire to simplify. Societies are highly complex organisms. Wars are complex endeavors. Military systems and armed services have millions of moving parts, as do modern economies. They also operate at many levels. Thus while critical factors such as leadership, morale, and national and ideological fervor in wartime are definitely important, they are intangible rather than measurable, and any discussion of the "how" factor cannot remain at that level. And each year, amazingly, there appear new archival sources, alongside fresh insights, controversies, and approaches, that should compel us to rethink earlier assumptions.

A lot of scholars tend to divide themselves into either "lumpers" or "splitters"—those who see a single overriding cause for what goes on, and those who see only confusion or multitudinous parts.[6] The second usually appears in some general or textbook form (that is, everything is included), the former in some tunnel-vision work with a sensationalist title ("the weapon that won the war"). Neither is very satisfactory. Surely it is possible to offer an analysis of how the tide of the Second World War turned in the Allies' favor without either claiming a single reason or offering a sanitized narrative in which everything is mentioned. It makes sense to think that some select parts of the tale have greater explanatory power than others. In this analysis of the Allies' gaining dominance in the war during the central eighteen months between January 1943 and June-July 1944, the selective process has been pursued. Not all contributions to the Allied victory count as equal. The claims of certain new methods, battles, individuals, and organizations as having been vital for victory have to be demonstrated.

The conclusions below are not assembled in a rigid hierarchy, although the order of the themes and reflections does advance toward a small number of key factors that explain how this particular war was won. That moves other explanations into a position that helps the reader see where they were important but also where they were not. I did not anticipate some of this ordering of the causes of the Allied victories when I began this book four years ago. The key determinant throughout, however, was a simple one: did this new device, organization, or new form of weaponry really *work* in winning battles, and can that be shown in practice? If scholars assumed afterward that a new invention was decisive simply because it was up and running, that was not enough.

A good example of this concerns the part played by intelligence in the winning of the war, especially in view of the massive attention given to that theme over the past four decades. So voluminous nowadays is the literature on intelligence in the Second World War that I originally assumed it was proper for the draft table of contents of this book to list a chapter titled "Chapter 6: How to Win the Intelligence War." Then two things happened. The first was that it gradually became apparent that it made no more sense to isolate intelligence into a single chapter than it did to isolate logistics or science and technology. All claim an important role in the eventual Allied victory in World War II, but the only way of measuring their true significance—as opposed to swallowing the assertions of their backers—is to integrate them into the story and analysis of each of the five parallel campaigns that make up this book. Claiming an astounding "intelligence breakthrough" that changed the course of the struggle without showing where it actually worked in the fighting fields is as sloppy as claiming that a new weapon transformed a campaign without proving that it did.*

---

*One might think of this as the litmus test for the P-51 Mustang versus the Me 262. Both were aviation breakthroughs, but one had a terrific impact on the war's outcome and the other didn't. Thus, arguments about the significance of this or that spy ring in the Second World War have to go through the same sort of litmus test: did they really help the war to be won, and where, specifically?

Once one attempts to integrate the hidden dimension of intelligence into the story of how the tide turned, a very mixed picture—more bluntly, a confusing one—emerges. For the fact is that the usefulness of intelligence varied from theater to theater, from service to service, and from year to year. Intelligence had a value in one encounter but no particular value in another. There is no way of amalgamating the many case studies into some complete rendering of account, and it remains a continual puzzle to this author that such a professional scholar as Sir Harry Hinsley could venture the opinion that Ultra probably shortened the war by as much as three years.[7] It is not provable.

One thing that emerges from the study of intelligence in the 1939–45 war, frankly, is the preponderance of intelligence *failures*. The litany probably begins in 1939–40, with the Red Army's crass underestimation of Finnish resistance capacities, the British intelligence failures regarding the Norwegian campaign, and the complete French ignorance of what was about to burst out of the Ardennes in May 1940. Add Stalin's utter refusal to listen to *all* the intelligence warnings about Operation Barbarossa and, in a different category, the American blindness to understand an attack was coming upon Pearl Harbor, and it may be that the year 1941 claims the accolade for having the most intelligence failures of the war. Then came the horribly expensive mistakes by Wehrmacht intelligence in not foreseeing the Red Army's encirclement at Stalingrad in 1942, the reverse pincer movement at Kursk in 1943, and the central punch at Bagration in 1944—a three-loss record without any equivalent.

The list goes on. MacArthur's prejudice against intelligence from naval sources was bound to affect the potential of the enemy messages that had been cracked by the brilliant Magic and Purple code breakers.[8] There was no realistic Allied assessment of how forcefully the German army would counterattack at Salerno and Anzio or at the Battle of the Bulge—it was as if no one in Anglo-American military intelligence had figured out that having to fight a Waffen-SS division was a very tough proposition. Brian Urquhart's warning that the Arnhem paratroop units would drop where seasoned German regiments were stationed was ignored. This incompetence was widely shared. Imperial General Headquarters in Tokyo was repeatedly blindsided by the multiple routes of American advances across the Pacific, yet surely anyone who

has played a complex board game could perceive that the counteroffensive might drive along various, changing routes. The final point may
be the most important of all: there were of course fabulous technological advances made during the war in this dimension, especially in decryption and signals intelligence, yet that did not automatically give an
advantage leading to victory in a battle or a campaign. Even the Ultra
system was imperfect; far, far better to have it than not, but by the very
nature of the intelligence war with Germany it could only produce
mixed results, not miracles.

By contrast, the successes—that is, the breakthroughs that had
provable battlefield victories that shortened the course of the war, which
is *the* litmus test—are on a short-order menu.[9] The Battle of Midway—
knowing where the enemy carriers are located, hiding one's own location from the enemy, and then being able to arrange one's forces
accordingly—certainly passes the test. But what else does? Several of
the Royal Navy's victories over the Italian fleet in Mediterranean encounters occurred because the British knew the location of their enemy,
and the sinking of the *Scharnhorst* far to the north in December 1943
was a clear triumph for British detection systems. The ability by 1943
to pick up and then sink distant U-boats was potentially a smart use of
intelligence, provided the high early success record didn't compromise
the detection system itself, as was threatened by the U.S. Navy's over-
zealous attacks upon U-boat meeting places in the Atlantic. Soviet military intelligence got better and better by 1944, but so did the entire
Red Army and Air Force. So, once again, one is driven back to the
blunt question: where can prior knowledge of the enemy's forces and
intentions be shown to have had military impacts? On the whole, and
even if one can readily concede that the Allied record on intelligence
was far better than that of the Axis, it is easier to demonstrate where
smooth logistics helped win the war than to show where intelligence
led to victory.[10]

A more positive surprise was the repeated evidence of the sheer in-
terconnectedness of the five narratives. Although the introduction suggested that an Allied success or failure in one campaign would have
consequences elsewhere, one of my early assumptions was that this
book would consist of five essentially self-contained, parallel narratives,
picking up from time to time the connections between, say, the cam-

paign against the U-boats and the Sicily landings, or between Operation Bagration and Normandy. Moreover, histories of the war in any single theater, even brilliant ones such as Atkinson's books on the U.S. Army in the Mediterranean, draw the reader's attention closer in, giving us even less perspective by which to understand what is happening elsewhere in the same season of the war.

Only after several chapters were drafted did it become clear to me that, for example, one would need to track the Wehrmacht and Luftwaffe's frantic juggling acts of moving divisions and air groups from France to Russia to the Mediterranean and back again, as explained in chapters 2, 3, and 4; to see better how the British fight for the Atlantic and Arctic sea-lanes, together with the strategic bombing campaign against Germany, dovetailed with the Red Army's defensive and offensive battles in the east; and to detail how the Japanese high command's excessive allocations of army divisions to central and southern China so undermined their capacity to hold ground in the Pacific. But campaign historians rarely connect the parts, and especially not across and between the five major theaters of war covered here. One comes away from this comparative exercise with a better understanding of the enormous tasks facing governments that have to conduct a multidimensional major conflict, and with a much higher appreciation of how the leaders in earlier wars had been involved in the same sort of juggling act.*

Some of the wartime Great Powers had less need to connect the parts than others; some had a need but failed to achieve balance and coherence. The Soviets did not have this problem of prioritizing; refusing to enter the Far Eastern war, they concentrated almost exclusively upon defeating the Nazi invader, and everything else was sensibly and terrifyingly subordinated to that end. The United States did face its classic Europe-versus-Pacific dilemma regarding priorities and resources, but by 1943–44 it had the capacity, remarkably, to surge for-

---

*Consider, among others, the Romans, holding their many fronts for many centuries; the Elder Pitt in the Seven Years' War, juggling Europe, Canada, and India; and Clemenceau and Lloyd George in the First World War. Consider, by contrast, Philip II of Spain's inability to focus, and Napoleon's double distraction in both Spain and Russia. To govern is indeed to choose. The implications for the American government in our troubled early twenty-first-century world are obvious.

ward in both theaters. Japan trapped itself by striking in too many directions—China, Burma, Southeast Asia, New Guinea, the Central Pacific—without the capacity to do so, and paid the eventual price for such lack of concentration, as well as for its shrinking resources. Germany practiced a multidimensional strategy, but Hitler's regime simply could not at the same time prevail in the great war in the Atlantic, the gigantic aerial battles over Europe, and the very large land campaigns in the Mediterranean, North Africa, the Balkans, and (especially) Russia, as well as preparing for the future Anglo-American invasion of France. Here, if anywhere, was the prime example of overstretch.

If the British did it rather better, it is perhaps because they had been juggling their global obligations for more than two hundred years and their decision-making systems had become reasonably good at it, but during the low point of 1941–42—Greece/Crete, Tobruk, Singapore, the Atlantic—it must have seemed to many that the old, battered empire was fading. Throughout the war British leaders *knew* they were overstretched—one reason the Alanbrooke diaries are so insightful— and that in turn explains why the Whitehall planners strove so intensively to make savings, to find new technologies that would reduce the future losses of ships, aircraft, and especially men, and welcomed volunteer fighting units and individuals from at least a dozen other nations into their own armed forces. It also explains why they attached such importance to getting their American allies to see things the British way, and were alarmed at the occasions when their more powerful and confident (and sometimes more ignorant) cousins brushed them away.* As it was, Britain ended the war with the German-Italian Axis indeed completely defeated and Japan crushed, but it also overstrained itself and its empire as a result—as we can now see, fatally so.[11]

It is striking that while these five major players were so differently constituted, with each fighting its own war and along the different di-

---

* Consider King's refusal to accept Royal Navy advice to convoy merchantmen steaming up the East Coast of the United States during 1942 (those were British oil tankers from Trinidad, supposedly under USN protection), Echolls's obstinacy to accept the Merlin-powered P-51, and Bradley's lack of interest in any British ideas regarding landing techniques, and compare these with accounts of American scientists' ready acceptance of the cavity magnetron, the shared intelligence work in the Bletchley huts, and the sharing of shipbuilding and ship-repair plans.

mensions of air, sea, and land, the end result here—that is, the 1944 result—was so uniform, with the Axis hard crust crumpling at the same time across four different theaters. Mark Clark may have justly complained that the arrival of his armies in Rome on June 4, 1944, was hardly noticed by the world's press, but in practical terms and effect the Allied victory in central Italy was far less significant for the final resolution of the war than the other three, that is, the almost simultaneous invasion of Normandy, the seizure of the Marianas, and the powerful blow of Operation Bagration.

Was this, then, just a new form of imperial fatigue? Was it simply that the German and Japanese war machines, which had fought with astonishing ferocity and efficiency for so long, were geographically overextended by their leaderships and ultimately collapsed in the face of the accumulated material power of the British Empire, the United States, and the USSR? The raw statistics certainly suggest that: by 1942 the Grand Alliance was equal to the Axis in terms of productive power, by 1943 it was surging well ahead, and by 1944 it was dominant.[12] Take any measure—oil supplies, steel output, or aircraft production—and the Allies were well in front.

Clearly it would be silly to deny the massive importance of the Allies' productive superiority by 1943–44. If instead it had been the Axis that possessed a productive power equaling $62.5 billion in 1943, with the Grand Alliance at only $18.3 billion, we would certainly not assume the defeat of Germany and Japan. But the argument that emerges from the chapters above is that these crude productive disparities could be, and very much were, affected by two other variables: namely, the role of geography (and its greater or lesser appreciation of that by the planners, designers, and decision makers on either side); and, perhaps the most important variable of all, the creation of war-making systems that contained impressive feedback loops, flexibility, a capacity to learn from mistakes, and a "culture of encouragement" (of which more in a moment) that permitted the middlemen in this grinding conflict the freedom to experiment, to offer ideas and opinions, and to cross traditional institutional boundaries.

If that is true, this work offers a different interpretation than the broad historiography of the Second World War, which tends to assume that the war was essentially won by the time of the battles of Midway, El Alamein, and Stalingrad. It attempts to trim the crude economic determinist explanation of the war's outcome and offers something more subtle. One is reminded here of Churchill's point that the war would be, and had to be, won by the "proper application" of force. Sheer numbers were not enough. Giving the RAF another several thousand Lancasters made little sense (as the prime minister famously complained in September 1941) if Bomber Command couldn't locate its targets. Building thousands of merchant ships couldn't win the war if ways could not be found to defeat Doenitz's great U-boat wolf packs in the mid-Atlantic. Quintupling the size of the U.S. Marine Corps didn't matter if they couldn't figure out a way of getting onto and past enemy-held beaches. Ten thousand T-34 tanks were simply a vast assemblage of steel and wire until someone figured out how to supply them with gas, oil, and munitions. One is tempted to modify Churchill's phrase to suggest that the Second World War was won by the "intelligent application" of superior force.

This point about the understanding of the factors affecting this global conflict becomes more important when one returns to consider the critical role played by geography. Size, distance, and topography attend all of these five narratives, just as they constrained the weary fighters who had to grapple with those spaces. Chapter 5 explicitly proclaims that it is about conquering the "tyranny of distance" across the broad Pacific reaches, but in their own ways each of the campaigns covered in the other four chapters can be viewed as the human struggle to gain control of great stretches of the air, sea, and land. Geography informed strategy. Clearly, in the American campaign against Japan, however lengthy the return route from Hawaii via the Gilberts, Marshalls, Carolines, Marianas, Iwo Jima, and Okinawa to Nihon's outer shores, the fact was that it remained the swiftest route to reach Japanese home waters, and that MacArthur's drive along the New Guinea coastline was, geographically, a far less obvious path to Tokyo Bay; for the same reason, it was impossible for the British exertions in the Burmese jungles, extraordinary as they were, to deliver a knockout blow.

Elsewhere also, geography shaped outcomes. Geography not only made the Torch landings feasible but very likely shaped their success:

because of the superior facility of movement across water, Anglo-American amphibious divisions could get to the shores of North Africa far more easily from the Clyde and Norfolk, Virginia, than could the Wehrmacht's heavy central-reserve divisions from southern Germany or western France, let alone the Ukraine. All that was needed was naval protection for the fast Allied troop convoys, which was provided. There were no U-boat-free routes, of course, until the great tussle for command of the oceanic center had been resolved. The mid-Atlantic air gap was a spatial problem that for three years lacked a military-technological solution; by June 1943 it had been found. The struggle to find a long-range fighter that would escort the B-17s all the way to Prague and back took somewhat longer; but by February 1944 even that solution had been found, thanks not to brute force but to human ingenuity and sympathetic initiative. Geography shaped the Ostfeldzug in a different way: the German army went too far into inhospitable lands and paid the price in a campaign where primitive material strength did help to counter military sophistication. And at El Alamein the fabled Afrika Korps came up against a vastly entrenched British imperial garrison at a point when Rommel was a very long way from his supplies, fighting against a foe who had been working on maintaining extended lines of communications to Egypt for many decades. The conclusion is clear: distance was a killer—or a winner—in this massive war. The Allied war-fighting systems improved dramatically when they fully took that into account.

One is therefore struck by the Axis powers' failure to see the Second World War as a gigantic geopolitical chessboard, and thus to appreciate the strategic significance of a small number of critical positions (bases) that gave the owner a disproportionate operational advantage.

The fact that Imperial General Headquarters in Tokyo seems to have given up attempting to seize Hawaii after the Battle of Midway, when U.S. forces in the Pacific were still so relatively weak, is amazing. Compared with the Hawaiian Islands, acquisitions such as New Guinea and Burma were mere bagatelles; they would in any case have fallen into Japan's lap as a consequence of Tokyo's having first taken the most vital strategic place in the entire Pacific. Hitler's failure to get his hands on Gibraltar—or, at the very least, to persuade Franco to neutralize it—was another major deficiency, explained perhaps by his obsession with the drive to the east. So also was the Italian-German inability to crush the British air and naval bases on Malta. Had the Pillars of Hercules been blocked, with Algeria staying in sympathetic Vichy hands and Malta transformed into a giant Luftwaffe base, how long would it have been before Egypt itself fell?

In all, the Axis squandered its enormous early advantages by driving into secondary theaters—the Balkans, Burma, southern China—and paying far less attention to the really critical targets. Hitler's strategy on the Eastern Front may be the most egregious example of all. Going at the same time for Leningrad in the north, Moscow in the middle, and Stalingrad, the Donbas, and the Caucasus in the south was militarily silly. Moscow, surely, was the supreme prize, even in a negative sense; that is, while a full occupation of that city might not have guaranteed the hoped-for swift victory—certainly the Politburo, even without Stalin, could have relocated eastward to continue the fight—the fact is that the German failure to seize the Soviet capital ensured that the war would go on. The student of all this gains a sense that both the Wehrmacht leadership and the Prussian-influenced Japanese generals had, ironically, forgotten Clausewitz's stress upon the importance of focusing upon the enemy's *Schwerpunkte* (centers of gravity, or key points) and paid the ultimate price for that forgetfulness. Many years ago Correlli Barnett suggested that the results of military battles, and of wars as a whole, could be regarded as the "great auditor of institutions."[13] The present study affirms that notion.

The final point that emerges from this investigation of how the war was won concerns the tricky, perhaps intangible issue of what might be termed the "culture of encouragement," or the culture of innovation. All of these massive and orchestrated war machines—in Britain, the United States, Germany, Japan, and the Soviet Union—naturally strove both to defend their own positions and to advance in order to defeat their enemies. That is a truism. The more interesting question, for this book, has been the *how* question: how did some of these politico-military systems do it more effectively than others? A good portion of the answer has to be that the successful systems were so because they possessed smarter feedback loops between top, middle, and bottom;

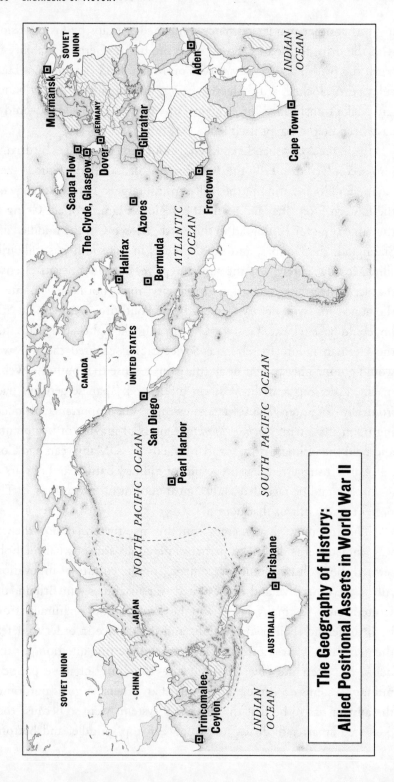

The Geography of History:
Allied Positional Assets in World War II

because they stimulated initiative, innovation, and ingenuity; and because they encouraged problem solvers to tackle large, apparently intractable problems.

Of course the men at the top made a difference. The individual leaders, or individual leaderships, revealed very marked alterations in style. Here the Japanese were the worst. With a god-leader kept from serious strategic decision making and authority devolved to the army and the navy, both of those services (once Yamamoto had been shot down in April 1943) displayed a rigidity and old-boys-solidarity that could in no way handle the imaginative American counterattacks that were unfolding across the Pacific. Hitler and Stalin were, as many historians have pointed out, very similar in their obsessions about control, with the critical difference that Stalin began to relax his iron grasp once he understood that he had a team of first-class generals working for him, whereas Hitler became ever more megalomaniacal and paranoid, giving his experienced generals in the field little room for operational maneuver.[14] Generally, the further the distance a commander was from OKW—Guderian along the Seine, Rommel in the desert, Kesselring in southern Italy—the freer he was to exercise his own undoubted battlefront skills.

Roosevelt and Churchill were equally different. The president appears to have had such unshakeable faith in America's power that he rarely interfered regarding either appointments or operational planning. Once the U.S. military leadership had been shaken out at the entry into war, he placed enormous faith in Arnold (air force), King (navy), and especially Marshall (army) to do the right thing, with his own trusted Admiral Leahy acting as titular head of the Joint Chiefs. The navy always had a special place in Roosevelt's affections, and his appointment of Ben Moreell to lead the new Seabees was a brilliant

THE GEOGRAPHY OF HISTORY: ALLIED POSITIONAL ASSETS IN WORLD WAR II
The 1937/1939–1945 global conflict was the World War of all History to date. Apart from the Russo-German land war, victory depended upon control of certain key maritime routes and of the ports that held the navies and aircraft that controlled the waters. While the British Empire and the United States had massive advantages here, both the Germans and Japanese were in a much weaker geopolitical situation. The map above highlights the most significant of the Allied naval bases during the war.

stroke. But otherwise FDR does not seem to have played an active or interfering role in service appointments. Why not let George Marshall do the tough job of clearing out unsuccessful generals at the beginning of the fighting, after the Kasserine Pass, and in the midst of the Anzio falterings?

It would have been temperamentally impossible for the prime minister to exercise that sort of hands-off approach—his personal notepaper was famously headed "Action This Day!" and it was with difficulty (including, finally, intervention by the king) that he was prevented from going off to the Normandy beaches on D-Day itself. At times—and there were many of those, including Norway, Dakar, and naval operations off Crete and Singapore—he drove Alanbrooke and the other British chiefs very close to desperation and resignation. And he certainly replaced a lot of generals in his search for effective battle-front leadership. Yet even the dyspeptic Alanbrooke frequently admitted that the prime minister's imagination, drive, and rhetoric were indispensable to the war effort.

There was another level of this worldwide conflict in which Churchill's enthusiasms and encouragements were invaluable: in recognizing talent, initiative, and, frankly, unorthodoxy in people and giving them a chance to prove themselves. His reproof about neglecting Major General Percy Hobart's talents because he "excited controversy" was quoted in chapter 4, and indeed it is not easy to see how Hobart could have gotten his weird tank notions turned into an entire armored division in the conservative British Army without the prime minister. It was Churchill who pulled Admiral Ramsay out of retirement to lead the empire's amphibious planning, who sent Orde Wingate to Burma to create the special-ops Chindits, and who arranged for the return of Sir Wilfrid Freeman to administer aircraft design and production at the Air Ministry. Who else but the prime minister, after witnessing in person the Dodgers and Wheezers' successful experiments with their forward-firing Hedgehogs, would have sent the First Sea Lord by early the next morning an instruction that resources be made available for this new weapon? That list could go on, but the plain fact is that there probably was never another war leader with his talent-spotting skills and his capacity to inspire and encourage. The man was sui generis, but the lesson he leaves with us is not: that without bold and worthy leadership, a large enterprise is likely to falter.

Yet there is more to this concept of a "culture of encouragement" than the personal tastes and whims of powerful leaders. The post-1919 U.S. Marine Corps, though much diminished in resources, was given enough freedom to develop its ideas on advanced naval bases in Micronesia. Stalin's tank (Koshkin) and aircraft (Yakovlev) designers, although terrified of the "boss," knew they would be supported if they were developing weapons that could kill lots of Germans—and thus got lots of resources. Someone in the RAF weapons development system was able to call Ronnie Harker down from Rolls-Royce in Derby to Duxford in April 1942 to test and report on the P-51 fighter; someone at Rolls-Royce knew how to reach Sir Wilfrid Freeman at the Air Ministry and get wheels turning. And a fighter commander such as Donald Blakeslee knew he could go to his air force superior, General Doolittle, and Doolittle in his turn knew he could go to Arnold and Lovett and get a faulty USAAF policy regarding the Mustang fighter's fate changed, to massive effect. Two young postdocs at the University of Birmingham were encouraged to figure out the problem of building a miniaturized radar set that did not collapse under the sheer energy of its pulsing beam, and when they did it, they were not turned back. After that, first their professor and then Whitehall, Tizard, Vannevar Bush, Bell Labs, and MIT took it further. Finally, this new technology helped to sink large numbers of U-boats. This wasn't a fluke; it was the result of a superior system.

But it had to start somewhere. And that somewhere was a space, a military-political culture, that allowed problem solving to go ahead. For reasons still unfathomable, Japan seems to have failed miserably on this scorecard: its impressive weapons systems such as the Zero fighter, *Akagi*-type carriers, and *Yamato*-class battleships were all essentially creations and designs of the late 1930s, and then the nation's innovative capacities appear to have faded. There were no latter-day equivalents to the Lancasters, the Mustangs, Walker's terrifying anti-U-boat squadrons, miniaturized radar, or decrypting machines, let alone the atomic bomb. The Third Reich went a different way, converting Germany's pre-Nazi technological strengths into "superweapons" such as the V-1 and V-2 rockets, the jet fighter, and the schnorkel U-boat; but none of those stupendous instruments could be effective in the war because Germany had already conceded command of the sea and then command of the air. Nobody in Nazi Germany—no midlevel organization, not even Speer himself—had worked out the problem that the new superefficient submarines being assembled in Bremen would not go to sea if their diesel engines could not be transported in safety from the Ruhr because Allied bombers had blasted all the rail lines.

In this particular measure, of the sheer efficiency of getting fighting equipment and fighting men from A to B, the British probably again take the palm, certainly not because of some innate cleverness, but because of their long organizational experiences and their acute sense after 1940 of fighting against the odds, together with the prospect of losing. Necessity was indeed here the mother of invention. They had to defend their cities, transport troops around Africa to Egypt, support the Greeks, hold the frontiers of India, get the United States into the war, and then bring that massive continental American potential to the European theater.[15] Here was another problem solver's task. How exactly did one get two million Americans soldiers, once arrived in the Clyde, to bases in southern England in preparation for the assault upon Normandy, when most of British Rail was concentrated on hauling coal wagons to the iron and steel mills that could not cease production? As it was, an organization staffed by people who had grown up memorizing Bradshaw's railway timetables as a hobby did it, while the high commanders took that all for granted because they were confident in their middle-level managers' capabilities. As Churchill famously said about the artificial harbors, don't worry too much about the problems, for the problems will sort themselves out—that is, a way will be found, step by step.

There is another way of thinking about this story of incremental problem solving, and it comes from a very contemporary example. In November 2011, as all the posthumous tributes were pouring in to the "genius" leader of Apple, Inc., Steve Jobs, an intriguing article appeared in the New Yorker. In it the author, Malcolm Gladwell, argued that Jobs was not an inventor of a machine or an insight that changed the world; few beings ever are (excepting perhaps Leonardo and Edison). Instead, he was a brilliant adopter of other people's early, clumsy inventions and partial insights, which he built upon, modified, and constantly improved. He was, to use today's parlance, a "tweaker," and his true genius was to push for ever-greater increases in the effectiveness of his company's products.

The story of Steve Jobs's success, however, was not new. The coming of the Industrial Revolution in eighteenth-century Britain—arguably, the greatest revolution that explains the rise of the West—came about precisely because that country possessed a plethora of tweakers in a national culture that encouraged progress:

> In 1779, Samuel Crompton, a retiring genius from Lancashire, invented the spinning mule, which made possible the mechanization of cotton manufacture. Yet England's real advantage was that it had Henry Stokes, of Horwich, who added metal rollers to the mule; and James Hargreaves, of Tottington, who figured out how to smooth the acceleration and deceleration of the spinning wheel; and William Kelly, of Glasgow, who worked out how to add water power to the draw stroke; and John Kennedy, of Manchester, who adapted the wheel to turn out fine counts; and, finally, Richard Roberts, also of Manchester, a master of precision machine tooling—and the tweaker's tweaker. He created the "automatic" spinning mule: an exacting, high-speed, reliable rethinking of Crompton's original creation.[16]

The reader who has followed this analogy in the five preceding chapters will be struck by the similarity. For was not the story of the evolution of the Soviets' T-34 tank from a badly designed, underpowered hunk of metal to an assured, fast-moving, deadly weapon of war a story of continuous tweaking (chapter 3)? Was that not also the case for the great American bomber, the B-29, at one stage so mired in difficulties that its cancellation was proposed until the Boeing teams fixed those problems (chapter 5)? And what about the miraculous tales of the P-51 Mustang (chapter 2), Percy Hobart's Funnies (chapter 4), and a powerful radar system so small that it could be inserted into the nose of a long-range patrol aircraft and turn the tide in the Battle of the Atlantic (chapter 1)? It all fits, once one brings the scattered pieces together. But all these projects needed time and support.

If these remarks about the culture of innovation are valid, the five peculiar and parallel tales narrated above carry a significant transferable message into other fields, other disciplines, other great contestations. This does not just mean transferability to such large-scale military conflicts as the Napoleonic War or the American Civil War, although they surely would serve as well, and indeed, each of those conflicts has its own splendid historians of how the war was won.[17] The argument in any of the five preceding chapters of this book can feed into the enormous literature and debates upon what has come to be called "military innovation studies."[18] The case histories also, I hope, contribute to the many writings on the operational level of war, and to the popularized "genius of design" television programs. And they relate (with full acknowledgment in the endnotes) to the transformative studies of military effectiveness.

By extension, then, any smart middle manager or management consultant in today's business world—or a CEO who reads widely— can see the lessons that emerge from these tales. Indeed, the managers of even the world's greatest companies can only presumably marvel at, say, Admiral Ramsay's planning and orchestration of the five simultaneous D-Day landings and wish they could achieve one-tenth of what he did. In a certain way, the ghost of the late, great Peter Drucker hangs around this present book, since all of his many works were about how to manage and organize. And no one, it is fair to argue, ever did it better than Ramsay and his team on the morning of June 6, 1944.

In sum, the winning of great wars always requires superior organization, and that in turn requires people who can run those organizations, not in a blinkered way but most competently and in a fashion that will allow outsiders to feed fresh ideas into the pursuit of victory. None of this can be done by the chiefs alone, however great their genius, however massive their energy. There has to be a support system, a culture of encouragement, efficient feedback loops, a capacity to learn from setbacks, an ability to get things done. And all this must be done in a fashion that is better than the enemy's. That is how wars are won.

Thus, the young, inquiring German worker of Brecht's poem cited at the very beginning of this book was right in his puzzled questions. Who indeed did make things work for Alexander and Caesar? Was it not the middle managers and the problem solvers? With all due respect to the great contributions made by the masters and commanders, on the one hand, and those made by the soldiers and sailors who had to cross the beaches and patrol the dangerous seas during the 1939–45 war, on the other, this book argues so. Without the middle personnel and the systems they managed, victory would remain out of grasp. It remains a puzzle that we have given the problem solvers of the Second World War such relatively small recognition.

By contrast, Frederick the Great, ever the canny manager of things, rewarded his generals, colonels, and middle managers with titles, honors, and lands (some of their heirs, bearing the same aristocratic titles, fought in the Ostfeldzug 180 years later, under a former Austrian corporal they despised). Alexander's team of fighters bore him back as far as they could, buried him, then returned to due honor in Macedonia. Philip II of Spain was not mean in handing out rewards to those who had served him faithfully and were willing to keep on fighting for his divine cause despite the Armada's failure. What happened to Caesar's cook is lost to history, but, assuredly, he played a role.

The same recognitions are deserved, surely, for the middle people who turned the Second World War from being the blunting of Axis aggressions in 1942 into the irreversible Allied advances of 1943–44, and thus the crushing of Germany and Japan. True, some of these individuals, weaponry, and organizations are recognized—such as the achievements of those who created the P-51 long-range fighter, the specialized tanks, miniaturized radar, the Fighting Seabees—but usually in a spotty and popularized manner, by fans of a certain aspect of the war. Rarely if at all have these individual threads been woven together to show how those advances then affected the many campaigns that swung in the Allies' favor during the middle parts of this global conflict. Even more rare has been an understanding of how the work of these various problem solvers also has to be joined by a fuller appreciation of the importance of having a "culture of encouragement," to ensure that the mere declarations and strategic intentions of great leaders get turned into reality and do not wither in the storms of war. If so, as has been argued throughout in the pages above, then we have lived with a large gap in our understanding of how World War II was won in its critical years. Perhaps the present work will help to close that gap a bit.

Acknowledgments

I first conceived of this book in 2007, a year after the publication of my study of the United Nations, *The Parliament of Man*. With the backing of Random House in New York and Penguin UK in London, I moved into the serious writing of chapters 1, 2, and 4 during six months of research leave in Cambridge in 2008. My return to full-time teaching, advising, and fund-raising at Yale left me little opportunity for further drafting, but I was able to take another stint of research leave at Cambridge in 2010, where I drafted the remaining chapters, 3 and 5. The introduction and conclusion were drafted in the fall of 2010, after my return to Yale. In the period following I sought to incorporate, or at least mull over, the comments and criticisms of many academic colleagues and my chief editors.

There is no way that senior, overcommitted academics these days can engage in major scholarly research and writing without the assistance of enlightened universities and foundations who are willing to support such endeavours. My thanks must therefore go in the first place to my home university, Yale, which not only provided an intellectual base, scholarly resources, and remarkable, sustaining collegiality but also freed me from time to time for serious research and writing abroad. In that latter regard, the welcome and hospitality I received from academic colleagues and colleges at Cambridge were both moving and essential. Above all, I am indebted to the Master and Fellows of St. John's College, who provided me with assistance and accommodation for three lengthy stays; without the college's generosity, I doubt that this book could have been written. I am also obliged to the IDEAS department of the London School of Economics and Political Science, which made me their first Philippe Roman Visiting Professor in 2008, thus allowing me the chance both to teach and to continue my research and writing. The generosity of Roger Hertog permitted me to put all those pieces together.

My colleagues at Yale, above all John Gaddis and Charles Hill, have tolerated my discourses on the nuts and bolts of war for many years now, and permitted me to inflict some of the ideas present in this book upon our co-taught class in grand strategy. The same tolerance and encouragement was shown by the genial Arne Westad and Michael Cox at the London School of Economics. As to the encouragement given by Cambridge scholars and officers, well, I don't know where to start, and I feel sure I will have missed a name or two: Zara Steiner, Christopher Andrew, David Reynolds, Jonathan Haslam, Richard Drayton (now at London), Adam Tooze (now at Yale), and Brendan Sims all made me think and think again. Allan Packwood at the Churchill College Archive was an immense help. At St. John's College, two successive masters, Richard Perham and Christopher Dobson, along with presidents Jane Heal and Mark Nicholls and, perhaps especially, that naval connoisseur cum domestic bursar John Harris, were always supportive.

As the sheer size of the published and unpublished materials for this project grew and grew, I in turn became indebted to a small group of wonderful research assistants: Will Chou, Will Owen, Evan Wilson, Joyce Arnold, Elisabeth Leake, Gabriel Perlman, Isabel Marin, and Daniel Hornung have helped me in gathering library materials, searching out obscure sources, and reading draft sections. Elisabeth was a particularly cruel and effective copy editor as well. Elizabeth Ralph nudged me into a rescrutiny of chapter 2 on the air war, and Igor Biryukov's researches on chapter 3 (Eastern Front) were invaluable. During the academic year 2011-12, Daniel Hornung joined Igor in weekly support of this project. In the intensive work stages of last summer, Isabel Marin and Sigrid von Wendel joined Igor and myself in the final "push." Checking texts for the last time, and assembling, under pressure, the most suitable maps, tables, and illustrations is no easy task. This was perfect teamwork, and accomplished everything that was required. It gave me great pleasure to work with all of the above and learn from them, something that constantly reminded me of my own role as Sir Basil Liddell Hart's research assistant a mere forty-plus years ago, where my task was to prepare first drafts for the chapters on the Pacific War, the Battle of the Atlantic, the strategic bombing campaign, and the latter half of the fighting in Italy.

A number of academic colleagues were kind enough to put aside their own work to read and critique various chapters, or, as a local alternative, to join me in the Granta pub in Cambridge for sustained discussions about where this work was going. Among them were Kathy Barbier, Tami Biddle, Michael Coles, John Harris, Jonathan Haslam, John Hattendorf, Milan Hauner, David Kahn, Rich Muller, Geoffrey Parker, Andrew Preston, John Reeve, Nicholas Rodger, and John Thompson. I am sure that I have inadvertently missed a name or two.

I was also lucky enough to be able to present some early ideas about this book in various stimulating academic environments: Yale itself, Cambridge, the Ohio State University, King's College London (the Annual War Studies Lecture), the Naval War College, Duke University, the Council on Foreign Relations, the National Defense University, and many other places over the past three years. The Society for Military History invited me to present my ideas in its January 2009 lecture at the annual conference of the American Historical Association, under the title of "History from the Middle," which was later published in their flagship journal, the *Journal of Military History*. Those ideas were also presented, in a variant form organized by the enterprising Martin Lawrence, as the first Lady Lucy Houston Lecture at Cambridge in March 2010, a nice tribute to the formidable lady whose largesse rescued the future Spitfire from oblivion.

This book makes no singular claim to arguing the importance of the nuts and bolts of war; it is certainly not the first to do so. Like other works with an ambitious theme, it stands on the shoulders of redoubtable scholars who have previously explored the same ground: more generally and distantly, Martin van Creveld and the incomparable Geoffrey Parker; and, with respect to World War II specifically, Rick Atkinson, Bruce Ellis, David Glantz, Richard Overy, Allan Millett and Williamson Murray, Marc Milner, Ronald Spector, Paddy Willmott, and the late and remarkable John Erickson, an engineer turned historian. Likewise, I owe a huge debt to the scholars who contributed to the American, British, Canadian, and West German official histories of the war; many of those volumes are fifty or sixty years old, yet their quality remains undimmed.

The marine paintings included among the photographs in this volume were all done by the incomparable artist Ian Marshall. They represent a prelude to his future illustrated book *Fighting Warships of World War II*. I am indebted to Ian for his great generosity here.

It has been almost forty years since I was introduced to Bruce Hunter, my literary agent, whose wise judgment and integrity have guided me on so many matters, and whose role has now been taken over by the able Andrew Gordon. Phyllis Westberg (Harold Ober) steered me through the New York side of things, and Ania Corless through all international contracts. Mika Kasuga of Random House really helped us in the final months. My editors Will Murphy (Random House, New York) and Stuart Proffitt (Penguin, United Kingdom) have been exemplary guides: supportive, firm, and very, very patient. Stuart sent me back to the writing board many times; I hope it shows. My debt to all these is not measurable.

Nor is my debt to my family, whose greatest virtue has been to be even more patient. I refer especially to Jim, John, and Matthew; to Sophia; to Cinnamon, Catherine, and Olivia; and to my grandson, Charlie Parker Kennedy, who was not even around to make life interesting when this work was begun.

My greatest debt is to the partner who has marched, laughed, and cried with me for well over a decade now, and to whom this book is dedicated.

Paul Kennedy
New Haven, 2012

Notes

INTRODUCTION

1. After our UN report group initially suggested major constitutional changes at the very top of the world organization—that is, amending the United Nations Charter to admit new permanent veto members to the Security Council—and discovered the great political roadblocks to proposals of that magnitude, it became obvious that the best way the UN could help itself was by ensuring that it was effective in the middle, in on-the-ground peacekeeping, development, and human rights work. Compare the larger agenda suggested in P. Kennedy and B. Russett, "Reforming the United Nations," *Foreign Affairs* 74, no. 5 (Sept.-Oct. 1995): 56–71, with the more cautious formulations in P. Kennedy, *The Parliament of Man: The Past, Present and Future of the United Nations* (New York: Random House, 2006), ch. 8.
2. The syllabus, for any reader interested, is available at http://iss.yale.edu/grand-strategy-program.
3. For some exceptions to this generalization, see many of the essays in W. Murray, M. Knox, and A. Bernstein, *The Making of Strategy: Rulers, States and War* (Cambridge: Cambridge University Press, 1994)—a very deliberate counterpoise to an older classic, E. M. Earle with G. A. Craig and F. Gilbert, eds., *Makers of Modern Strategy: Military Thought from Machiavelli to Hitler* (Princeton: Princeton University Press, 1971), with its emphasis upon strategic writings and thought.
4. See, respectively, G. Parker, *The Army of Flanders and the Spanish Road: The Logistics of Spanish Victory and Defeat in the Low Countries War* (Cambridge: Cambridge University Press, 1972); P. Padfield, *Guns at Sea* (London: Evelyn, 1973), ch. 10 and 11; J. C. Riley, *International Government Finance and the Amsterdam Capital Market 1740–1815* (Cambridge: Cambridge University Press, 1980); P. M. Kennedy, "Imperial Cable Communications and Strategy 1870–1914," *English Historical Review* 86, no. 341 (1972): 728–52.
5. These operational directives are most sensibly summarized by State Department historian Herbert Feis in *Churchill, Roosevelt, Stalin: The War*

They Waged and the Peace They Sought (Princeton: Princeton University Press, 1957), 105–8.

6. The best way to test this remark is to examine four of the most useful histories of the Second World War that I quote repeatedly in my own text and check their descriptions (or lack of mention) of such jigsaw puzzle pieces as the Merlin-powered P-51 Mustang, the cavity magnetron, the Hedgehog, and Hobart's Funnies. See B. Ellis, *Brute Force: Allied Strategy and Tactics in the Second World War* (New York: Viking, 1990); R. Overy, *Why the Allies Won* (London: Jonathan Cape, 1995); W. Murray and A. R. Millett, *A War to Be Won: Fighting the Second World War* (Cambridge, MA: Harvard University Press, 2000); B. H. Liddell Hart, *History of the Second World War* (London: Cassell's, 1970). Another useful point of comparison would be with two recent and much acclaimed works on World War II. The first is Ian Kershaw's *Fateful Choices: Ten Decisions That Changed the World 1940–41* (London: Penguin, 2007), a wonderful read, though deliberately constructed as a set of top-down stories—see ch. 2, "Hitler Decides to Attack the Soviet Union," ch. 7, "Roosevelt Decides to Wage Undeclared War," and so on. The second is Andrew Roberts's gripping book *Masters and Commanders: How Roosevelt, Churchill, Marshall and Alanbrooke Won the War in the West* (London: Allen Lane/Penguin, 2008), which to an amazing extent can be seen as the top-level complement to the middle-level thrust of my book.

CHAPTER ONE: HOW TO GET CONVOYS SAFELY ACROSS THE ATLANTIC

1. B. H. Liddell Hart, *History of the Second World War* (London: Cassell's, 1970), 316–17. For Alanbrooke's comments on the travel delays, see the pertinent pages in his *War Diaries, 1939–1945: Field Marshal Lord Alanbrooke*, ed. Alex Danchev and Daniel Todman (London: Weidenfeld and Nicolson, 2001).

2. Liddell Hart, *History*, 386. The actual math calculations from Liddell Hart's figures would suggest that Doenitz possessed 204 operational submarines (out of a total of 366) by the end of 1942, but the difference is insignificant. Grand Admiral Karl Doenitz was of course the remarkable C in C of the German submarine arm, then the Kriegsmarine itself. There have been so many books written on this campaign over the past sixty years that it is difficult to know which ones to list. Readers might begin with Marc Milner's fine summary, *Battle of the Atlantic* (Stroud, UK: Tempus, 2005), and take things from there, that is, from his brief bibliography on 265–67. The U.K. official history is S. W. Roskill, *The War at Sea, 1939–1945*, 3 vols. (London: HMSO, 1954–61). The U.S. official history is S. E. Morison, *History of United States Naval Operations in World War II*, 15 vols. (Boston: Little, Brown, 1947–62), vols. 1 and 10 being those most relevant, because they are specifically on the Battle of the At-

lantic. Milner's additional value is that he brings in the increasingly important role of the Royal Canadian Navy and Air Force. There are useful Navy
Records Society volumes by the late David Syrett and by Eric Grove and a
great amount of further information in the many publications of the German Military History Research Office.

3. The quote is from Roskill, *War at Sea*, 2:367.
4. For the general strategic theory here, see John Winton, *Convoy: The Defence of Sea Trade, 1890–1990* (London: Michael Joseph, 1983); S. W.
   Roskill, *The Strategy of Sea Power* (London: Collins, 1962); and Sir Julian
   Corbett, *Some Principles of Maritime Strategy* (London: Collins, 1962).
5. A.R. Millet and W. Murray, *Military Effectiveness* (Allen & Unwin, Sydney, 1989).
6. The most available tables are in Roskill, *War at Sea*, vol. 3, part II, various
   appendices, where the naval war as a whole is summed up.
7. The fall of France? Perhaps, but only for the French. Stalingrad? Perhaps,
   though the German army was advancing eastward again in the spring of
   1943. Midway? But it marked the limit of Japan's expansion in the Central
   Pacific, not the start of Nimitz's great counteroffensive, which was well
   over a year away.
8. There are lots of fine works on the critical months of the Battle of the Atlantic, including Roskill, *War at Sea;* Milner, *Battle of the Atlantic;* and
   Correlli Barnett, *Engage the Enemy More Closely: The Royal Navy in the
   Second World War* (London: Hodder and Stoughton, 1991). But perhaps
   the most remarkable of them all is German historian (and former naval
   officer) Juergen Rohwer's *The Critical Convoy Battles of 1943* (London: Ian
   Allen, 1977), a feat of historical reconstruction. Also very impressive:
   Martin Middlebrook, *Convoy* (London: William Morrow, 1986).
9. Rohwer, *Critical Convoy Battles*, 55.
10. Ibid., 211, for technical data on the U-boat types.
11. Barnett, *Engage the Enemy*, 599.
12. On this theme more generally, see D. Howse, *Radar at Sea: The Royal Navy
    in the Second World War* (London: Macmillan, 1993), an impeccable study;
    Sir Arthur Hezlet, *The Electron and Sea Power* (London: Stein and Day,
    1976); and the excellent study (one of many by the same author) by G.
    Hartcup, *The Effect of Science on the Second World War* (Basingstoke: St.
    Martin's, 2000), especially ch. 3–4.
13. Rohwer, *Critical Convoy Battles*, 113.
14. Ibid., 121.
15. For Alanbrooke's many concerns in this critical period, see his *War Diaries*,
    ca. 330–425.
16. See, for example, Vice Admiral Sir Peter Gretton's analysis in the final
    chapter of his book *Crisis Convoy: The Story of HX 231* (London: P. Davies,
    1974), "Why Did the Germans Lose the Battle of the Atlantic in the

Spring of 1943?" which feints and ducks around that very question. Gretton is by no means the only author to do so, and he is someone who witnessed and played a major role in the turn of the tide.

17. Roskill, *War at Sea,* vol. 2, map 38 (opposite p. 365).

18. Milner, *Battle of the Atlantic,* 127.

19. See again Roskill's *War at Sea,* vol. 2, ch. 5, and the scatter-map of the lost merchantmen of Convoy PQ 17, opposite p. 141.

20. Barnett, *Engage the Enemy,* 600.

21. Assessing the significance of HX 231's contribution to the Battle of the Atlantic is something of a puzzle. Roskill, in *War at Sea,* vol. 2, does not mention it at all. Perhaps this is what prompted its escort commander, Peter Gretton, to write *Crisis Convoy,* a self-important work, though with some interesting tidbits on morale and training, plus details of what the merchantmen were carrying, the number of tankers, and so on. Milner, *Battle of the Atlantic,* gives this clash a mere sentence on p. 158, mentioning Doenitz's disappointment that only six ships were lost from the convoy. R. Overy, *Why the Allies Won,* 2nd ed. (London: Pimlico, 2006), 69, writes, "Convoy HX 231 from Newfoundland fought its way through four days of gale-force winds against a pack of seventeen submarines. Four U-boats were sunk for almost no loss." But Roskill's fastidious compilation of German U-boat losses (Appendix J, p. 470) shows only two U-boats sunk in the North Atlantic in these days.

22. Gretton, P. *Crisis Convoy: The Story of the HX 231* (London: Naval Institute Press, 1974), 157.

23. Ibid., 173.

24. The warships' performances—and group photos of the remarkably young commanders of each vessel—are in Sir Peter Gretton's *Convoy Escort Commander* (London: Cassell, 1964); the narrative is on 149–62.

25. Barnett, *Engage the Enemy,* 610.

26. For Doenitz's political character, see P. Padfield, *Doenitz: The Last Fuehrer* (New York: Harper and Row, 1984). For his mid-May report to Hitler, see Liddell Hart, *History,* 389–90. For his assessment of the role of Allied airpower in blunting the attacks upon convoys HX 239 and SC 130, see Barnett, *Engage the Enemy,* 611.

27. K. Doenitz, *Memoirs: Ten Years and Twenty Days* (London: Weidenfeld and Nicolson, 1959), 326.

28. Bomber Harris's dismissive memo is recorded in M. E. Howard, *Grand Strategy* (London: HMSO, 1972), 4:21, while Doenitz's sober assessment is reprinted in Barnett, *Engage the Enemy,* 611. The dismal story on the British side has been recently confirmed by the research of Duncan Redford in "Inter- and Intra-Service Rivalries in the Battle of the Atlantic," *Journal of Strategic Studies* 32, no. 6 (Dec. 2009): 899–928. Gretton's amazing claims about his personal air-sea cooperation experiences are in

his introduction to R. Seth, *The Fiercest Battle: The Story of North Atlantic Convoy ONS 5, 22nd April–7th May 1943* (London: Hutchinson, 1961), 14–15. Gretton also reports that the Liverpool Tactical School had "joint classes" for Royal Navy and Coastal Command officers, which would be quite remarkable.

29. Also known, under another code name, as the Mark 24 mine. See "Mark 24 Mine," Wikipedia, http://wikipedia.org/wiki/Mark_24_Mine; Kathleen Williams, "See Fido Run: A Tale of the First Anti-U-boat Acoustic Torpedo," paper presented at the U.S. Naval Academy Naval History Symposium, 2009.

30. H. P. Willmott, *The Great Crusade: A New Complete History of the Second World War* (London: Michael Joseph, 1989), 266–267.

31. For illustrations of a depth charge crew at work, see Roskill, *War at Sea*, vol. 3, part I, opposite p. 257; on improved sonar, see Hartcup, *Effect of Science*, 60–69.

32. Hartcup, *Effect of Science*, 72–74. There is a photo of a Hedgehog as illustration 9.

33. G. Pawle, *The Secret War 1939–1945* (London: White Lion Press, 1956), ch. 12. However, as a fine corrective to the "British eccentrics" interpretation of how the war was won, see D. Edgerton, *Britain's War Machine: Weapons, Resources and Experts in the Second World War* (London: Allen Lane, 2011), and my discussion in "Reflections" at the end of this book.

34. Rohwer, *Critical Convoy Battles*, 198. Hartcup, *Effect of Science*, 47–49, explains how it works.

35. For the best, brief summary, see Hartcup, *Effect of Science*, 24–31.

36. Liddell Hart, *History*, 389.

37. E. G. Bowen, *Radar Days* (Bristol: Adam Hilger, 1987), ch. 9–12, explains the technology, and his role in the transfer to the Rad Lab. John Burchard's *Q.E.D: M.I.T. in World War II* (New York: Wiley, 1948) observes that "the British achievement of the cavity magnetron was perhaps the most important single contribution to technical development of the first years of the war" (219). See also Howse, *Radar at Sea*, 67–68, 156.

38. There is a photo of a Leigh Light in Barnett, *Engage the Enemy*, between 588 and 589. Earlier, on 258–59, he details the interminable delays. In general Barnett is very critical of the shortcomings of British industry and authorities to get the right weapons to the fronts. My own feeling is that the beleaguered island state performed rather well in the strained circumstances of total war, a view now much reinforced by Edgerton, *Britain's War Machine*.

39. Barnett, *Engage the Enemy*, 609. There is a balanced (actually, rather cool) assessment of Ultra's contribution to the Allies' overall victory in Hartcup, *Effect of Science*, ch. 5. Rohwer, *Critical Convoy Battles*, 229–44, explains naval code breaking. The founding father of code-breaking history, David

Kahn, also is cautious about ascribing too much importance to Ultra or indeed other deciphering systems; see his "Intelligence in World War II: A Survey," *Journal of Intelligence History* 1, no. 1 (Summer 2001): 1–20.

40. Herbert E. Werner, *Iron Coffins: A Personal Account of the German U-boat Battles of World War II* (London: Arthur Barker, 1969), is a grim and fascinating account, with rather wonderful illustrations, and possibly was the inspiration for the great movie *Das Boot.*

41. Roskill, *War at Sea,* vol. 3, part 1, ch. 2–3; Morison, *History,* has a great spreadsheet map of the U-boat kills in the Bay of Biscay, 10:97. The glider bombs are discussed in Milner's fine *Battle of the Atlantic,* 193–94.

42. On the Polish Mosquitos (and other nationalities in the squadrons), see Milner, *Battle of the Atlantic,* 189.

43. Barnett, *Engage the Enemy,* 606.

44. Most of the more detailed accounts of the Battle of the Atlantic mention Walker's role—how could they not?—but great data, including the quotation, can also be found at an individual website, captainwalker.info/

45. Barnett, *Engage the Enemy,* 605; more generally on operations research, see Hartcup, *Effect of Science,* ch. 6.

46. Roskill, *War at Sea,* vol. 3, part 1, p. 395; and Morison, *History,* vol. 10, ch. 8.

47. Werner, *Iron Coffins,* 213, recounts he and fellow commanders received the orders to attack the D-Day craft "with the final objective of destroying enemy ships by ramming." See also Morison, *History,* 10:324–25, on the massive Allied naval and aerial screen (and the loss of *Pink*).

48. This final period of the struggle is covered well in Roskill, *War at Sea,* vol. 3, parts 1 and 2, and in Barnett, *Engage the Enemy,* 852–58; Milner, *Battle of the Atlantic,* ch. 9, shows how tough those later battles were.

49. Willmott, *Great Crusade,* 273. Ellis, *Brute Force,* 160–61, also makes a strong case for numbers and production ultimately being key.

CHAPTER TWO: HOW TO WIN COMMAND OF THE AIR

1. E. Bendiger, *The Fall of Fortresses* (New York: Putnam, 1980); quotes are from 219–21.

2. Ibid., 236, 225.

3. The literature here is vast. B. H. Liddell Hart, *History of the Second World War* (London: Cassell's, 1970), ch. 23, and R. Overy, *Why the Allies Won* (London: Jonathan Cape, 1995), ch. 4, provide succinct single-chapter overviews. I found M. Hastings, *Bomber Command* (London: Michael Joseph, 1979) perhaps the single best volume, very critical but also discriminating. The four-volume British official history by C. Webster and N. Frankland, *The Strategic Air Offensive Against Germany* (London: HMSO, 1961), is a model of its kind.

4. N. Longmate, *The Bombers: The RAF Offensive Against Germany 1939–1945* (London: Hutchinson, 1983), 298.

5. W. Murray and A. R. Millett, *A War to Be Won: Fighting the Second World War* (Cambridge, MA: Harvard University Press, 2000), 317; also, more generally, see J. Terraine, *The Right of the Line* (London: Pen and Sword, 2010), which examines the RAF's role in the European war from beginning to end.

6. Overy, *Why the Allies Won*, 147; Hastings, *Bomber Command*, 318; Longmate, *The Bombers*.

7. I. F. Clarke, *Voice Prophesying War 1763–1984* (Oxford: Oxford University Press, 1966).

8. See A. Gollin, *No Longer an Island: Britain and the Wright Brothers 1902–1909* (London: Heinemann, 1984), especially the later chapters. For Amery's 1904 contention, see Paul Kennedy, *Strategy and Diplomacy 1870–1945: Eight Studies* (London: Allen and Unwin, 1983), 47.

9. See Overy's classic comparative analysis *The Air War 1939–1945* (London: Europa, 1980). There are fabulous photographs in R. Higham, *Air Power: A Concise History* (Yuma, KS: Sunflower Press, 1984).

10. Compare, for example, the hugely critical accounts by Longmate in *The Bombers* and by Bendiger, *Fall of Fortresses,* with that by a staunch defender of RAF Bomber Command's policies, Dudley Saward, in *Victory Denied: The Rise of Air Power and the Defeat of Germany 1920–1945* (London: Buchan and Enright, 1985). Saward is also the author of the authorized biography of Bomber Harris, whose own account, *Bomber Offensive* (London: Collins, 1947) was published very shortly after the war and well captures his own strong opinions.

11. Quoted from Longmate, *The Bombers,* 21–22, though the italics are mine. Chapter 1 of Hastings's *Bomber Command* has a fine, brief survey of the RAF between 1917 and 1940, and vol. 1 of Webster and Frankland's *Strategic Air Offensive* is invaluable.

12. On the theories of air warfare, see ch. 20 in E. M. Earle, ed., *Makers of Modern Strategy* (Princeton: Princeton University Press, 1971); on Trenchard, see Hastings, *Bomber Command,* ch. 1. Much of my analysis here derives from Tami Biddle, *Rhetoric and Reality: The Evolution of British and American Ideas About Strategic Bombing, 1914-1945* (Princeton: Princeton University Press, 2001).

13. Trenchard's remarkable statement, and the equally remarkable replies of the chief of the Imperial General Staff and the First Sea Lord, are extensively quoted in Longmate, *The Bombers,* 43–47.

14. U. Bialer, *The Shadow of the Bomber: The Fear of Air Attack and British Politics 1932–1939* (London: Royal Historical Society, 1980). The Baldwin quotation is in Hastings, *Bomber Command,* 50.

15. B. Collier, *The Defence of the United Kingdom* (London: HMSO, 1957), sets the larger scene, as does Overy, *Air War*, ch. 2.

16. W. Murray, *Luftwaffe* (Baltimore: Nautical and Aviation Publishing Company of America, 1985), 43–61, is superb here.

17. See Collier, *Defence*, and T. C. G. Jones, *The Battle of Britain* (London: Frank Cass, 2000).

18. See Robert Wohl, *A Passion for Wings: Aviation and the Western Imagination 1908–1918* (New Haven, CT: Yale University Press, 1994).

19. Liddell Hart, *History*, 91.

20. For the best brief account, see G. Hartcup's superb *The Effect of Science on the Second World War* (London: Macmillan, 2000), ch. 2–3.

21. Quoted in B. Schwarz, "Black Saturday," *Atlantic*, April 2008, 85, which is a review of Peter Stansky's *The First Day of the Blitz* (New Haven, CT: Yale University Press, 2007).

22. Murray, *Luftwaffe*, 60, Table XI. The comparative aircraft production rates are in Overy, *Air War*, 33.

23. Murray, *Luftwaffe*, 60, 10.

24. Figures from Overy, *Air War*, 150.

25. Quotes from Liddell Hart, *History*, 595–96; see also Webster and Frankland, *Strategic Air Offensive*, 1:178.

26. Webster and Frankland, ibid, 233

27. A. Tooze, *The Wages of Destruction: The Making and Breaking of the Nazi Economy* (London: Penguin, 2006), 596–602, a brilliant, revisionist analysis.

28. See Hastings, *Bomber Command*, 246, for the damage statistics; see W. Murray, *Strategy for Defeat: The Luftwaffe 1939–1945* (Maxwell, AL: Air University Press, 1983), 169, for Speer and Hitler. Note that this is a different, slightly earlier book than Murray's *Luftwaffe*, though it uses a lot of the same data.

29. See Longmate, *The Bombers*, ch. 21, "The Biggest Chop Night Ever"; Hastings, *Bomber Command*, 319, gives the Harris quotes, and on 320 quotes the official history.

30. R. Weigley, *The American Way of Warfare: A History of United States Military Strategy and Policy* (Bloomington: Indiana University Press, 1973); ch. 14 has his own comments on the USAAF aerial offensives.

31. W. F. Craven and J. L. Cate, eds., *The Army Air Forces in World War II* (Chicago: University of Chicago Press, 1948–1958), 7 vols., is revealing here: see vol. 2, ch. 9, Arthur B. Ferguson's "The Casablanca Directive."

32. Bendiger, *Fall of Fortresses*, 232–34. Bendiger's language here is withering.

33. These figures come from A. Furse, *Wilfrid Freeman: The Genius Behind Allied Survival and Air Supremacy 1939 to 1945* (Staplehurst, UK: Spellmount Press, 1999), 234; they are slightly different in Liddell Hart, *History*, 603.

34. Craven and Cate, eds., *Army Air Forces*, 2:702–3; the actual description is in chapter 20, "Pointblank," by Arthur B. Ferguson.

35. Craven and Cate, eds., *Army Air Forces*, 2:706, 705.

36. Karl Mendelssohn, *Science and Western Domination* (London: Thames and Hudson, 1976). The story has been developed in Daniel Headrick's *Power over Peoples: Technology, Environments, and Western Imperialism, 1400 to the Present* (Princeton: Princeton University Press, 2010).

37. A. Harvey-Bailey, *The Merlin in Perspective* (Derby, UK: Rolls-Royce Heritage Trust, 1983), is very technical but also important in stressing the key role of Rolls-Royce's managing director E. W. Hives after Royce's death. See also H. Glancy, *Spitfire: The Biography* (London: Atlantic Books, 2006), ch. 1.

38. Harvey-Bailey, *Merlin*.

39. Glancy's *Spitfire* is only the most obvious. See also Alfred Price's *The Spitfire Story* (London: Arms and Armour, 1995) and Len Deighton's remarkable *Fighter: The True Story of the Battle of Britain* (London: Pimlico, 1996).

40. *The Story of the Spitfire*, DVD (Pegasus, 2001).

41. Harvey-Bailey, *Merlin*, is excellent on the steady enhancement of the engine's power. The equally impressive efforts by Packard engineers to mass-produce Merlin 61 engines in the U.S.A. is nicely covered in Herman Arthur, *Freedom's Forge: How American Business Produced Victory in World War II* (New York: Random House, 2012), 103-105.

42. See Furse, *Wilfrid Freeman*.

43. D. Birch, *Rolls-Royce and the Mustang* (Derby, UK: Rolls-Royce Historical Trust, 1987), 10; the photo of the original plane is on the same page.

44. Furse, *Wilfrid Freeman*, 226–29, presents remarkable detective work.

45. Paul A. Ludwig, *P-51 Mustang: Development of the Long-Range Escort Fighter* (Surrey, UK: Ian Allen, 2003), esp. ch. 5, on the resistance of Echols.

46. Birch, *Rolls-Royce and the Mustang* reproduces Hitchcock's letter in full on 37–39; see also 147–48. The quote by the official historians is in Craven and Cate, eds., *Army Air Services*, 4:217–18.

47. Craven and Cate, eds., *Army Air Services*, 3:8.

48. Lovett's report, and Arnold's response, are best covered in Ludwig, *P-51 Mustang*, 143–45, 148.

49. For two of them, see Murray, *Luftwaffe*, and N. Frankland, *The Bombing Offensive Against Germany* (London: Faber, 1965).

50. Furse, *Wilfrid Freeman*, 234–35.

51. See www.cebudanderson.com/droptanks.html—an unusual source (accessed May 2008), the memoir of Donald W. Marner, a U.S. mechanic serving a Mustang squadron based in Suffolk in 1944–45, whose chief task was to get his hands on enough of them from his RAF buddies. Bendiger

also mentions the American fliers' gratitude for these quaint papier-mâché drop tanks. For confirmation of the immense significance of the drop tanks (especially the paper version) in the air war, see Ludwig, *P-51 Mustang*, 168–70.

52. Craven and Cate, eds., *Army Air Forces*, vol. 3, ch. 3, on "Big Week"; and Murray, *Luftwaffe*, 223ff. Ludwig, *P-51 Mustang*, 204, has comparative figures of P-38, P-47, and P-51 kill ratios. To some degree, then, as G. E. Cross points out, the P-47 Thunderbolts became overshadowed by the Mustangs, rather the way the Hurricanes were overshadowed by the Spitfires in the Battle of Britain, while in reality all four aircraft types played a vital role. See Cross's *Jonah's Feet Are Dry: The Experience of the 353rd Fighter Group During World War Two* (Ipswich, UK: Thunderbolt, 2001).

53. Furse, *Wilfrid Freeman*, 239–41.

54. See Craven and Cate, eds., *Army Air Forces*, 3:63, on air battles doing "more to defeat the Luftwaffe than did the destruction of the aircraft factories." There is a vast German-language literature, perhaps best summarized in English in the seventh volume of the German official history: Horst Boog et al., *The Strategic Air War in Europe, 1943–1944/45*, vol. 7 of *Germany and the Second World War* (Oxford: Oxford University Press, 2006).

55. Overy, *Why the Allies Won*, 152. The figures in the preceding paragraph come from Murray and Millett, *A War to Be Won*, 324–25. The Kocke anecdote is from E. R. Hooton, *Eagle in Flames: The Fall of the Luftwaffe* (London: Arms and Armour, 1997), 270–71, also with names of fellow aces killed at that time. (This is a fine, almost intimidating statistical analysis of the air war in Europe.)

56. Frankland, *The Bombing Offensive*, 86, offers a really crisp account of how the coming of the American long-range fighters redounded to the secondary advantage of Bomber Command.

57. Murray and Millett, *A War to Be Won*, 413. Craven and Cate, eds., *Army Air Forces*, vol. 3, is excellent throughout. See also W. Hays Parks, " 'Precision' and 'Area' Bombing: Who Did Which and When?" *Journal of Strategic Studies* 18, no. 1 (March 1995): 145–74. Also, personal communication of June 30, 2008, to author from Professor Tami Biddle, whose own writings (including *Rhetoric and Reality*) are compelling scholars into a serious reconsideration of the challenges Harris faced from mid-1942 onward.

58. Hastings, *Bomber Command*, 342–43.

59. J. Scutts, *Mustang Aces of the Eighth Air Force* (Oxford: Osprey Military Series, 1994), 56–60, on the coming of the Me 262s.

60. On the diminishing aviation fuel figures, see M. Cooper, *The German Air Force, 1933–1945: An Anatomy of Failure* (London: Jane's, 1981), 348–49, 360.

61. Hastings, *Bomber Command*, 422–23.

62. Saward, *Victory Denied;* Harris, *Bomber Offensive.*

63. All these works have been cited above. It will be obvious how much I am indebted to the works of Hastings, Murray, Biddle, and Overy, and how much their conclusions make sense to me. See Hastings, *Bomber Command,* ch. 14–15; Murray, *Luftwaffe,* ch. 7–8; Biddle, *Rhetoric and Reality,* ch. 5 and conclusion; Overy, *Why the Allies Won,* 149–63. But perhaps the prize goes to Webster and Frankland, *Strategic Air Offensive,* vol. 3, and Craven and Cate, eds., *Army Air Forces,* vol. 3, passim, as models of scholarship, objectivity, and insight.

64. The calculation about V-rocket costs versus aircraft figures is in Overy, *Why the Allies Won,* 294. Hitler's bizarre demands about the Messerschmitt Me 262 are neatly covered in D. Irving, *The Rise and Fall of the Luftwaffe: The Life of Luftwaffe Marshall Erhard Milch* (London: Weidenfeld and Nicolson, 1973), ch. 21.

65. The later volumes of Craven and Cate, eds., *Army Air Forces,* on the Pacific War, are best here, but there is also a great survey in Murray and Millet, *A War to Be Won,* ch. 17–18.

66. The infamous 9/11 attacks on Manhattan and the Pentagon claimed almost 3,000 lives.

67. The best obituary of Harker is that of the *Times* (London) on June 14, 1999, describing him as "the man who put the Merlin in the Mustang." The obituarist clearly has no idea of the later opposition to the Merlin-Mustang and talks of it as being greeted "like manna in heaven in Washington." But he gets Harker right, at least.

CHAPTER THREE: HOW TO STOP A BLITZKRIEG

1. The quotation is from R. Atkinson, *An Army at Dawn: The War in North Africa, 1942–1943* (New York: Henry Holt, 2003), 350; the main battle is covered on 359–92. See also S. W. Mitcham, *Blitzkrieg No Longer: The German Wehrmacht in Battle, 1943* (Barnsley, UK: Pen and Sword Books, 2010), 66ff.

2. Vividly described in Atkinson, *Army at Dawn,* 212–13.

3. Williamson Murray, *German Military Effectiveness* (Baltimore: Nautical and Aviation Publishing, 1992), esp. ch. 1. B. H. Liddell Hart, *History of the Second World War* (London: Cassell's, 1970), is brief but good on the Polish campaign (ch. 3) and the defeat of France (ch. 7). On how surprising the latter result was, see E. R. May's great revisionist book, *Strange Victory: Hitler's Conquest of France* (New York: Hill and Wang, 2000).

4. See Waugh's classic Sword of Honour trilogy, especially the middle volume, *Officers and Gentlemen,* where he graphically describes his fictional "Royal Halbardiers" regiment being routed by the Germans in Greece and Crete. Only the New Zealanders seem to have stood up to the invaders, man for man, but at very severe cost.

5. E. L. Jones, *The European Miracle* (Cambridge: Cambridge University Press, 1981), for this thesis, and very much followed in P. Kennedy, *The Rise and Fall of the Great Powers* (New York: Random House, 1987), ch. 1.

6. There are nice, clear details and good maps in Archer Jones, *The Art of War in the Western World* (Urbana: University of Illinois Press, 1987).

7. T. Lupfer, *The Dynamics of Doctrine: The Changes in German Tactical Doctrine During the First World War* (Leavenworth, KS: Combat Studies Institute, 1981), and, more generally, T. N. Dupuy, *A Genius for War: The German Army and the General Staff* (Fairfax, VA: Hero Books, 1984). See also the running commentary in R. M. Citino, *Death of the Wehrmacht: The German Campaigns of 1942* (Lawrence: University Press of Kansas, 2007).

8. See M. Boot's fine distillation of this offense versus defense spiral in his *War Made New* (New York: Gotham Books, 2006).

9. Some fine maps covering this campaign were lovingly put together for Liddell Hart's *History*, on 110–11, 282, 292, and 300, but see also the maps in C. Messenger, *World War Two: Chronological Atlas* (London: Bloomsbury, 1989).

10. A simple but most useful summary of all the moves in the North African Campaign is accessible in Messenger, *World War Two*, 46–55, 88–93, 116–23, 134–35.

11. There is a fine article by L. Ceva, "The North African Campaign 1940–43: A Reconsideration," *Journal of Strategic Studies* 13, no. 1 (March 1990): 84–104 (part of a special issue, edited by J. Gooch, called "Decisive Campaigns of the Second World War"), which among other things reminds the reader of the very significant role played by Italian forces in this campaign.

12. Quoted in *The Rommel Papers*, ed. B. H. Liddell Hart (London: Collins, 1953), 249. See also Rommel's amazingly candid letters home to his wife in the surrounding pages. The critical importance of fuel shortages is stressed again and again in B. Ellis, *Brute Force: Allied Strategy and Tactics in the Second World War* (New York: Viking, 1990), ch. 5.

13. A nice summation of this evolution is S. Bidwell, *Gunners at War: A Tactical Study of the Royal Artillery in the Twentieth Century* (New York: Arrow Books, 1972).

14. Liddell Hart, *History*, 296.

15. The best (and almost the only) authority here is M. Kroll, *The History of Landmines* (London: Leo Cooper, 1998). Clearly it is an unappealing topic, even for military historians themselves.

16. For Hobart's flail tanks (actually invented by a South African captain, Abraham du Toit), see chapter 4, and the "Mine Flail" article in Wikipedia, http://en.wikipedia.org/wiki/Mine_flail. For the mine detector, see "Polish Mine Detector," Wikipedia, http://en.wikipedia.org/wiki/Polish_mine_detector (both accessed June 2010).

17. It is unsurprising, therefore, that Coningham also commanded the tactical air forces in both later major campaigns. On Dawson's quietly outstanding organizational skills, see Ellis's approving remarks in *Brute Force,* 266–67; for the larger story of the RAF in the North African campaign at this time, see D. Richards and H. St. G. Saunders, *Royal Air Force 1939–1945* (London: HMSO, 1954), 2:160ff. The earlier, sad tale is in D. I. Hall, *Strategy for Victory: The Development of British Tactical Air Power, 1919–1943* (Westport, CT: Praeger, 2008).

18. All the general World War II books referred to in this volume— H. P. Willmott, *The Great Crusade: A New Complete History of the Second World War* (London: Michael Joseph, 1989); Messenger, *World War Two;* W. Murray and A. R. Millett, *A War to Be Won: Fighting the Second World War* (Cambridge, MA: Harvard University Press, 2000); R. Overy, *Why the Allies Won* (London: Jonathan Cape, 1995); Ellis, *Brute Force;* J. Keegan, *The Second World War* (New York: Penguin, 1990); Liddell Hart, *History;* and so—naturally cover El Alamein and point to the usual aspects: the constrained geographical limits, the importance of supplies, the British superiority in numbers, the importance of minefields and artillery, and the Wehrmacht's fighting skills. Nothing has emerged in recent writings to change this overall outline.

19. Mitcham, *Blitzkrieg No Longer,* ch. 4, is excellent on the Arnim-Rommel tensions.

20. The second volume of Atkinson's trilogy (*Day of Battle*), on the Sicilian and Italian campaigns, offers an excellent analysis, plus an introduction to an enormous body of further literature, such as C. D'Este's *World War Two in the Mediterranean (1942–1945)* (Chapel Hill, NC: Algonquin Books, 1990). For the casualties claim, see Keegan, *Second World War,* 368.

21. T. N. Dupuy, *Numbers, Prediction and War: Using History to Evaluate Combat Factors and Predict the Outcome of Battles* (Fairfax, VA: Hero Books, 1985), has masses of statistics. One doesn't need to follow the predictive part of this exercise to find the historical statistics interesting.

22. http://en.wikipedia.org/wiki/Eastern_Front_(World_War_II), 2nd paragraph—accessed May 2010.

23. Messenger, *World War Two,* 63–64; David M. Glantz and Jonathan M. House, *When Titans Clashed: How the Red Army Stopped Hitler* (Lawrence: University Press of Kansas, 1995) has many other good maps.

24. Requoted in Liddell Hart, *History,* 169.

25. There are fuller details in the overlapping final chapters of J. Erickson's *The Road to Stalingrad* (London: Weidenfeld and Nicolson, 1975) and the first chapters of the successor volume, *The Road to Berlin* (London: Weidenfeld and Nicolson, 1983). Really, with Erickson as the master, but so many other Anglo-American historians such as Earl F. Ziemke, the prodigious David M. Glantz, Ian Bellamy, Malcolm MacIntosh, Albert Seaton,

and the many excellent German experts on this topic, it is difficult to stop turning the endnote apparatus on the Russo-German War into something larger than the text. For Liddell Hart's approval of the Stavka-orchestrated advances around the greater Stalingrad area, see *History,* 481.

26. R. Forczyk, *Erich Von Manstein* (Oxford: Osprey Press, 2010), 36–42 (it has good illustrations); Erickson, *Road to Berlin,* 51ff.

27. M. K. Barbier, *Kursk: The Greatest Tank Battle, 1943* (St. Paul, MN: MBI Publishing, 2002); M. Healy, *Kursk 1943* (Oxford: Osprey Press, 1992), for remarkable detail; and Lloyd Clarke, *The Battle of the Tanks: Kursk, 1943* (New York: Atlantic Monthly Press, 2011).

28. A. Nagorski, *The Greatest Battle* (New York: Simon and Schuster, 2007).

29. Erickson, *Road to Stalingrad;* Erickson, *Road to Berlin.* See also the reflections in Citino, "Death of the Wehrmacht," esp. 14–19.

30. B. Wegner, "The Road to Defeat: The German Campaigns in Russia, 1941–1943," *Journal of Strategic Studies* 13, no. 1 (March 1990): 122–23. A most intriguing article.

31. See J. E. Forster, "The Dynamics of Volksgemeinschaft: The Effectiveness of the German Military Establishment in the Second World War," in A. Millett and W. Murray, eds., *Military Effectiveness* (London: Allen and Unwin, 1988), 3:201–2.

32. P. Carell, *Hitler's War on Russia,* trans. Ewald Osers (London: Corgi, 1966), 623. Carell (actually, Paul Karl Schmidt) was an early Nazi and a leading wartime propagandist who managed to escape the Nuremberg dragnet and transform himself into a highly successful writer of military histories— works that were always informative, but with dodgy judgments.

33. Cited again from Wegner, "The Road to Defeat," 122–23.

34. Email communication to author by Mr. Igor Biryukov, June 7, 2010.

35. The titles give this away: Ellis, *Brute Force;* Glantz and House, *When Titans Clashed;* and R. Overy's fine *Russia's War: Blood upon the Snow* (New York: TV Books, 1997).

36. D. Orgill, *T-34: Russian Armor* (New York: Ballantine Books, 1971), is full of such quotes.

37. Carell, *Hitler's War,* 75–76; see also the fine Wikipedia article "T-34," http://wikipedia.org/wiki/T-34 (accessed May 2010), with a wonderful bibliography.

38. The Mellethin, von Kleist, and Guderian quotations come from Orgill, *T-34.* The amazing postwar sales of the T-34 across the globe are detailed in the Wikipedia article "T-34."

39. Albeit in a backhanded way, by describing the post-1942 improvements; see Orgill, *T-34,* 73ff.

40. "T-34," Wikipedia.

41. Mary R. Habeck, *Storm of Steel: The Development of Armor Doctrine in Germany and the Soviet Union* (Ithaca: Cornell University Press, 2003),

has many interesting comments on the mutual "borrowings" of various of the interwar armored services. See also "J. Walker Christie," Wikipedia, http://wikipedia.org/Wiki/J._Walter_Christie (accessed May 2011).

42. Brief details in Orgill, *T-34*.

43. M. Bariatinsky, "Srednii Tank T-34-85," *Istoria Sozdania* (accessed May 26, 2011, from http://www.cardarmy.ru/armor/articles/t3485.htm). I am grateful to Professor Jonathan Haslam (Cambridge) for drawing my attention to this source.

44. A very informative piece, despite its aggressive title, is A. Isaev, "Against the T-34 the German Tanks Were Crap," in A. Drabkin and O. Sheremet, eds., *T-34 in Action* (Mechanicsburg, PA: Stackpole Books, 2008), ch. 2.

45. "An Evaluation of the T-34 and KV Tanks by Workers of the Aberdeen Testing Grounds of the U.S., Submitted by Firms, Officers and Members of Military Commissions Responsible for Testing Tanks," available at http://www.battlefield.ru/en/documents/80-armor-andequipment/300-t34 -kv1-aberdeen-evaluation.htr. I am grateful to Professor Jonathan Haslam (Cambridge) for drawing my attention to this source.

46. Ibid. Carell, *Hitler's War*, also frequently notes the need for the T-34 commander to have a sledgehammer nearby, and their lack of a decent radio. It is amazing that they didn't do much worse in the early years.

47. Healey, *Kursk 1943*, 31; Mitcham, *Blitzkrieg No Longer*, 132.

48. I found the best general source here to be G. L. Rottman, *World War II Anti-Tank-Tactics* (Oxford: Osprey Publishing, 2005), 45ff.

49. The Keegan quote is from his *Second World War*, 407. And see the confirmation in David M. Glantz, *Colossus Reborn* (Lawrence: University Press of Kansas, 2005), 29.

50. Barbier, *Kursk*, 55. There are similar figures in the valuable work by W. S. Dunn Jr., *The Soviet Economy and the Red Army 1930–1945* (Westport, CT: Praeger, 1995), 179.

51. Dunn, *Soviet Economy*, has impressive figures.

52. Glantz, *Colossus Reborn*, 355ff.

53. Ibid. The creation of these massive pontoon-bridge parks, containing Lego-like bridges of various lengths and load-carrying capacities, sounds very similar to the story of the Seabees (see chapter 5), but I have not yet found a Soviet equivalent to Admiral Ben Moreell.

54. Glantz, *Colossus Reborn*, 361–62; Barbier, *Kursk*, 58.

55. Barbier, *Kursk*, 58; see Mitcham, *Blitzkrieg No Longer*, 138, on the partisans' efforts at Kursk. Stone's observation is from *A Military History of Russia* (Westport, CT: Praeger, 2006), 212–13.

56. http://www.dupuyinstitute.org/ubb/Forum 4/HTML/000052.html.

57. Both W. Murray, *Luftwaffe* (Baltimore: Nautical and Aviation Publishing Company of America, 1985), and R. Muller, *The German Air War in Russia* (Baltimore: Nautical and Aviation Publishing, 1992), show the tre-

mendous effects that the RAF and USAAF strategic bombing campaigns had in pulling away the German air from the Eastern Front, leaving behind chiefly planes for supporting the ground forces. See Richard J. Evans, *The Third Reich at War* (New York: Penguin, 2008), 461, for the particular statistic.

58. Mitcham's figures, which are based upon Niepold's older work, are in *The German Defeat in the East 1944–45* (Mechanicsburg, PA: Stackpole Books, 2001), 16, 36. Hardesty, V. *Red Phoenix: The Rise of Soviet Power* (Minnetonka, MN: Olympic Marketing Corp, 1982) is best here.

59. W. Murray, *Luftwaffe* and Muller, *German Air War,* are good introductions here.

60. For details of the Sturmoviks, see "Ilyushin Il-2," Wikipedia, http://en.wikipedia.org/wiki/Ilyushin_Il-2; also described in A. Brookes, *Air War over Russia* (Hersham, Surrey: Ian Allen, 2003), 63.

61. Erickson, *Road to Berlin,* ch. 5; Mitcham, *German Defeat in the East;* Glantz and House, *When Titans Clashed,* ch. 13.

62. See "Operation Bagration," Wikipedia, http://en.wikipedia.org/wiki/Operation-Bagration; Mitcham, *The German Defeat in the East,* ch. 1; and the fuller account in S. Zaloga, *Bagration 1944: The Destruction of Army Group Centre* (Oxford: Osprey Press, 1996).

63. See in particular Habeck, *Storm of Steel,* 232–33. The irony of the Soviet armies becoming more flexible just when the German armies were seizing up is covered in the middle chapters of Glantz and House's fine work *When Titans Clashed,* ch. 9–13; Stone discusses this in *A Military History of Russia,* 202ff.

64. Mitcham, S. *Blitzkrieg No Longer,* passim.

65. Ibid.; Erickson, *Road to Berlin,* ch. 11–16, provides enormous detail.

66. See the horrifying details in N. Fergusson, *The War of the World: Twentieth-Century Conflict and the Descent of the West* (New York: Penguin, 2004); now supplemented by I. Kershaw, *The End: Hitler's Germany, 1944–45* (London: Allen Lane, 2011).

67. Mitcham, *Blitzkrieg No Longer,* 215–16. Zeitzler's reflective piece appeared in the April 1962 issue of *Military Review,* with the intriguing title "Men and Space in War: A German Problem in World War II." It is so little known, and so worth recovering. Willmott also stresses this point, in a nifty comparative way, in *The Great Crusade,* ch. 5, "Time, Space, and Doctrine."

68. O. P. Chaney, *Zhukov,* rev. ed. (Norman: University of Oklahoma Press, 1996), provides fine detail.

CHAPTER FOUR: HOW TO SEIZE AN ENEMY-HELD SHORE

1. See, for example, D. J. B. Trim and M. C. Fissel, eds., *Amphibious Warfare 1000–1700* (Leiden: Brill, 2006); and, for those reliant upon the elec-

tronic media, a rather good Wikipedia piece is http:// en.wikipedia.org/ wiki/Amphibious_warfare (accessed May 1, 2008). There is the impressive work by D. Abulafia, *The Great Sea: A Human History of the Mediterranean* (London: Allen Lowe, 2010), with great coverage of campaigns in that sea.

2. Accounts of such raids are in Bernard Fergusson's classic *The Watery Maze: The Story of Combined Operations* (London: Collins, 1961), along with chapters on full invasions themselves. But Fergusson is so enthused about any actions taken against the enemy that the operational distinction is not made clear.

3. Ibid., 47.

4. http://en.wikipedia.org/wiki/Amphibious_warfare, the "16th century" portion, is the quickest way to get to this tale. Scholars interested in most details can check on the "Terceras Landing."

5. See B. H. Liddell Hart, *The British Way in Warfare* (London: Faber and Faber, 1932), especially ch. 1; and the more modern treatment by M. E. Howard, *The Continental Commitment* (London: Maurice Temple Smith, 1972).

6. For the Tanga fiasco, see Fergusson, *Watery Maze*, 24–29.

7. See A. Millett and W. Murray, eds., *Military Effectiveness* (London: Allen and Unwin, 1988), especially the introduction and the conclusion to vol. 1.

8. It is hard to know where to start (or stop) with references to Gallipoli. The military account is Alan Moorhead's *Gallipoli* (London: Hamish Hamilton, 1956), with R. Rhodes James, *Gallipoli* (London: Batsford, 1965) best on the political side; and a very fine recent survey by L. A. Carlyon, *Gallipoli* (London: Doubleday, 2002). Those far from a good library can find a fair summary, with a useful ANZAC angle, in "Gallipoli Campaign," Wikipedia, http://en.wikipedia.org/wiki/Battle_of_Gallipoli.

9. See Correlli Barnett, *Engage the Enemy More Closely: The Royal Navy in the Second World War* (London: Hodder and Stoughton, 1991), 540–43; Fergusson, *Watery Maze*, 36–43; and the important memoir by the ISTCD's first director, L. E. H. Maund, *Assault from the Sea* (London: Methuen, 1949).

10. S. W. Roskill, *The War at Sea, 1939–1945* (London: HMSO, 1954–61), vol. 1; Barnett, *Engage the Enemy*, ch. 3–13.

11. The two great historians of the twentieth-century Royal Navy, Arthur Marder and Stephen Roskill, disagreed on many issues. On Churchill's interference and poor performance in the Norwegian campaign, however, there was remarkable overlap: see A. J. Marder, "Winston Is Back!" *English Historical Review*, supp. 5 (1972); S. W. Roskill, *Churchill and the Admirals* (London: Collins, 1977), ch. 8 and appendix, 283–99.

12. B. H. Liddell Hart, *History of the Second World War* (London: Cassell's, 1970), 226. For the Crete campaign, see Barnett, *Engage the Enemy*, ch.

11–12, and especially C. MacDonald's *The Lost Battle: Crete 1941* (New York: Macmillan, 1993), especially the truly scary chapter 10, "Ordeal at Sea."

13. W. Murray and A. R. Millett, *A War to Be Won: Fighting the Second World War* (Cambridge, MA: Harvard University Press, 2000), 106. The Norway airpower statistics are in Liddell Hart, *History*, 59, with the *Prince of Wales/Repulse* statistics on 226.

14. Barnett, *Engage the Enemy*, 203–6, is withering; Fergusson, *Watery Maze*, 59–69, is eye-opening. See also A. J. Marder's detailed study *Operation "Menace"* (Oxford: Oxford University Press, 1976).

15. Fergusson, *Watery Maze*, 166; Barnett, *Engage the Enemy*, 864–68.

16. On Combined Operations' learning curve, see Barnett, *Engage the Enemy*, 545–46, and ch. 11–15 of P. Ziegler's *Mountbatten: A Biography* (New York: Knopf, 1985).

17. The literature on the Dieppe Raid is itself a minefield. Fergusson, *Watery Maze*, 175–85, is unrepentant about its utility. See, by contrast, T. Robertson, *The Shame and the Glory* (Toronto: McLelland and Stewart, 1967), and Denis and Shelagh Whitaker, *Dieppe: Tragedy to Triumph* (Whitby, ON: McGraw-Hill, 1967), 293–304, which is very critical but ultimately comes down on the benefits of the operation for the later D-Day successes.

18. Fergusson, *Watery Maze*, 185; Churchill's language was rather more circumspect—"Their sacrifice was not in vain"—when he later wrote his *History*. But it is clear from D. Reynolds's illuminating study, *In Command of History: Churchill Writing and Fighting the Second World War* (London: Penguin, 2004), 345–48, that Churchill, Ismay, Mountbatten, and others in the British high command were embarrassed about how to explain the operation after the war.

19. On Anglo-American "jointness," especially between 1942 and 1944, there is nothing quite like H. Feis's classic *Churchill Roosevelt Stalin: The War They Waged and the Peace They Fought* (Princeton: Princeton University Press, 1957), 37–324. The more military aspect is covered by M. Matloff, *Strategic Planning for Coalition War, 1943–1944* (Washington, DC: Center of Military History, 1994), and M. E. Howard's superb *Grand Strategy*, vol. 4: *August 1942–September 1943* (London: HMSO, 1972).

20. Barnett, *Engage the Enemy*, 554.

21. Ibid., 563. A wonderful account of all this chaos is R. Atkinson, *An Army at Dawn: The War in North Africa, 1942–1943* (New York: Holt, 2002).

22. See M. E. Howard's judicious *The Mediterranean Strategy in the Second World War* (London: Weidenfeld and Nicolson, 1967), as well as his official history, *Grand Strategy*, vol. 4. Barnett's unrelenting criticism here of Mediterranean "blue water" strategy in *Engage the Enemy*, ch. 17–18, 20–22, seems to me less balanced. S. E. Morison gives an American perspective in *Strategy and Compromise* (Boston: Little, Brown, 1958) as well

as his *History of United States Naval Operations in World War II* (Boston: Little, Brown, 1947–62), vol. 9.

23. Morison, *History*, vol. 9, is marvelously thorough. Liddell Hart, *History*, ch. 27 and 30, is nicely succinct. Atkinson's *The Day of Battle: The War in Sicily and Italy, 1943–1944* (New York: Holt, 2007), is epic at ground level.

24. Liddell Hart, *History*, 445; see also Barnett, *Engage the Enemy*, 627–650.

25. For the above, see Liddell Hart, *History*, 460–65; Morison, *History*, vol. 9, part III; Atkinson, *Day of Battle*, part Two.

26. Liddell Hart, *History*, 526–32; Atkinson, *Day of Battle*, part Three; and Morison, *History*, vol. 9, part IV.

27. Note also the titles of the major parts of Liddell Hart's *History*, part V, "The Turn," part VI, "The Ebb," part VII, "Full Ebb," and part VIII, "Finale."

28. The text is probably most easily found in David Eisenhower's biography of his grandfather, *Eisenhower at War 1943–1945* (New York: Random House, 1986), 252.

29. There are thousands of books on D-Day and the Normandy campaign, including some superb official histories (British, Canadian, U.S.) on their air forces, armies, navies, and intelligence. I thought the best single-volume works to be M. Hastings, *Overlord* (London: Michael Joseph, 1984); S. Ambrose, *D-Day, June 6th, 1944* (New York: Simon and Schuster, 1994); and C. Ryan, *The Longest Day* (New York: Simon and Schuster, 1959). There are also some marvelously good maps and illustrations in *Purnell's History of the Second World War*, 5:1793–942.

30. Barnett, *Engage the Enemy*, ch. 24–25, gives a very fine summary of the planning and organization, as does *Purnell's History*, 5:1794–5, 1870–5.

31. "Bertram Ramsay," Wikipedia, http://en.wikipedia.org/wiki/Bertram _Ramsay.

32. R. Overy, *Why the Allies Won* (London: Jonathan Cape, 1995), 183.

33. C. Gross and M. Postlethwaite, *War in the Air: The World War Two Aviation Paintings of Mark Postlethwaite* (Marlborough, UK: Crowood Press, 2004), 78.

34. M. K. Barbier's *D-Day Deception: Operation Fortitude and the Normandy Invasion* (Westport, CT: Praeger, 2007) is a really important study on this topic, and nicely supplements C. Cruikshank's *Deception in World War II* (Oxford: Oxford University Press, 1979), which has remarkable, sometimes hilarious photographs. But the serious student of this deception should also consult a more cautious work: C. Bickell, "Operation Fortitude South: An Analysis of Its Influence upon German Dispositions and Conduct of Operations in 1944," *War and Society* 18, no.1 (May 2000): 91–122. The article has an excellent bibliography, although obviously not including the findings in Barbier and other later works.

35. F. H. Hinsley et al., *British Intelligence in the Second World War,* vol. 3, part 2 (New York: Cambridge University Press, 1988), section 13, *Overlord,* is impressive in describing both deception and intelligence aspects of the Normandy operation. See also two other esteemed works, D. Kahn, *Hitler's Spies: German Military Intelligence in World War Two* (New York: Macmillan, 1978), and M. E. Howard, *Strategic Deception* (London: HMSO, 1990), which is also vol. 5 of *British Intelligence in the Second World War.*

36. F. H. Hinsley et al. *British Intelligence in the Second World War,* vol. 3, part 2, 107n, 127, 153, on Resistance attacks. Quite amazing details of SOE cooperation with the French are in M. R. D. Foot, *SOE: An Outline History of the Special Operations Executive, 1940–1945* (London: BBC Publications, 1984), esp. 222–29.

37. Quoted in Overy, *Why the Allies Won,* 195. As usual, a superb, brief summary.

38. This is probably better captured in the great scene in the 1962 movie version of Ryan's book *The Longest Day* (with Curt Juergens playing Blumentritt) than in any written account.

39. There is an unsurpassed analysis (with excellent maps and tables) in Roskill, *War at Sea,* vol. 3, part 2, 5–74; but see also Barnett, *Engage the Enemy,* ch. 24–25, for another version; and Ambrose, *D-Day,* ch. 5–9.

40. There is a remarkable photo of four Beaufighters, coming in from all directions to attack German minesweepers and destroyers inside a Channel port in June 1944. Photo is in C. Bekker's *The German Navy 1939-1945* (Hamlyn: London/New York, 1972), 179.

41. Roskill, *War at Sea,* vol. 3, part 2, 53–59, details the containment and defeat of the U-boats as well as the Allied losses.

42. Hobart's career has attracted a number of studies, including a nice biography by K. Macksay, *Armored Crusader* (London: Hutchison, 1967), and an extremely lively article by T. J. Constable, "The Little-Known Story of Percy Hobart," *Journal of Historical Review* 18, no. 1 (Jan.-Feb. 1999), also at ihr.org/jhr/v18/v18n1p-2_Constable.html (accessed Feb. 20, 2008). The illustrations of these weird contraptions in *Purnell's History,* 5:1834–5, 1919, are worth savoring. Churchill's directive about reemploying Hobart is in Constable's electronic version of this remarkable story.

43. Quoted in Ambrose, *D-Day,* 551, a characteristically generous acknowledgment of what was happening on beaches other than Omaha and Utah.

44. Ibid., 323; he devotes nine chapters to the Omaha Beach story. See also Hastings, *Overlord,* 105–21; and Murray and Millett, *A War to Be Won,* 417–23, which is extremely critical of Bradley, the U.S. Navy, and the whole Omaha operation. I am also obliged to Professor Tami Biddle for bringing me to a better understanding of the panoply of difficulties facing the Omaha planners and commanders.

45. Ambrose, *D-Day*, 576; Roskill, *History*, vol. 3, part 2, 53, give an unusually exact total of 132,715 men landed. It is not clear when these various tallies were taken—at dusk, at midnight, or at dawn next day—or whether the airborne forces are included. It hardly matters.

46. D. Eisenhower, *Eisenhower at War*, 425–26.

47. Barnett, *Engage the Enemy*, 843–51; and, less tartly, Roskill, *History*, vol. II, part 2, 142–53.

48. See its use in *Purnell's History*, 5:1793.

49. There is a rather nice Anglo-American synergy here, spotted by J. A. Isley and P. A. Crowl, *The U.S. Marines and Amphibious War: Its Theory, and Its Practice in the Pacific* (Princeton: Princeton University Press, 1951), 583–84: the larger landing craft ship and infantry craft were of British design, the LVTs and DUKWs were an American idea. Together the match was perfect.

50. Ibid., 581–82. Their ch. 12, "Amphibious Progress, 1941–1945," is a fine reflection, with some cross-references to European amphibious operations as well.

CHAPTER FIVE: HOW TO DEFEAT THE "TYRANNY OF DISTANCE"

1. G. Blainey, *The Tyranny of Distance* (London: Macmillan, 1968), a book chiefly about how vast distances shaped Australia's history, but with implications for the whole history of the Pacific Ocean as well.

2. For what follows, see generally R. Storry, *Japan and the Decline of the West in Asia 1894–1943* (London: Longman, 1979); P. Kennedy, *The Rise and Fall of the Great Powers* (New York: Random House, 1987), 206–9, 298–302.

3. See R. Hackett, *Yamagata Arimoto and the Rise of Modern Japan 1838–1922* (Cambridge, MA: Harvard University Press, 1971).

4. Apart from Storry, *Japan*, see also R. Myers and M. Peattie, *The Japanese Colonial Empire 1895–1945* (Princeton: Princeton University Press, 1984).

5. The best succinct coverage of the East Asian crises of the 1930s (there are many older, wonderful works) is in A. Iriye, *The Origins of the Second World War in Asia and the Pacific* (London: Longman, 1987). The economic impulses to Japan's outward thrust is handled brilliantly in M. A. Barnhart, *Japan Prepares for Total War: The Search for Economic Security, 1919–1941* (Ithaca, NY: Cornell University Press, 1988). Also useful is B. H. Liddell Hart, *History of the Second World War* (London: Cassell's, 1970), ch. 16; and H. P. Willmott, *The Great Crusade: A New Complete History of the Second World War* (London: Michael Joseph, 1989).

6. There is a summary of the Japanese military position in R. Spector, *Eagle Against the Sun: The American War with Japan* (New York: Free Press,

1985), ch. 2; but above all, A. Coox, "The Effectiveness of the Japanese Establishment in the Second World War," in A. Millett and W. Murray, *Military Effectiveness,* 3:1–44 (London: Allen and Unwin, 1988).

7. See a nice speculative essay by J. Black, "Midway and the Indian Ocean," *Naval War College Review* 62, no. 4 (Autumn 2009): 131–40.

8. Best discussed in A. Danchev's intellectual biography of Liddell Hart, *The Alchemist of War: The Life of Basil Liddell Hart* (London: Weidenfeld and Nicolson, 1988).

9. Willmott, *Great Crusade,* 314ff. For a rather similar discussion, see Liddell Hart, *History,* ch. 29.

10. Apart from Willmott, *Great Crusade,* see P. Kennedy, *Strategy and Diplomacy 1870–1945* (London: Fontana, 1983), ch. 7, "Japanese Strategic Decisions, 1939–45," for a development of this argument.

11. The best brief recent analysis is by W. Tao, "The Chinese Theatre and the Pacific War," in S. Dockrill, ed., *From Pearl Harbor to Hiroshima* (London: Macmillan, 1994). There is also the vast library of books on Stilwell in China, the most entertaining being B. Tuchman, *Sand Against the Wind: Stilwell and the American Experience in China 1911–1945* (New York: Macmillan, 1970).

12. The immensely difficult struggle by British Empire forces in the India-Burma theater is analyzed in vast detail in the official history, S. Woodburn Kirby et al., *The War Against Japan,* 5 vols. (London: HMSO, 1957–69); and a later rendition in C. Bayly and T. Harper, *Forgotten Armies* (London: Penguin & Allen Lane, 2004). As a compensation, it also gave cause for the best single-volume memoir by a general of the entire war, namely, Slim's *Defeat into Victory* (London: Cassell, 1956).

13. J. Masters, *The Road Past Mandalay* (London: Michael Joseph, 1961). The pun on the title of Kipling's poem/song is obvious. The clearest and most balanced book of all: L. Allen, *Burma: The Longest War, 1941–45* (New York: St. Martin's, 1984).

14. MacArthur's driving nature and his strategic opinions are covered in W. Manchester's *American Caesar: Douglas MacArthur 1880–1964* (Boston: Little, Brown, 1978). There are also acute running comments in Spector, *Eagle.*

15. For Eichelberger (and the "don't come back alive") instruction, see Spector, *Eagle,* 216; for the Marines, see the fine "1st Marine Division (United States)," Wikipedia, http://en.wikipedia.org/wiki/1st_Marine_Division (United_States) (accessed June 2010).

16. Liddell Hart, *History,* 620.

17. Ibid., 617; and, in very good detail, S. E. Morison, *History of United States Naval Operations in World War II* (Boston: Little, Brown, 1947–62), vol. 8, ch. 9.

18. L. Allen, "The Campaigns in Asia and the Pacific," *Journal of Strategic*

Studies 13, no. 1 (March 1990): 175. This is an extraordinarily rich source and summation, especially for the Japanese side.

19. Ibid., 165.

20. E. S. Miller, *War Plan Orange: The U.S. Strategy to Defeat Japan 1897–1945* (Annapolis, MD: U.S. Naval Institute Press, 1991), has the full story.

21. The best reminder of this important point is in M. van Creveld's ingenious work *Supplying War: Logistics from Wallenstein to Patton* (Cambridge: Cambridge University Press, 1977).

22. For massive detail, nothing will beat the official *History of U.S. Marine Corps Operations in World War II*, 5 vols. (Washington, DC: U.S. Marine Corps, 1958–68). The single-volume classics are A. R. Millett, *Semper Fidelis: The History of the United States Marine Corps*, 2nd ed. (New York: Free Press, 1991), ch. 12; and J. A. Isley and P. A. Crowl, *The U.S. Marines and Amphibious War: Its Theory, and Its Practice in the Pacific* (Princeton: Princeton University Press, 1951), ch. 1–3.

23. The "Success . . . Failure" quotation is from Isley and Crowl, *U.S. Marines*, 14–21.

24. Millett, *Semper Fidelis*, 320.

25. Isley and Crowl, *U.S. Marines*, 26.

26. For a full treatment, see D. A. Ballendorf and M. Bartlett, *Pete Ellis: Amphibious Warfare Prophet 1880–1923* (Annapolis, MD: Naval Institute Press, 1997).

27. See Millett, *Semper Fidelis*, 327; there is also a lively account in Isley and Crowl, *U.S. Marines*, 30–31.

28. Millett, *Semper Fidelis*, 336.

29. Both Morison's official naval history volumes, especially vols. 5–8 and 12–14, and the U.S. Army's official history volumes (dozens of them) cover their respective service's record in the Pacific theater. For a smooth-running commentary on the marines, the army, and amphibious warfare, see Spector, *Eagle*.

30. Quotation from C. G. Reynolds, *The Fast Carriers: The Forging of an Air Navy* (New York: McGraw-Hill, 1968), 1.

31. A nice summary is in ibid., 4–13; for technical data, see M. Stille, *Imperial Japanese Navy Aircraft Carriers 1921–1945* (London: Osprey, 2005).

32. Reynolds, *Fast Carriers*, ch. 3, "Weapon of Expediency, 1942–1943," gives the context in the critical period of the Pacific War. S. W. Roskill, *The War at Sea, 1939–1945* (London: HMSO, 1954–61), vol. 2, discusses HMS *Victorious*'s unusual experience.

33. Extremely useful details are in "Essex Class Aircraft Carrier," Wikipedia, http://en.wikipedia.org/Essex_class_aircraft _carrier (accessed May 2010). The author's notations and further references are the best I have seen.

34. These exploratory missions are covered in Reynolds, *Fast Carriers*, ch. 2, and Morison, *History*, vol. 7.

35. Spector, *Eagle,* 257.

36. For the Rabaul attacks, see Morison, *History,* vol. 6, Part 4, 369ff; see also Reynolds, *Fast Carriers,* 96ff.

37. The number of books and articles on the legendary Hellcat come close to the total for the equally legendary Spitfire. The best starting place may be with another one of those remarkably scholarly Wikipedia articles on aspects of the Pacific War: "F6F Hellcat," http://en.wikipedia.org/wiki/F6F _Hellcat (accessed May 2010).

38. Ibid.; Reynolds, *Fast Carriers,* 57, and passim.

39. Most historians of the war in the Central Pacific realize that there was something of a hiatus in the fighting—at least in the significant fighting—between November 1943 (Tarawa) and June 1944 (Marianas, Rabaul), so they tend to devote less space to operations in those months and more to the arrival of the newer weapons systems, the coming of radar, and so on. Morison, being the official naval historian, fills this gap in *History,* vol. 7.

40. "The Great Marianas Turkey Shoot"—apart from Midway, everyone's favorite aerial clash of the Pacific War. Reynolds, *Fast Carriers,* 190–204, is as good as any. Morison, *History,* has terrific details on 8:257–321.

41. Reynolds, *Fast Carriers,* makes the strongest (in my view, overly forced) argument about the Jutland analogy on 163–65, 209–10, followed by Spector, *Eagle,* 312.

42. Reynolds, *Fast Carriers,* has a rather generous ch. 9 on the performance of the British Pacific Fleet; Correlli Barnett, *Engage the Enemy More Closely: The Royal Navy in the Second World War* (London: Hodder and Stoughton, 1991), ch. 28, is a gloomy and almost dismissive account.

43. Two excellent introductions: C. Berger, *B29: The Superfortress* (New York: Ballantine, 1970); and an impressive Wikipedia entry, "B-29 Superfortress," http://en.wikipedia.org/wiki/B-29_Superfortress (accessed May 2010). Both these works have fine lists for further reading.

44. All of these details are in Wikipedia, "B-29 Superfortress."

45. Berger, *B29,* has a wonderful section on "The Battle of Kansas," 48–59. The "urgent struggle for airspeed" is a neat phrase from the Wikipedia article. Also excellent on the problem solvers of the B-29's many defects is Herman, *Freedom's Forge,* 297–322.

46. Berger, *B29,* 60–107.

47. John Toland, *The Rising Sun: The Decline and Fall of the Japanese Empire 1936–1945* (London: Cassell & Company, 1971) 676, 745. For the larger issue, see the powerful reflections of M. Sherry, *The Rise of American Air Power: The Creation of Armageddon* (New Haven: Yale University Press, 1987).

48. I could not find a satisfying study of Moreell, but there are some basic biographical details in "Ben Moreell," Wikipedia, http://en.wikipedia.org/ wiki/Ben_Moreell (accessed spring 2010).

49. Almost all that follows is taken from "Seabees in World War II," another very thorough Wikipedia entry on aspects of the war in the Pacific and Far East, http://en.wikipedia.org/wiki/Seabees_in_World_War_II (accessed spring 2010).

50. Ibid.

51. Spector, *Eagle,* 318–19.

52. The best insight into this tale of independence and resourcefulness comes from reading the memoirs of the American submariners themselves, of which there are many. For a taste, try Richard H. O'Kane, *Clear the Bridge!* (New York: Bantam, 1981); James F. Calvert, *Silent Running: My Years on an Attack Submarine* (New York: John Wiley, 1995)—withering in his comments on the Naval Ordnance Bureau; and Edward Beach, *Submarine!* (New York: Bantam, 1952).

53. Giving much detail is Clay Blair, *Silent Victory,* 2 vols. (New York: Lippincott, 1975); good comparative comments are in P. Padfield, *War Beneath the Sea: Submarine Conflict 1939–1945* (London: John Murray, 1995), especially ch. 9.

54. Edwin P. Hoyt, *The Destroyer Killer* (New York: Pocket Books, 1989).

55. S. E. Morison, *Two Ocean War* (Boston: Little, Brown & Co., 1963), 510–11.

56. Exact statistics for Japanese losses in the Pacific are (as for so many other conflicts) virtually impossible to arrive at. For example, a heavy explosion might convince a submariner that his target had been destroyed, but it might only be damaged—or the torpedo might have exploded prematurely. And in a hectic action, an aircraft and a sub might claim to have sunk the same ship. Wartime medals were awarded on the basis of what appeared to be substantive proof of kills. But at the end of the war a Joint Army-Navy Assessment Committee (JANAC) was set up to compare all claims with Japan's own records. In almost all cases—including the overall totals—the figures were strongly reduced, yet without altering the overall picture. For the figures above, see Padfield, *War Beneath the Sea,* 476, and Morison, *Two Ocean War,* 511.

57. The story is told in every general account (and almost all memoirs) of the Pacific War. The clearest explanation, even though containing much technical detail, is a five-part article by Frederick J. Milford in the *Submarine Review,* appearing between April 1996 and October 1997. See in particular Part Two (October 1996), "The Great Torpedo Scandal, 1941–1943."

58. The quotation and statistics following come from Roskill, *History,* vol. 3, part 2, 367.

59. Calculated from Mackenzie J. Gregory, "Top Ten US Navy Submarine Captains in WW2 by Number of Confirmed Ships Sunk," at http://ahoy .tk-jk.net/macslog/TopTenUSNavySubmarineCap/html (accessed March 2010).

60. Quoted in Morison, *Two Ocean War*, 486.

61. C. Boyd and A. Yoshida, *Japanese Submarine Forces in World War Two* (Annapolis: Naval Institute Press, 1995). See also the various comparisons made in Padfield, *War Beneath the Sea*.

62. Morison, *Two Ocean War*, 486.

63. Padfield, *War Beneath the Sea*, ch. 9, is, as ever, reliable here.

64. R. Spector, "American Seizure of Japan's Strategic Points, Summer 1942–44," in S. Dockrill, ed., *From Pearl Harbor to Hiroshima: The Second World War in Asia and the Pacific, 1941–45* (London: Macmillan, 1994), ch. 4.

65. The best brief, and rather sardonic, account of the Aleutian Islands is in Spector, *Eagle*, 178–82.

66. Liddell Hart, *History*, 356–62.

67. Morison, *History*, vol. 8, is the most detailed.

68. Millett, *Semper Fidelis*, 410–19, is excellent on the Marianas campaign.

69. Morison, *History*, 8:162.

70. J. B. Wood, *Japanese Military Strategy in the Pacific War: Was Defeat Inevitable?* (Lanham, MD: Rowman and Littlefield, 2007).

CONCLUSION: PROBLEM SOLVING IN HISTORY

1. A. Bryant, *The Turn of the Tide, 1939–1943: Based on the Diaries of Field Marshall Viscount Alanbrooke* (London: Collins, 1957); A. Bryant, *Triumph in the West 1943–1945* (London: Collins, 1959).

2. See Alanbrooke, *War Diaries, 1939–1945: Field Marshal Lord Alanbrooke*, ed. Alex Danchev and Daniel Todman (London: Weidenfeld and Nicolson, 2001). This later edition is a model of its kind. It not only includes many of the more candid entries that Bryant had felt it prudent to omit while Churchill and other key personages were still alive, but it also distinguishes between Alanbrooke's original uncensored entries, Alanbrooke's later notes, and Bryant's own variants (see "Note on the Text," xxxi–xxxiv).

3. See Andrew Roberts's clever use of the Alanbrooke diaries in *Masters and Commanders: How Roosevelt, Churchill, Marshall and Alanbrooke Won the War in the West* (London: Allen Lane/Penguin, 2008).

4. Alanbrooke, *War Diaries*, 433.

5. Ibid., 557. "It was a wonderful moment to find myself re-entering France almost 4 years after being thrown out" (entry of June 12, 1944).

6. The phrase seems to have been invented by the great historian of Stuart Britain, J. H. Hexter, in "The Burden of Proof," *Times Literary Supplement*, October 24, 1974—part of the raging debate in those years on the causes of the English Civil War.

7. The claim comes in the penultimate paragraph of Hinsley's 1988 Harmon Memorial Lecture to the U.S. Air Force Association, "The Intelligence Revolution: A Historical Perspective." It is actually a wonderful piece,

showing due skepticism of the many popular works of the 1970s and 1980s on spy rings, decrypting geniuses, and intelligence breakthroughs. So it is odd that he put his neck so far out with this nonprovable estimate.

8. R. Spector, *Eagle Against the Sun: The American War with Japan* (New York: Free Press, 1985), 457. Chapter 20, "Behind the Lines," is an impressive survey on many aspects of the intelligence war—and the limitations.

9. I had already composed these paragraphs before David Kahn sent me his extremely important article, "An Historical Theory of Intelligence," *Intelligence and National Security* 16, no. 3 (Autumn 2001): 79–92, which I had quite missed earlier. Note especially 85–86: "Intelligence is necessary to the defense, it is only contingent to the offense."

10. D. Kahn, "Intelligence in World War II: A Survey," *Journal of Intelligence History* 1, no. 1 (Summer 2001): 1–20, a fine summation because it repeatedly asks for the proof that intelligence worked.

11. It comes as something of a relief to this author that the most powerful criticism of certain U.S. commanders toward British ideas and inventions are made by American historians themselves: see Paul A. Ludwig, *P-51 Mustang: Development of the Long-Range Escort Fighter* (Surrey, UK: Ian Allen, 2003), on Echolls's opposition to the P-51; and W. Murray and A. R. Millett, *A War to Be Won: Fighting the Second World War* (Cambridge, MA: Harvard University Press, 2000), 249–50, on the slaughters of the chiefly U.K. merchantmen along the eastern seaboard ("It was Admiral King at his worst; he was simply not going to learn anything from the British, whatever the costs"); and ibid., 418–19, about Bradley's unwillingness to learn anything about the "tactical problems confronted by an amphibious assault on prepared defenses." Compare this American confidence of their own sheer muscle power with Churchill's insistence that it would not be by vast numbers of men and shells but by devising newer weapons and by scientific leadership "that we shall best cope with the enemy's superior strength," a key refrain in P. Delaforce's *Churchill's Secret Weapons* (London: Robert Hale, 1998).

12. P. Kennedy, *The Rise and Fall of the Great Powers* (New York: Random House, 1987), esp. 355, Table 35.

13. C. Barnett, *The Swordbearers* (London: Eyre and Spottiswood, 1963), 11.

14. See, among others, A. Bullock, *Hitler and Stalin: Parallel Lives* (London: Harper Collins, 1991); and the contrasting S. Bialer, ed., *Stalin and His Generals* (London: Souvenir Press, 1970), and H. Heiber, ed., *Hitler and His Generals* (New York: Enigman Press, 2003).

15. See D. Edgerton, *Britain's War Machine: Weapons, Resources and Experts in the Second World War* (London: Allen Lane, 2011), a sharp contrast with Barnett's *The Audit of War: The Illusion and Reality of Britain as a Great Nation* (London: Macmillan, 1986).

16. Malcolm Gladwell, "The Tweaker: The Real Genius of Steve Jobs," *New Yorker,* November 14, 2011. In a rather wonderful way, Herman's book on American innovation and productivity in WWII, *Freedom's Forge,* passim, is simply an extended version of this story of constant improvement of an initial design to get a satisfying final product.

17. One thinks here of that brilliant work by H. Hattaway and A. Jones, *How the North Won: A Military History of the Civil War* (Urbana: University of Illinois Press, 1983).

18. Neatly summarized in A. Grisson, "The Future of Military Innovation Studies," *Journal of Strategic Studies* 29, no. 5 (October 2006): 905–34, paying due tribute to Barry Posen, Eliot Cohen, Williamson Murray, MacGregor Knox, Timothy Lupfer, and other notable figures in this field. My own brief venture here was in my 2009 George Marshall Memorial Lecture, published as P. Kennedy, "History from the Middle: The Case of the Second World War," *Journal of Military History* 74, no. 1 (January 2010): 35–51. On *The Genius of Design,* see http://www.bbc.co.uk/programmes/b00sjIfg. The Millett and Murray "military effectiveness" concepts run through this present text, and many an endnote.

Bibliography

SPECIAL REMARKS

The alphabetical bibliography that follows below is chiefly of the standard type, but I would like to make several comments upon sources used for this book. The first is in regard to the well-known electronic database Wikipedia and others like it. Many university professors worry about the incomplete or nonverifiable aspects of entries, and about an undue reliance of their students upon easy-to-access electronic sources rather than the wonderful experience of slowly perusing books on musty library shelves and discovering works that have escaped the imperfect electronic catalogs. I understand that very well. But I have to confess to being mightily impressed by certain of the lengthy and detailed and scholarly (and anonymous) Wikipedia entries to which reference is made here, in particular to those which relate to aspects of the Pacific War. They are substantive and very well documented, and I would like to pay tribute to their authors (or the single author, since many suggest that the same craftsman was at work). Many other Wikipedia items are, as is alleged, rather embarrassing to peruse.

I should also like to acknowledge a heavy debt to the authors of works that are often referred to scornfully by professional (Ph.D.) historians as being written solely for "military history buffs." Actually, I do not think this present book could have been written—certainly not to the depth it was—had it not been for my reliance upon numerous titles made available by Osprey Press, Pen and Sword, and other notable publishers of military-technological history. Many years ago Professor Lawrence Stone, generously reviewing a book far removed from his own tastes, observed that in Clio's great mansion there are many corridors and many apartments; and that there is room for all. I like that.

Finally, I am indebted to the extraordinary quality of so many of the official histories put out by national governments or their armed services upon the various aspects of World War Two, military and civilian. Those composed by the American, British, and Dominion historians, some of them a full sixty years ago, are a tribute to their profession; well-written, balanced, critical (for example, the British and American histories of their own strategic bombing campaigns), and models of detached analysis. Then there is the impressive, more recent German official history, *Das Deutsche Reich und der Zweite Welt-krieg,* taking scholarship to new levels. This bibliography only contains titles expressly contained in the notes, but I did want to acknowledge my broader debts here.

BOOKS

Alanbrooke, Field Marshal Lord. *War Diaries 1939–1945.* Edited by A. Danchev and D. Todman. London: Weidenfeld and Nicolson, 2001.

Allen, L. *Burma: The Longest War, 1941–1945.* New York: St. Martin's, 1984.

Ambrose, S. *D-Day, June 6th, 1944.* New York: Simon and Schuster, 1994.

Atkinson, R., *An Army at Dawn: The War in North Africa, 1942–1943.* New York: Holt, 2002.

———. *The Day of Battle: The War in Sicily and Italy, 1943–1944.* New York: Henry Holt, 2007.

Ballendorf, D.A., and M. Bartlett. *Pete Ellis: Amphibious Warfare Prophet 1880–1923.* Annapolis, MD: Naval Institute Press, 1997.

Barbier, M. K. *D-Day Deception: Operation Fortitude and the Normandy Invasion.* Westport, CT: Praeger, 2007.

———. *Kursk: The Greatest Tank Battle, 1943.* Leicestershire: Ian Allan Publishing, 2002.

Barnett, C. *The Audit of War: The Illusion and Reality of Britain as a Great Nation.* London: Macmillan, 1986.

———. *Engage the Enemy More Closely: The Royal Navy in the Second World War.* London: Hodder and Stoughton, 1991.

———. *The Swordbearers.* London: Eyre and Spottiswood, 1963.

Barnhart, M. A. *Japan Prepares for Total War: The Search for Economic Security, 1919–1941.* Ithaca, NY: Cornell University Press, 1988.

Beach, E. *Submarine!* New York: Bantam, 1952.

Bekker, C. *The German Navy 1939–1945.* London: Hamlyn, 1972.

Bendiger, E. *The Fall of Fortress.* New York: Putnam, 1980.

Berger, C. *B29: The Superfortress.* New York: Ballantine, 1970.

Bernstein, A., M. Knox, and W. Murray. *The Making of Strategy: Rulers, States and War.* Cambridge: Cambridge University Press, 1994.

Bialer, S., ed. *Stalin and His Generals.* London: Souvenir Press, 1970.

Bialer, U. *The Shadow of the Bomber: The Fear of Air Attack and British Politics 1932–1939.* London: Royal Historical Society, 1980.

Biddle, Tami. *Rhetoric and Reality in Air Warfare: The Evolution of British and American Ideas About Strategic Bombing, 1914–1945.* Princeton: Princeton University Press, 2001.

Bidwell, S. *Gunners at War: A Tactical Study of the Royal Artillery in the Twentieth Century.* New York: Arrow Books, 1972.

Birch, D. *Rolls-Royce and the Mustang.* Derby, UK: Rolls-Royce Heritage Trust, 1987.

Blair, C. *Silent Victory.* 2 vols. New York: Lippincott, 1975.

Brookes, A. *Air War over Russia.* Hersham, Surrey: Ian Allan, 2003.

Bryant, A. *The Turn of the Tide: 1939–1943.* London: Doubleday, 1957.

———. *Triumph in the West, 1943–1946.* London: Collins, 1959.

Bullock, A. *Hitler and Stalin: Parallel Lives.* London: HarperCollins, 1991.

Burchard, John. *Q.E.D.: MIT in World War II.* New York: Wiley, 1948.

Calvert, J. F. *Silent Running: My Years on an Attack Submarine.* New York: John Wiley, 1995.

Carell, P. *Hitler's War on Russia: The Story of the German Defeat in the East.* Sheridan, CO: Aberdeen Books, 2002.

Carlyon, L. A. *Gallipoli.* London: Batsford, 1965.

Citino, R. *Death of the Wehrmacht: The German Campaigns of 1942.* Lawrence: University Press of Kansas, 2007.

Clarke, I. F. *Voice Prophesizing War 1763–1984.* Oxford: Oxford University Press, 1966.

Collier, B. *The Defence of the United Kingdom.* London: HMSO, 1957.

Cooper, M. *The German Air Force 1933–1945: An Anatomy of Failure.* London: Jane's, 1981.

Corbett, J. S. *Some Principles of Maritime Strategy.* London, 1911.

Craven, W. F., and J. L. Cate, eds. *The Army Air Forces in World War II.* Chicago: University of Chicago Press, 1948–1958.

Cross, G. E. *Jonah's Feet Are Dry: The Experience of the 353rd Fighter Group During World War II.* Ipswich, Suffolk: Thunderbolt, 2001.

Cruikshank, C. *Deception in World War II.* Oxford: Oxford University Press, 1979.

D'Este, C. *World War II in the Mediterranean, 1942–1945.* Chapel Hill, NC: Algonquin Books, 1990.

Danchev, A. *The Alchemist of War: The Life of Basil Liddell Hart.* London: Weidenfeld and Nicolson, 1988.

Deighton, L. *Fighter: The True Story of the Battle of Britain.* London: Pimlico, 1996.

Dupuy, T. N. *Numbers, Prediction, and War: Using History to Evaluate Combat Factors and Predict the Outcome of Battles.* Fairfax, VA: Hero Books, 1985.

———. *A Genius for War: The German Army and General Staff, 1807–1945.* Fairfax, VA: Hero Books, 1984.

Earle, E. M., with G. A. Craig and F. Gilbert. *Makers of Modern Strategy: Military Thought from Machiavelli to Hitler.* Princeton: Princeton University Press, 1971.

Edgerton, D. *Britain's War Machine: Weapons, Resources and Experts in the Second World War.* London: Allen Lane, 2011.

Eisenhower, D. *Eisenhower at War 1943–1945*. New York: Random House, 1986.

Ellis, B. *Brute Force: Allied Strategy and Tactics in the Second World War*. New York: Viking, 1990.

Erickson, J. *The Road to Berlin*. London: Weidenfeld and Nicolson, 1983.

———. *The Road to Stalingrad: Stalin's War with Germany*. London: Weidenfeld and Nicolson, 1975.

Evans, R. *The Third Reich at War*. New York: Penguin, 2008.

Fergusson, B. *The Watery Maze: The Story of Combined Operations*. London: Collins, 1961.

Fissel, M. C., and D. J. B. Trim, eds. *Amphibious Warfare 1000–1700: Commerce, State Formation and European Expansion*. Leiden, Netherlands: Brill, 2005.

Foot, M. R. D. *SOE: An Outline History of the Special Operations Executive, 1940–1945*. London: BBC Publications, 1984.

Forczyk, R. *Erich von Manstein*. Oxford: Osprey, 2010.

Frankland, N. *The Bombing Offensive Against Germany*. London: Faber, 1965.

Furse, A. *Wilfrid Freeman: The Genius Behind Allied Survival and Air Supremacy 1939–1945*. Staplehurst, UK: Spellmount Press, 1999.

Glancy, H. *Spitfire: The Biography*. London: Atlantic Books, 2006.

Glantz, D. *Colossus Reborn: The Red Army at War, 1941–1943*. Westport, CT: Praeger, 1995.

Glantz, D., and J. House. *The Battle of Kursk*. Lawrence: University Press of Kansas, 2004.

———. *When Titans Clashed: How the Red Army Stopped Hitler*. Lawrence: University Press of Kansas, 1998.

Gollin, A. *No Longer an Island: Britain and the Wright Brothers 1902–1909*. London: Heinemann, 1984.

Gretton, P. *Convoy Escort Commander*. London: Cassell, 1964.

———. *Crisis Convoy: The Story of the HX 231*. London: Naval Institute Press, 1974.

Gross, C., and M. Postlethwaite. *War in the Air: The World War Two Aviation Paintings of Mark Postlethwaite*. Marlborough, UK: Crowood Press, 2004.

Habeck, M. R. *Storm of Steel: The Development of Armor Doctrine in Germany and the Soviet Union, 1919–1939*. Ithaca, NY: Cornell University Press, 2003.

Hackett, R. *Yamagata Arimoto and the Rise of Modern Japan 1838–1922*. Cambridge, MA: Harvard University Press, 1971.

Hall, D. I. *Strategy for Victory: The Development of British Tactical Air Power, 1919–1943*. Westport, CT: Praeger, 2008.

Hardesty, *Red Phoenix: The Rise of Soviet Power*. Minnetonka, MN: Olympic Marketing Corp, 1982.

Harris, A. *Bomber Offensive*. London: Collins, 1947.

Hartcup, G. *The Effect of Science on the Second World War.* London: Macmillan, 2000.

Harvey-Bailey, A. *The Merlin in Perspective.* Derby, UK: Rolls-Royce Heritage Trust, 1983.

Hastings, M. *Overlord.* London: Michael Joseph, 1984.

Hattaway, H., and A. Jones. *How the North Won: A Military History of the Civil War.* Champaign: University of Illinois Press, 1983.

Headrick, D. *Power over Peoples: Technology, Environments, and Western Imperialism, 1400 to the Present.* Princeton: Princeton University Press, 2010.

Healy, M. *Kursk 1943: The Tide Turns in the East.* Oxford: Osprey Press, 1992.

Heiber, H., ed. *Hitler and His Generals.* New York: Enigma Press, 2003.

Herman, Arthur. *Freedom's Forge: How American Business Produced Victory in World War II.* New York: Random House, 2012.

Hezlet, A. *The Electron and Sea Power.* London: C. Davies, 1975.

Higham, R. *Air Power: A Concise History.* Yuma, KS: Sunflower Press, 1984.

Hinsley, F. H. et al. *British Intelligence in the Second World War,* 5 vols. New York: Cambridge University Press, 1979–1990.

Hooton, E. R. *Eagle in Flames: The Fall of the Luftwaffe.* London: Arms and Armour Press, 1997.

Howard, M. E. *Grand Strategy.* London: HMSO, 1970.

———. *The Continental Commitment.* London: Maurice Temple Smith, 1972.

———. *The Mediterranean Strategy in World War Two.* London: Weidenfeld and Nicolson, 1967.

———. *Strategic Deception.* London: HMSO, 1990.

Howse, D. *Radar at Sea: The Royal Navy in World War 2.* Annapolis: US Naval Institute Press, 1993.

Hoyt, E. P. *The Destroyer Killer.* New York: Pocket Books, 1989.

Iriye, A. *The Origins of the Second World War in Asia and the Pacific.* London: Longman, 1987.

Irving, D. *The Rise and Fall of the Luftwaffe: The Life of Luftwaffe Marshal Erhard Milch.* London: Weidenfeld and Nicolson, 1973.

Isley, J. A., and P. A. Crowl. *The U.S. Marines and Amphibious War: Its Theory, and its Pratice in the Pacific.* Princeton: Princeton University Press, 1951.

Jones, E. L. *The European Miracle.* Cambridge: Cambridge University Press, 1981.

Jones, T. C. G. *The Battle of Britain.* London: Frank Cass, 2000.

Keegan, J. *The Second World War.* New York: Penguin, 1990.

Kennedy, P. M. *Strategy and Diplomacy 1870–1945: Eight Studies.* London: Allen and Unwin, 1983.

———. *The Rise and Fall of the Great Powers.* New York: Random House, 1987.

Kershaw, I. *Fateful Choices: Ten Decisions That Changed the World, 1940–1941.* London: Penguin, 2007.

Kirby, S. W. *The War Against Japan.* London: HMSO, 1957–69.

Liddell Hart, Basil H. *History of the Second World War.* London: Cassell's, 1970.

————. *The British Way in Warfare.* London: Faber and Faber, 1932.

Ludwig, P. *P-51 Mustang: Development of the Long-Range Escort Fighter.* Surrey, UK: Ian Allen, 2003.

Lupfer, T. *The Dynamics of Doctrine: The Changes in German Tactical Doctrine During the First World War.* Leavenworth, KS: Combat Studies Institute, 1981.

MacDonald, C. *The Lost Battle: Crete 1941.* New York: Macmillan, 1993.

Macksay, K. *Armored Crusader.* London: Hutchinson, 1967.

Manchester, W. *American Caesar: Douglas MacArthur 1880–1964.* Boston: Little, Brown, 1978.

Marder, A. J. *Operation "Menace."* Oxford: Oxford University Press, 1976.

Masters, J. *The Road Past Mandalay.* London: Michael Joseph, 1961.

Matloff, M., and E. M. Snell. *Strategic Planning for Coalition Warfare, 1941–1942.* Washington, DC: Office of the Chief of Military History, 1953.

Maund, L. E. H. *Assault from the Sea.* London: Methuen, 1949.

Mendelssohn, K. *Science and Western Domination.* London: Thames and Hudson, 1977.

Messenger, C. *World War Two: Chronological Atlas.* London: Bloomsbury, 1989.

Middlebrook, M. *Convoy: The Greatest U-boat Battle of the War.* London: Phoenix Books, 2003.

Miller, E. S. *War Plan Orange: The U.S. Strategy to Defeat Japan 1897–1945.* Annapolis, MD: U.S. Naval Institute Press, 1991.

Millett, A. R. *Semper Fidelis: The History of the United States Marine Corps.* New York: Free Press, 1991.

Millett, A. R., and W. Murray. *Military Effectiveness.* Sydney: Allen and Unwin, 1989.

Milner, M. *Battle of the Atlantic.* Stroud, Gloucester: Tempus, 2005.

Mitcham, S. *Blitzkrieg No Longer: The German Wehrmacht in Battle, 1943.* Mechanicsburg, PA: Stackpole Books, 2010.

Moorhead, A. *Gallipoli.* London: Hamish Hamilton, 1956.

Morison, S. E. *History of United States Naval Operations in World War II.* Boston: Little, Brown, 1947–1962.

Murray, W. *German Military Effectiveness.* Baltimore: Nautical and Aviation Publishing, 1992.

————. *Luftwaffe.* Baltimore: Nautical and Aviation Publishing, 1985.

————. *Strategy for Defeat: The Luftwaffe 1939–1945.* Maxwell, AL: Air University Press, 1983.

Murray, W., and A. R. Millett. *A War to Be Won: Fighting the Second World War.* Cambridge, MA: Harvard University Press, 2000.

Myers, R., and M. Peattie. *The Japanese Colonial Empire 1895–1945.* Princeton: Princeton University Press, 1984.

Nagorski, A. *The Greatest Battle: Stalin, Hitler, and the Desperate Struggle for Moscow That Changed the Course of World War II.* New York: Simon and Schuster, 2007.

O'Kane, R. *Clear the Bridge!* New York: Bantam, 1981.

Orgill, D. *T-34: Russian Armor.* New York: Ballantine Books, 1971.

Overy, R. *The Air War 1939–1945.* London: Europa, 1980.

———. *Russia's War: Blood upon the Snow 1939–1945.* New York: TV Books, 1997.

———. *Why the Allies Won.* London: Jonathan Cape, 1995.

Padfield, P. *Guns at Sea.* London: Evelyn, 1973.

———. *War Beneath the Sea: Submarine Conflict 1939–1945.* London: John Murray, 1995.

Parker, G. *The Army of Flanders and the Spanish Road: The Logistics of Spanish Victory and Defeat in the Low Countries War.* Cambridge: Cambridge University Press, 1972.

Pawle, G. *The Secret War 1939–1945.* London: White Lion Press, 1956.

Price, A. *The Spitfire Story.* London: Arms and Armour, 1995.

Reynolds, C. G. *The Fast Carriers: The Forging of an Air Navy.* New York: McGraw-Hill, 1968.

Reynolds, D. *In Command of History: Churchill Writing and Fighting the Second World War.* London: Penguin, 2004.

Richards, D., and H. St. G. Saunders. *Royal Air Force 1939–1945.* London: HMSO, 1954.

Riley, J. C. *International Government Finance and the Amsterdam Capital Market 1740–1815.* Cambridge: Cambridge University Press, 1980.

Roberts, A. *Masters and Commanders: How Roosevelt, Churchill, Marshall and Alanbrooke Won the War in the West.* London: Allen Lane, 2008.

Robertson, T. *Dieppe: The Shame and Glory.* Toronto, ON: McLelland and Stewart, 1967.

Rohwer, J. *The Critical Convoy Battles of 1943.* London: Ian Allen, 1977.

Rommel, E. *The Rommel Papers.* Ed. B. H. Liddell Hart. London: Collins, 1953.

Roskill, S. W. *Churchill and the Admirals.* London: Longman, 1972.

———. *The Strategy of Sea Power.* London: John Goodchild, 1986.

———. *The War at Sea, 1939–1945.* London: HMSO, 1954–61.

Rottman, G. L. *World War II Infantry Anti-Tank Tactics.* Oxford: Osprey, 2005.

Ryan, C. *The Longest Day.* New York: Simon and Schuster, 1959.

Saward, D. *Victory Denied: The Rise of Air Power and the Defeat of Germany 1920–1945.* London: Buchan and Enright, 1985.

Scutts, J. *Mustang Aces of the Eighth Air Force*. Oxford: Osprey, 1994.

Seth, R. *The Fiercest Battle: The Story of North Atlantic Convoy ONS 5, 22nd April–7th May 1943*. London: Hutchinson, 1961.

Sherry, M. *The Rise of American Air Power: The Creation of Armageddon*. New Haven: Yale University Press, 1987.

Slim, W., *Defeat Into Victory: Battling Japan in Burma and India, 1942–1945*. London: Cassell, 1956.

Sobel, D. *Longitude*. London: Penguin, 1996.

Spector, R. *Eagle Against the Sun: The American War with Japan*. New York: Free Press, 1985.

Stille, M. *Imperial Japanese Navy Aircraft Carriers 1921–1945*. London: Osprey, 2005.

Stone, D. *A Military History of Russia: From Ivan the Terrible to the War in Chechnya*. Westport, CT: Praeger, 2006.

Storry, R. *Japan and the Decline of the West in Asia 1894–1943*. London: Longman, 1979.

Terraine, J. *The Right of the Line: The Role of the RAF in World War Two*. London: Pen and Sword, 2010.

Tooze, A. *The Wages of Destruction: The Making and Breaking of the Nazi Economy*. London: Penguin, 2006.

Tuchman, B. *Sand Against the Wind: Stilwell and the American Experience in China 1911–1945*. New York: Macmillan, 1970.

Van Creveld, M. *Supplying War: Logistics from Wallenstein to Patton*. Cambridge: Cambridge University Press, 1977.

Waugh, E. *The Sword of Honour Trilogy*. New York: Everyman's Library, 1994.

Webster, C., and N. Frankland. *The Strategic Air Offensive Against Germany, 1939–1945*. London: HMSO, 1961.

Weigley, R. *The American Way of Warfare: A History of United States Military Strategy and Policy*. Bloomington: Indiana University Press, 1973.

Werner, H. *Iron Coffins: A Personal Account of the German U-boat Battles of World War II*. New York: Da Capo Press, 2002.

Whitaker, D., and S. Whitaker. *Dieppe: Tragedy to Triumph*. Whitby, ON: McGraw-Hill, 1967.

Willmott, H. P. *The Great Crusade: A New Complete History of the Second World War*. London: Michael Joseph, 1989.

Winton, J. *Convoy: Defence of Sea Trade, 1890–1990*. London: Michael Joseph, 1983.

Wohl, R. *A Passion for Wings: Aviation and the Western Imagination 1908–1918*. New Haven: Yale University Press, 1994.

Zaloga, S. *The Destruction of Army Group Centre*. Oxford: Osprey Press, 1996.

Zieger, P. *Mountbatten: A Biography*. New York: Knopf, 1985.

ARTICLES

Allen, L. "The Campaigns in Asia and the Pacific." *Journal of Strategic Studies* 13, no. 1 (March 1990).

Black, J. "Midway and the Indian Ocean." *Naval War College Review* 62, no. 4 (Autumn 2009).

Ceva, L. "The North African Campaign 1940–43: A Reconsideration." *Journal of Strategic Studies* 13, no. 1 (March 1990): 84–104.

Constable, T. J. "The Little-Known Story of Percy Hobart." *Journal of Historical Review* 18, no. 1 (Jan.-Feb. 1999).

Coox, A. "The Effectiveness of the Japanese Establishment in the Second World War." In A. Millett and W. Murray, eds., *Military Effectiveness*, vol. 3 (London: Allen and Unwin, 1988).

Forster, J. E. "The Dynamics of Volksgemeinschaft: The Effectiveness of the German Military Establishment in the Second World War." In A. Millett and W. Murray, eds., *Military Effectiveness*, vol. 3 (London: Allen and Unwin, 1988).

Grisson, A. "The Future of Military Innovation Studies." *Journal of Strategic Studies* 29, no. 5 (October 2006).

Isaev, A. "Against the T-34, the German Tanks Were Crap." In A. Drabkin and O. Sheremet, eds., *T-34 in Action*. Mechanicsburg, PA: Stackpole Books, 2008.

Kahn, D. "Intelligence in World War II: A Survey." *Journal of Intelligence History* 1, no. 1 (Summer 2001).

———. "An Historical Theory of Intelligence." *Intelligence and National Security* 16, no. 3 (Autumn, 2001).

Kennedy, P. M. "History from the Middle: The Case of the Second World War." *Journal of Military History* 74, no. 1 (January 2010).

———. "Imperial Cable Communications and Strategy 1870–1914." *English Historical Review* 86, 341 (1972): 728–52.

Marder, A. J. "Winston Is Back!" *English Historical Review*, supp. 5 (1972).

Schwarz, B. "Black Saturday." *Atlantic*, April 2008.

Spector, R. "America's Seizure of Japan's Strategic Points, Summer 1942–1944." In S. Dockrill, ed., *From Pearl Harbor to Hiroshima: The Second World War in Asia and the Pacific, 1941–1945*. London: Macmillan, 1994.

Tao, W. "The Chinese Theatre and the Pacific War." In S. Dockrill, ed, *From Pearl Harbor to Hiroshima*. London: Macmillan, 1994.

Wegner, B., "The Road to Defeat: The German Campaigns in Russia, 1941–43." *Journal of Strategic Studies* 13, no. 1 (March 1990): 122–23.

Zeitzler, K. "Men and Space in War: A German Problem in World War II." *Military Review*, April 1962.

INTERNET SOURCES

http://www.cebudanderson.com/droptanks.htm The memoir of Donald W. Marner, a mechanic serving a Mustang squadron based in Suffolk in 1944–45.

http://en.wikipedia.org/wiki/F6F_Hellcat

http://en.wikipedia.org/wiki/Essex_class_aircraft_carrier

http://en.wikipedia.org/wiki/B-29_Superfortress

http://en.wikipedia.org/wiki/Ben_Moreell

http://en.wikipedia.org/wiki/Operation_Bagration

http://en.wikipedia.org/wiki/Polish_mine_detector

http://en.wikipedia.org/wiki/Seabees_in_World_War_II

http://en.wikipedia.org/wiki/Battle_of_Gallipoli

http://wikipedia.org/wiki/T-34

http://en.wikipedia.org/wiki/Amphibious_warfare

http://www.dupuyinstitute.org/ubb/Forum4/HTML/000052.html

http://www.ihr.org/jhr/v 18/v 18n 1 p-2_Constable.html

http://www.cardarmy.ru/armor/articles/t3485.htm

http://en.wikipedia.org/wiki/Ilyushin_Il-2

Credits

Ronnie Harker (Rolls-Royce plc Archives)

Packard Merlin is lowered into a P-51 (Paul A. Ludwig, *Development of the P-51 Long-Range Escort Fighter Mustang.* Hersham, Surrey, Ian Allan Printing Ltd, 2003, p. 89)

Captain Pete Ellis (United States Marine Corps Archives)

Guadalcanal, September 1942 (Getty Images)

Major General Sir Percy Hobart (Imperial War Museum, London)

Hobart's Funnies (Imperial War Museum, London)

Admiral Ben Moreell (United States Naval Institute)

Mulberry harbor, Arromanches, June 1944 (Imperial War Museum, London)

Messerschmitt Me 262 (Getty Images)

Me 262 with bombs (Manfred Jurleit, *Strahljäger Me 262 im Einsatz.* Berlin Transpress, 1993, p. 90)

Montgomery and Zhukov in Berlin (Getty Images)

German prisoners of war marching (Russian State Documentary Film & Photo Archives at Krasnogorsk)

MAPS

"Location of Merchant Ships of the British Empire, November 1937" (original title "British Empire Shipping, 1937," image F0516) © National Maritime Museum, Greenwich, London. Reproduced by permission of Royal Museums Greenwich

"The North Atlantic Air Gap and Convoys" based on a map from *Convoy* by John Winton (London: Michael Joseph, 1983). Adapted by permission of The Estate of John Winton

"Fighter Command Control Network, Circa 1940" based on a map from *The Battle* by Richard Overy. Adapted by permission of W. W. Norton and Penguin Books UK

"Increasing Escort Fighter Range" based on a map from *Luftwaffe* by Williamson Murray. Adapted by permission of The Nautical & Aviation Publishing Company of America.

Anglo-American Armies Advance in Northern Africa and Southern Italy (U.S. Army Center of Military History)

The Rapid German Expansion in the East, July–December 1941 (Wikicommons)

Red Army Advances During Operation Bagration, June–August 1944 (Department of History, United States Military Academy)

British Landings in Madagascar, May 1942 (Sigrid von Wendel)

The Anglo-American Maritime Routes for Operation Torch, November 1942 (Sigrid von Wendel)

The D-Day Invasions, June 6, 1944 (Department of History, United States Military Academy)

The Japanese Empire's Expansion at Its Peak, 1942 (Department of History, United States Military Academy)

The Four Options for Allied Counterattack Against Tokyo After 1942–1943 (Sigrid von Wendel)

The Geography of History: Allied Positional Assets in World War II (Sigrid von Wendel)

Index

ABOUT THE AUTHOR

PAUL KENNEDY is internationally known for his writings and commentaries on global political, economic, and strategic issues. He earned his B.A. at Newcastle University and his doctorate at the University of Oxford. Since 1983, he has been the Dilworth Professor of History and director of international security studies at Yale University. He is on the editorial board of numerous scholarly journals and writes for *The New York Times, Los Angeles Times, The Atlantic,* and many foreign-language newspapers and magazines. Kennedy is the author and editor of nineteen books, including *The Rise and Fall of the Great Powers,* which has been translated into more than twenty languages, followed by *Preparing for the Twenty-first Century* (1993) and *The Parliament of Man* (2006).